Springer Series in
Electronics and Photonics 32

Edited by Walter Engl

Springer Series in Electronics and Photonics

Editors: I. P. Kaminov W. Engl T. Sugano

Managing Editor: H. K. V. Lotsch

This series was originally published under the title
Springer Series in Electrophysics
and has been renamed starting with Volume 22.

Volumes 1–21 are listed at the end of the book

J.-F. Luy P. Russer (Eds.)

Silicon-Based Millimeter-Wave Devices

Foreword by E. Kasper

With 202 Figures and 14 Tables

Springer-Verlag

Berlin Heidelberg New York
London Paris Tokyo
Hong Kong Barcelona
Budapest

Dr.-Ing. Johann-Friedrich Luy
Daimler-Benz AG, Forschung und Technik,
Wilhelm-Runge-Str. 11, 89081 Ulm, Germany

Professor Dr. techn. Peter Russer
Ferdinand-Braun-Institut für Höchstfrequenztechnik,
Rudower Chaussee 5, 12489 Berlin, Germany

Series Editors:
Dr. Ivan P. Kaminov
AT&T Bell Laboratories, P.O.Box 400, Holmdel, NY 07733-0400, USA

Professor Dr. Walter Engl
Kopernikusstr.16, 52074 Aachen, Germany

Professor Takuo Sugano, Ph. D.
Department of Electrical and Electronic Engineering, Toyo University
2100 Kujirai Kawagoe Saitama 350, Japan

Managing Editor: Dr. Helmut K.V. Lotsch
Springer-Verlag, Tiergartenstr. 17, 69121 Heidelberg, Germany

ISBN-13:978-3-642-79033-1   e-ISBN-13:978-3-642-79031-7
DOI: 10.1007/978-3-642-79031-7

Library of Congress Cataloging-in-Publication Data. Silicon-based millimeter-wave devices/
J.-F. Luy, P. Russer. p. cm. – (Springer series in electronics and photonics: 32) Includes
bibliographical references and index. ISBN 3-540-58047-6 (Berlin: acid-free paper). – ISBN 0-
387-58047-6 (New York: acid-free paper) 1. Millimeter wave devices. 2. Silicon crystals, I. Luy,
J.-F. (Johann-Friedrich), 1958– II. Russer, P. (Peter), 1943– . III. Series: Springer Series in
electronics and photonics; v. 32. TK7876.5.S55 1994 621.381'3–dc20 94-12946

© Springer-Verlag Berlin Heidelberg 1994
Softcover reprint of the hardcover 1st edition 1994

Typesetting: Macmillan India Ltd, Bangalore-25

SPIN: 10069501   54/3140/SPS - 5 4 3 2 1 0 - Printed on acid-free paper

Foreword

Today there is an increasing demand for monolithic integrated milli-meter-wave circuits for a variety of applications in communications and sensorics. At frequencies beyond 30 GHz also distributed planar circuit structures may be included in the monolithic circuits. Indeed, many efforts are now made to integrate the highly developed III-V high-frequency devices with passive networks to microwave integrated circuits. It is generally believed that similar circuits based on silicon will not be available because of the limited frequency regime of silicon transistors and because of the high loss of microstrip lines on usual silicon substrates.

Rapid progress in certain fields – only as example are given the speed push delivered by SiGe/Si heterostructures or the successful demonstration of low-loss microstrip and coplanar lines on high-resistivity silicon – requires to reconsider the use of silicon circuits for applications in the mm-wave region. Furthermore, attractive for all kinds of integrated circuits is the rather high value of the thermal conductivity of silicon.

At first look the entrance of a silicon-based technology in a field up to now completely dominated by the III-V technology may be con-sidered from the latter point of view unwanted and ruinous competi-tion in that small market segment. But there are good chances – as I believe – that the new technical solutions (e.g., integration of anten-nae and transmitter/receiver) and the extension to very economic solu-tions will accelerate the acceptance and broadening of this market with benefits for both.

The first steps in a new research direction are always only docu-mented by journal publications and conference talks. One might spec-ulate when the right moment for the publication of a book on a young topic is given. The editors of this volume made a successful attempt to find a good balance between basic contributions describing the underly-ing principles and applied chapters describing for the reader the actual state of realization.

I wish that this remarkable book will find readers from various disciplines. Although mainly written for researchers working in the area of microwave devices and circuits, the many facets of this well

balanced collection should gather interest from physicists, material sci-
entists, high-frequency engineers, and system engineers, and possibly
by managers searching for small-sized, economic solutions. Readers
looking only for specific aspects will acknowledge that every chapter
is written self-consistently.

Stuttgart   *Erich Kasper*
July 1994

Preface

Today silicon plays a dominating role as the basic semiconductor material for electronic components for frequencies up into the lower Gigahertz range. However, it is less noticed that silicon is also a viable material for many circuit applications up into the millimeter-wave range. In silicon circuits IMPATT diodes may be used as the active devices for power generation up to far beyond 100 GHz. With the silicon-germanium heterobipolar transistor a three-terminal active silicon based device is available, too. Schottky diodes allow the realization of sensitive mixers and detectors.

Monolithic integration of solid-state devices provides the possibility of low-cost production, improved reliability, small size and light weight, and easy assembly. Besides III-V compound based MMICs also Silicon Monolithic Millimeter-Wave Integrated Circuits (SIMMWICs) operating at frequencies up to above 100 GHz with growing interest. In the frequency region above 60 GHz meet SIMMWICs with dimensions of only a few millimeters may also include planar antenna structures. This opens the way for the realization of monolithic single-chip transmitters and single-chip receivers in the millimeter wave region.

The future availability of low-cost millimeter-wave components and Millimeter-wave Monolithic Integrated Circuits (MMICs) will stimulate the penetration of millimiter-wave technology into commercial and consumer electronics. The advantages of millimeter waves are high bandwidth for communication applications, high resolution for sensor applications, and high antenna gain with small antennas.

Chapter 1 outlines the fundamentals of different planar transmission lines, planar resonators, and gives a survey of possible modelling approaches. Special emphasis has been placed on planar antennas which may be integrated on high-resistivity silicon. Chapter 2 treats the family of transit-time devices – IMPATT diodes, BARITT diodes and related devices. The physics of Schottky contacts on silicon represent the topic of Chap. 3. Schottky-barrier models and electrical transport properties as well as epitaxial diodes on silicon and measurement techniques are described. The tremendous development of the high-frequency performance of SiGe HBTs is treated in Chap. 4, and a detailed discussion of operation principles, design constraints and technological aspects is presented. After this discussion of the fundamentals of silicon based millimeter-wave devices Chap. 5 provides a treatment of the silicon-based monolithic integrated millimeter-wave devices realized up to now. The self-mixing oscillator mode offers a possibility to come to low-cost radar-sensor solutions, and the properties of this operation mode are discussed with special consideration of transit-time diodes in Chap. 6. Chapter 7 is probably the first comprehensive representation of silicon millimeter-wave technology: Substrate properties, basic technologies and fabrication processes of monolithic integrated two- and three-terminal devices are treated. What are the prospects of silicon/silicon-germanium millimeter-wave devices? Chapter 8 deals with future devices which are now in an early stage of research. Chapter 9 treats applications of this millimeter-wave integration technology which represents now and in the near future the driving force for the research and development of SiGe/SIMMWICs.

This quasi-monograph gives a survey of the state-of-the-art of silicon and silicon-germanium millimeter-wave devices, integrated circuits, and their present and future applications.

Ulm   *J.-F. Luy*
Berlin   *P. Russer*
July 1994

Contents

3. Schottky Contacts on Silicon

Contributors

E. Biebl
   Lehrstuhl für Hochfrequenztechnik der Technischen
   Universität München, Arcisstr. 21, 80333 München, Germany

J. Buechler
   Daimler-Benz Research Center
   Wilhelm-Runge-Str. 11, 89081 Ulm, Germany

Q. Chen
   Department of Physics and Measurement Technology,
   Linköping University, S-581 83 Linköping, Sweden

M. Claassen
   Lehrstuhl für Allgemeine Elektrotechnik und Angewandte
   Elektronik, TU München, Arcisstr. 21, 80333 München, Germany

Y. Fu
   Department of Physics and Measurement Technology,
   Linköping University, S-581 83 Linköping, Sweden

A. Gruhle
   Daimler-Benz Research Center
   Wilhelm-Runge-Str. 11, 89081 Ulm, Germany

J. -F. Luy
   Daimler-Benz Research Center, Wilhelm-Runge-Str. 11,
   89081 Ulm, Germany

W. Menzel
   University of Ulm, Department of Microwave Techniques,
   89069 Ulm, Germany

U. Rau

    Max-Planck-Institut für Festkörperforschung,
    Heisenbergstrasse 1, 70569 Stuttgart, Germany

P. Russer

    Ferdinand-Braun-Institut für Hochfrequenztechnik,
    Rudower Chaussee 5, 12489 Berlin, Germany

K.M. Strohm

    Daimler-Benz Research Center, Wilhelm-Runge-Str. 11,
    89081 Ulm, Germany

J.H. Werner

    Max-Planck-Institut für Festkörperforschung,
    Heisenbergstrasse 1, 70569 Stuttgart, Germany

M. Willander

    Department of Physics and Measurement Technology,
    Linköping University, S-581 83 Linköping, Sweden

1 Fundamentals

P. Russer and E. Biebl

Ferdinand-Braun-Institut für Hochfrequenztechnik, 12489 Berlin, Germany
Lehrstuhl für Hochfrequenztechnik, Technische Universität München, 80290
München, Germany

The demands for future civil sensor and communication systems are stimulating the development of MilliMeter-wave Monolithic Integrated Circuits (M^3ICs) [1.1]. The advantages of millimeter waves are high bandwidth for communication applications, high resolution for sensor applications, and high antenna gain with small antennas. Besides III–V compound based devices [1.2,3] also SIlicon Monolithic Millimeter-Wave Integrated Circuits (SIMMWICs) were investigated [1.4–6]. The suitability of silicon as the substrate material for monolithic integrated millimeter-wave circuits has been successfully demonstrated in numerous examples [1.7–22]. Monolithic implementation of solid-state devices provides the possibility of low-cost production, improved reliability, small size and light weight, and easy assembly. If technically available high-resistance silicon material with a specific resistance of up to 10,000 Ωcm is used, the dominant loss contribution in the planar monolithic millimeter-wave circuits come from the skin-effect losses, and the circuit properties are not influenced by the ohmic losses in the semiconductor material. In silicon circuitry IMPact Avalanche Transit-Time (IMPATT) diodes may be used as the active devices whereas mixer and detector circuits may be realized with Schottky diodes. IMPATT diodes as well as Schottky diodes may be integrated in planar monolithic silicon technology. In the future with the silicon-germanium hetero bipolar transistor also a three-terminal active device will become available for integrated silicon millimeter-wave devices. In the frequency region above 60 GHz SIMMWICs with only dimensions of a few millimeters also include planar antenna structures. This opens the way for the realization of monolithic single-chip transmitters and single-chip receivers in the millimeter-wave region.

In the following the fundamentals of planar silicon monolithic millimeter-wave integrated circuits for frequencies up to above 100 GHz are discussed. We first consider the material properties of silicon and their influence on the feasibility of millimeter-wave integrated circuits. Thereafter, passive planar millimeter-wave structures on silicon are treated. Experimental results with different millimeter-wave integrated circuits are presented.

The future availability of easy to handle low-cost monolithic integrated millimeter-wave components will stimulate the penetration of millimeter-wave technology into the field of commercial and consumer electronics. SIMMWICs will meet with these requirements in an ideal manner. The main future applications will be communications, sensorics, and radio-location [1.23,24]. The millimeter-wave applications will take advantage of the high gain and good angular resolution

Springer Series in Electronics and Photonics, Vol. 32
Silicon-Based Millimeter-Wave Devices, Eds.: Luy et al.
© Springer-Verlag Berlin Heidelberg 1994

obtainable with very small antenna structures. The integration of the antenna structures allows the direct coupling of SIMMWIC to the radiation field. Complete receiver or transmitter circuits may be realized monolithically on chips with dimensions of a few millimeters. SIMMWIC transmitters and receivers may be compared to optoelectronic circuit elements like luminescent diodes, laser diodes, and photodiodes. The antenna gain may be increased, if necessary, by using lenses and mirrors. Base-band signal-processing circuits can be integrated on the SIMMWIC.

Radio-location applications will be in millimeter-wave sensors for short-distance precision velocity and position sensing. There will be numerous applications, especially in the short-distance region between a few millimeters and some ten meters. In traffic future applications will be the anticollision radar for vehicles, Doppler sensors for direct measurement of velocity over ground, and stationary sensors for analysis of traffic situations. For industrial applications millimeter-wave sensors are of interest in any situation where position or movement has to be measured without contact. In communications short-distance high-bandwidth radio links, specially for in-house communications will be of interest.

1.1 Silicon as the Base Material for MMICs

Early investigations of silicon-based planar circuits have been reported in [1.25,26]. *Rosen* et al. [1.4] have been the first suggesting the use of silicon as the base material for monolithic integrated microwave circuits. Table 1.1 offers a comparison of the data of Si and GaAs. The dielectric constant of both materials is comparable. Its high value is advantageous for the miniaturization of distributed microwave circuits.

Today silicon-substrate material with a specific resistance of 10,000 Ωcm is available. For this material in planar circuits the conductor losses due to the skin effect are the dominant material-loss contribution. The ohmic losses in the silicon give only a minor loss contribution. The dielectric-loss factor of Si and GaAs is

Table 1.1. Material parameters of Si and GaAs

	Si	GaAs
Dielectric constant	11.7	12.9
Specific resistance	$> 10^4$ Ωcm	$> 10^6$ Ωcm
Dielectric loss factor (90 GHz)	$1.3 \cdot 10^{-3}$	$0.7 \cdot 10^{-3}$
Thermal conductivity	1.5 W cm^{-1}K^{-1}	0.46 W cm^{-1}K^{-1}
Electron mobility	1500 cm^2/Vs	8500 cm^2/Vs
Hole mobility	450 cm^2/Vs	400 cm^2/Vs
High field drift velocity	$8 \cdot 10^6$cm s^{-1}	$4 \cdot 10^6$cm s^{-1}
Density	2.33 g cm^{-3}	5.32 g cm^{-3}

within the same order at 90 GHz. On the other hand, the electron mobility of GaAs is six times higher than the electron mobility of Si. This yields a correspondingly higher f_T value for bipolar transistors and field-effect transistors. The high-field drift velocities of Si and GaAs are in the same order of magnitude. For this reason silicon as well as GaAs is suited for IMPATT diodes, FETs with submicrometer channel length and for bipolar transistors with ultra-thin bases. The thermal conductivity of silicon is three times higher than that of GaAs. From this point of view silicon is advantageous for power circuits. Further advantages of silicon are the possibility of forming insulating layers by oxidation and the low costs of the material.

Experimental investigations have shown that the deposition of molecular beam epitaxial layers increases the attenuation of striplines fabricated on this material. However, if the epitaxial layers are removed by etching and planar structures are fabricated on this material, the microstrip lines will have the same low attenuation as lines on unprocessed material [1.8]. Considering this, it is possible to monolithically integrate active elements on the silicon substrate without degrading the material properties with respect to the passive planar circuit structures.

1.2 Linear Passive Planar Millimeter Wave Circuits on Silicon

The linear passive parts of SIMMWICs may be realized in planar circuit technology. Figure 1.1 shows the calculated attenuation coefficient of a gold microstrip line on a 7,000 Ωcm silicon substrate for a ratio of strip width to substrate thickness $w/h = 0.8$ [1.8]. Since for highly insulating silicon material for millimeter-wave

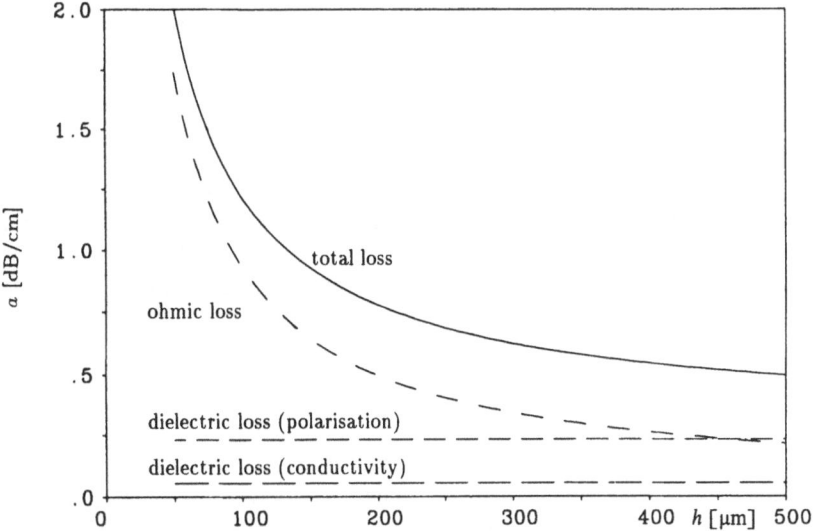

Fig. 1.1. Calculated attenuation coefficient of a 50 Ω gold microstrip line on a silicon substrate

frequencies the dielectric losses in the silicon play a minor role compared with the total losses, the use of silicon yields no drawback. Experimental investigations have shown that on substrates with 195 μm thickness microstrip lines with an attenuation as low as 0.6 dB/cm at 90 GHz can be realized [1.8]. Taking into consideration that with increasing frequency due to the decrease of the linear circuit dimensions a linear increase of the attenuation coefficient may be accepted, the logarithmic attenuation per unit of wavelength in silicon strip-line circuits above 90 GHz is only by a factor 2 higher than the logarithmic attenuation per unit of wavelength in a room temperature duroid micro-strip-line circuit at 10 GHz.

The basic linear passive structures of monolithic integrated millimeter-wave circuits are planar transmission lines. The required resonant or radiating structures are, in general, open or shorted transmission lines with a length of a half or a full wavelength. In this section transmission lines and resonators with small radiation losses are considered. The radiating structures, i.e. planar antennas, will be treated in the next section.

The commonly used types of planar transmission lines are summarized in Fig. 1.2. The behavior of these structures at microwave frequencies has been studied extensively in the past thirty or forty years. Details can be found in several textbooks [1.27,28]. The extension into the millimeter-wave region, however, is not simply a matter of wavelength scaling. Because of fabrication tolerances, the dimensions cannot be reduced proportionally to the wavelength. Thus, transmission-line discontinuities, such as bends or T-junctions, are no longer electrically small and produce considerable energy leakage. Moreover, the substrate is no longer electrically thin. Depending on the geometry surface-wave modes or parallel-plate waveguide modes will be excited. Surface waves trapped in the substrate may increase substantially the mutual coupling between the circuit elements compared to the coupling by the radiated fields. A way out of these problems is to avoid transmission lines and, in particular, discontinuities whenever it is practicable. This design rule is best met by integrated antennas, where the active elements are integrated in the radiating structure. However, particularly in substrates with high permittivity the excitation of surface waves may result in poor radiation efficiency.

The design of passive planar millimeter-wave circuits requires careful treatment of all the excited electromagnetic fields. In some cases, that can only be achieved by employing an expensive full wave analysis. In this text we give some fundamental data for the design of commonly used passive structures in the millimeter-wave region. Information about treatment by means of field-theoretic approaches can be found in [1.29–33].

1.2.1 Wave Propagation in Planar Structures

In Fig. 1.2 the desired direction of energy flow is perpendicular to the drawing plane along the z-axis. The fundamental mode is purely TEM for the stripline, quasi-TEM for the microstrip, coplanar and coplanar strip line, and quasi-TE for the slot line.

Fig. 1.2. Planar transmission lines

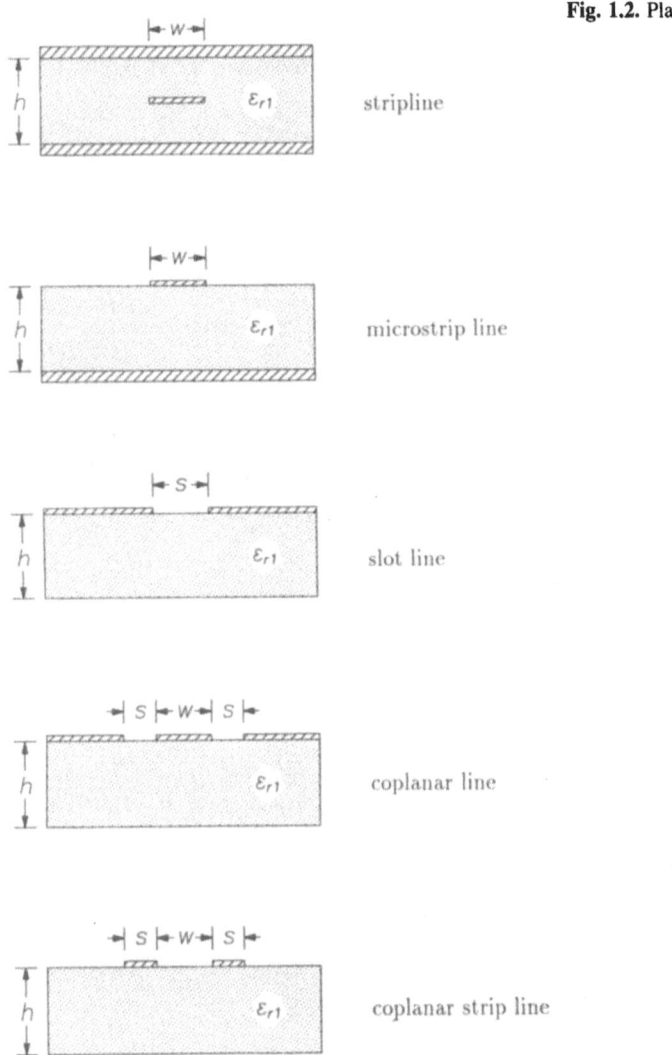

Besides the modes guided by the structured metallization some parasitic modes may propagate, which are supported by the dielectric layer itself [1.34]. In the strip-line structure the dielectric substrate is metallized on both sides providing a parallel-plate waveguide. In the microstrip, slot line, and coplanar line one side is metallized and guides also a trapped surface wave. Fixing a slot line or coplanar line on a metallic mount results in a parallel-plate waveguide. The coplanar strip-line structure provides a symmetric dielectric slab waveguide. In what follows the propagation characteristics of the parasitic modes will be discussed. Only one dielectric layer with $\varepsilon_1 = \varepsilon_0 \varepsilon_{r1}$ and height h is assumed. The structure is surrounded by free space.

Parallel-Plate Waveguide

The fundamental mode of the parallel-plate waveguide is a TEM mode. Consequently the propagation constant of this mode is

$$\beta_{\text{TEM}} = \sqrt{\varepsilon_{\text{r1}}} k_0 = \sqrt{\varepsilon_{\text{r1}}} \frac{2\pi f}{c_0} \tag{1.1}$$

with the frequency f and the free space light velocity c_0. The TEM mode has no cutoff frequency and is non-dispersive.

The higher-order modes are the TE_n and TM_n modes, respectively. The phase coefficients in a dielectric slab with thickness h

$$\beta_n = \sqrt{\varepsilon_{\text{r1}} k_0^2 - k_n^2} \tag{1.2}$$

with

$$k_n = \frac{n\pi}{h} \tag{1.3}$$

are equal for TE and TM modes. The cutoff frequency of the first higher-order modes ($n = 1$) is in silicon

$$f_{\text{c1}} = \frac{c_0}{2h\sqrt{\varepsilon_{\text{r1}}}} = \frac{43.9 \text{ GHz}}{h[\text{mm}]} \tag{1.4}$$

Dielectric Slab with Single Sided Metallization

The dielectric slab with metallization on one side supports trapped TE and TM surface waves. Let us suppose that the trapped surface wave is excited by an electrical line current on the surface of the slab. Then the dominant propagation direction of the TE and TM surface waves is normal and parallel to the line current, respectively [1.35]. The propagation constant β cannot be given in a closed-form expression. The solution of the Helmholtz equation taking into account Dirichlet's condition $E_t = 0$ at the metallized slab surface yields the eigenvalue equations

$$\begin{aligned} u \tan u &= \varepsilon_{\text{r1}} w \quad \text{TM modes,} \\ -u \cot u &= w \qquad \text{TE modes,} \end{aligned} \tag{1.5}$$

where $u = h\sqrt{\varepsilon_{\text{r1}} k_0^2 - \beta^2}$ and $w = h\sqrt{\beta^2 - k_0^2}$. The eigenvalue equations (1.5) can easily be solved, either numerically or graphically, employing the normalized frequency $v = \sqrt{u^2 + w^2} = k_0 h \sqrt{\varepsilon_{\text{r1}} - 1}$ [1.36].

The TM_0 mode has no cutoff frequency and large components of the electric fields normal to the slab surface. Thus, the TM_0 mode is very similar to a TEM mode and can easily be excited by a planar waveguide TEM or quasi-TEM mode. The normalized cutoff frequencies of the higher-order modes are

$$\begin{aligned} v_{\text{c,TM}_n} &= n\pi \qquad\qquad\quad \text{TM modes,} \\ v_{\text{c,TE}_n} &= \left(n + \frac{1}{2}\right)\pi \quad \text{TE modes.} \end{aligned} \tag{1.6}$$

The first higher-order mode is the TE_0 mode with the cutoff frequency in silicon

$$f_{c,TE_0} = \frac{c_0}{4h\sqrt{\varepsilon_{r1} - 1}} = \frac{22.9 \text{ GHz}}{h[\text{mm}]}. \tag{1.7}$$

Dielectric Slab Without Metallization

The dielectric slab without metallization can be treated in a manner analogous to the case with single sided metallization. The Helmholtz equation has to be solved taking into account the continuity of the transversal field components at the surfaces of the slab. This yields the eigenvalue equations

$$
\begin{aligned}
\frac{u}{2} \tan\left(\frac{u}{2}\right) &= \varepsilon_{r1} \frac{w}{2} && \text{even TM modes,} \\
-\frac{u}{2} \cot\left(\frac{u}{2}\right) &= \varepsilon_{r1} \frac{w}{2} && \text{odd TM modes,} \\
\frac{u}{2} \tan\left(\frac{u}{2}\right) &= \frac{w}{2} && \text{even TE modes,} \\
-\frac{u}{2} \cot\left(\frac{u}{2}\right) &= \frac{w}{2} && \text{odd TE modes.}
\end{aligned}
\tag{1.8}
$$

The solution with respect to β may be obtained numerically or graphically. Both, the TE_0 and the TM_0 mode have no cutoff frequency. The normalized cutoff frequencies for the higher-order modes

$$v_{cn} = n\pi \tag{1.9}$$

are identical for the TE_n and TM_n modes. The lowest higher-order modes are the TM_1 and the TE_1 mode with the cutoff frequency

$$f_{c,TM_1} = f_{c,TE_1} = \frac{c_0}{2h\sqrt{\varepsilon_{r1} - 1}} = \frac{45.8 \text{ GHz}}{h[\text{mm}]}. \tag{1.10}$$

Suppression of the Effects of Surface Waves

Interaction of surface waves and planar transmission-line modes is caused by coupling at discontinuities and by synchronous coupling. At discontinuities the incident guided wave is partially scattered into radiated waves and surface waves. Reflection of surface waves at the substrate edges results in standing substrate waves and causes ripple in the frequency response of the planar circuit. Radiation of surface waves from the substrate edges may result in pattern distortion of planar antennas. These effects can be reduced by placing absorbing material at the substrate edges and by application of absorbing layers on these parts of the substrate surface, which are not covered by the waveguiding structures.

Synchronous coupling occurs in the case of phase matching between the transmission-line mode and the surface-wave mode. Synchronous coupling is very strong and results in unacceptably high losses in the planar circuit. Due to the dispersive nature of the involved modes synchronism occurs above a certain frequency resulting in an upper limit for the frequency of operation, which will be given for the microstrip line in (1.15).

For the design of passive planar circuits information about the strength of the excitation of parasitic fields would be very useful. In general, for a given structure a quantitative determination requires a full wave analysis since the excitation depends on the current distribution in the conductors. However, the analysis of the radiation efficiency of microstrip antennas has shown that generally the excitation of substrate modes is not very much affected by either the antenna shape or its feed [1.37]. This implies that the radiation efficiency may be estimated from an assumed current distribution, or even an infinitesimal current source. *Alexopoulos* et al. [1.35] presented data for the excitation of trapped surface waves by line currents. They calculated the power radiated into surface-wave modes of a grounded substrate by elementary printed dipoles on the substrate and by slots in the ground plane. As a result one can state that for weak excitation of surface-wave modes the substrate height h should be as small as possible.

1.2.2 Planar Transmission Lines

In monolithic integrated circuits the conventional three-dimensional waveguides such as hollow or coaxial guides have to be replaced by planar two-dimensional guiding structures. Cross-sectional views of commonly used planar transmission lines are sketched in Fig. 1.2. For millimeter-wave monolithic integrated circuits the microstrip has been employed extensively. Since the reduction of the substrate height is limited for reasons of mechanical stability, the substrate becomes electrically thick at millimeter-wave frequencies. Thus, microstrip lines suffer from dispersion and from losses due to the excitation of surface waves. In order to overcome these problems, there is increasing interest in coplanar waveguides. Another advantage of this waveguide type is an easy realization of shunt as well as series connections of active and passive devices. Due to its high dispersion and radiation losses the slot line is almost exclusively used as a resonator and an antenna [1.38–40]. Due to its balanced nature the coplanar strip line is ideally suited as a feed line for printed dipole antennas [1.41]. Each of these types of planar transmission lines has some distinctive advantages over the others regarding impedance range, power handling capabilities, losses, dispersion, and technological aspects. In this subsection we will discuss these characteristics and give some fundamental design data.

Striplines

The stripline plays no role in monolithic integrated circuits. However, we give a brief discussion of its fundamental formulae since the characteristics of the transmission lines discussed in the following may be derived from the strip line.

Strip lines are homogeneous waveguides since no field lines leave the dielectric material. The fundamental mode is purely TEM and, thus, non-dispersive.

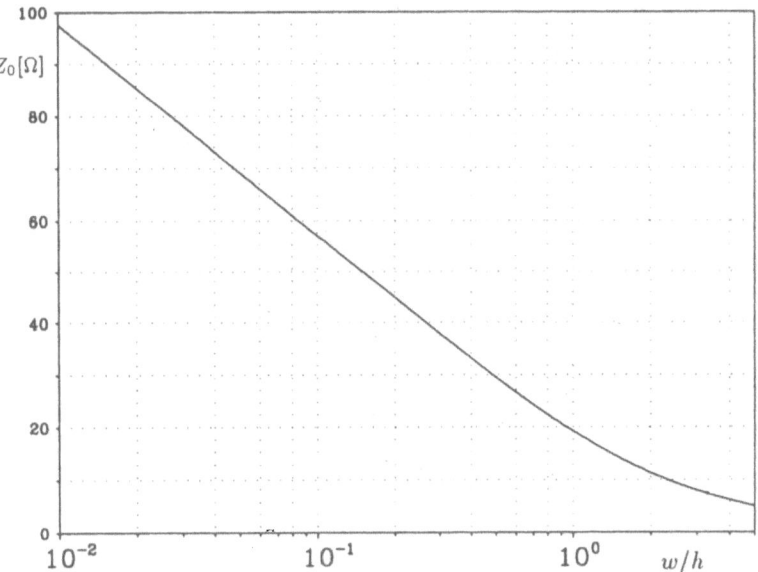

Fig. 1.3. Characteristic impedance of stripline

The characteristics can be calculated completely by electrostatic analysis such as conformal mapping. For an infinitesimally thin core conductor the characteristic impedance [1.42]

$$Z_0 = \frac{\eta_0}{4\sqrt{\varepsilon_{rl}}} \frac{K'(k)}{K(k)}, \quad k = \tanh\left(\frac{\pi w}{2h}\right) \tag{1.11}$$

is obtained, where $\eta_0 = \sqrt{\mu_0/\varepsilon_0}$ is the characteristic impedance of free space. For a silicon substrate Z_0 versus w/h is plotted in Fig. 1.3. K represents a complete elliptic function of the first kind and K' its complementary function [1.43]. A very accurate approximation for K/K' is given in the Sect. 1.5. Since the fundamental mode is a TEM mode, the propagation constant is given by (1.1). The first higher-order mode is the TE_1 parallel-plate waveguide mode. For $w \ll h$ its cutoff frequency is given by (1.4). The influence of the core conductor can be taken into account by a correction term yielding

$$f_c = \frac{c_0}{2h\sqrt{\varepsilon_{rl}}} \frac{1}{w/h + \pi/4}. \tag{1.12}$$

Microstrip Lines

Microstrip lines are most commonly used in monolithic integrated circuits. Due to the back side metallization the substrate can easily be fixed onto a metallic mount providing an efficient heat sink. The series connection of lumped elements is very easy. The parallel connection of lumped elements and fabrication of shorts requires via holes, which are technologically expensive in monolithic integrated circuits.

In microstrip lines the field is not confined entirely in the dielectric substrate. Thus, microstrip lines are transversely inhomogeneous lines and even the fundamental mode propagating along the microstrip is only quasi-TEM. However, the fundamental mode can be treated by a quasi-static approach, if the substrate thickness and the strip width are much smaller than the wavelength in the dielectric material. It should be noted that this condition might be violated in the millimeter-wave region since the reduction of the substrate height is technologically limited. In case of an electrically thick substrate the hybrid nature of the fundamental mode has to be taken into account by a dynamic analysis.

Hammerstad has presented design data [1.44,45], which have been obtained by a functional approximation to numerical calculations on the basis of a quasi-static approach, as included in Sect. 1.5. The characteristic impedance is represented as

$$Z_0 = \frac{Z_0(\varepsilon_{r1} = 1)}{\sqrt{\varepsilon_{r,\text{eff}}}} \tag{1.13}$$

where the effective dielectric constant $\varepsilon_{r,\text{eff}}$ takes into account the ratio of the field in the air to the field in the substrate. Microstrip lines on a silicon substrate ($\varepsilon_{r1} = 11.6$) exhibit characteristic impedances $15\Omega \leq Z_0 \leq 100\Omega$ and effective dielectric constants $7 \leq \varepsilon_{r,\text{eff}} \leq 9.5$ for normalized strip width $0.1 \leq w/h \leq 5$, as shown in Fig. 1.4.

In the millimeter-wave region the dispersive behavior of microstrip line has to be considered. In Fig. 1.5 Z_0 and $\varepsilon_{r,\text{eff}}$ versus the normalized frequency h/λ_0 have been plotted for various normalized strip widths w/λ_0. For typical microstrip lines on silicon substrates the quasi-static approach for the characteristic impedance and

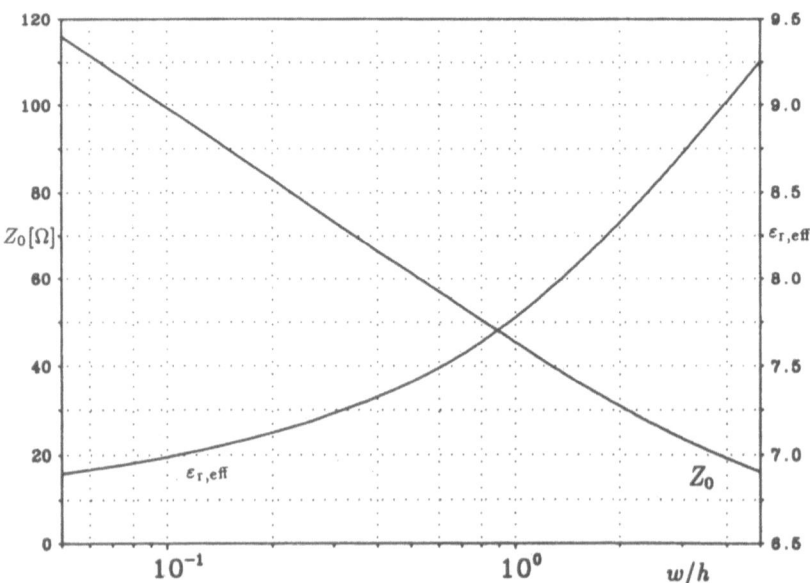

Fig. 1.4. Characteristic impedance and effective dielectric constant of microstrip line

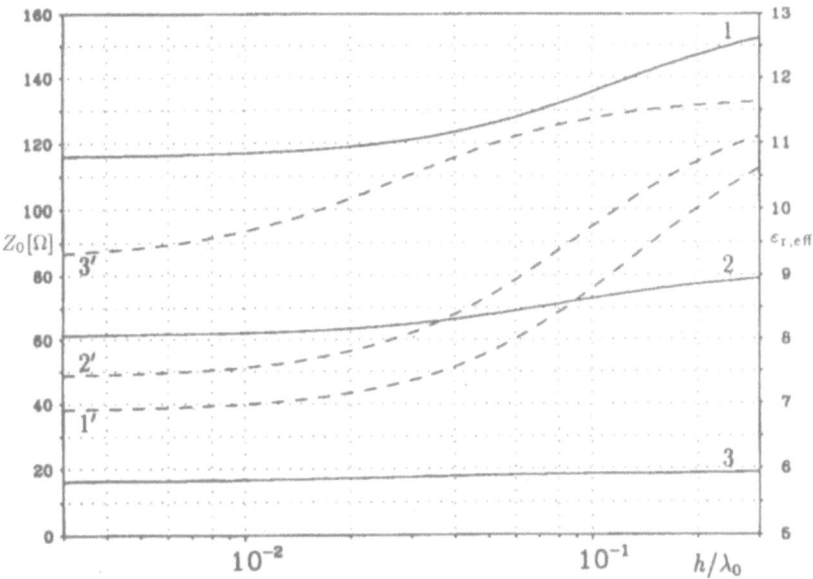

Fig. 1.5. Dispersive behavior of microstrip line on silicon, (———) Z_0, (-----) $\varepsilon_{r,eff}$: 1,1': $w/h = 0.05$; 2,2': $w/h = 0.5$; 3,3': $w/h = 5.0$

the effective permittivity holds for $h/\lambda_0 \leq 0.02$ and $h/\lambda_0 \leq 0.005$, respectively. Above these values the field concentrates within the substrate and the effective permittivity increases, until the permittivity of the substrate ε_{r1} is reached. In the same manner the characteristic impedance rises by a factor of about $\sqrt{\varepsilon_{r1}/\varepsilon_{r,eff}}$, where $\varepsilon_{r,eff}$ is the quasi-static effective permittivity.

The maximum frequency of operation of a microstrip is mainly limited by the excitation of spurious modes, namely higher-order hybrid microstrip modes, trapped surface waves, and radiating waves. A coarse approximation for the cutoff frequency of the first higher-order hybrid microstrip mode is given by [1.46]

$$f_{c,HE_1} \doteq \frac{c_0 Z_0}{2\eta_0 h}. \tag{1.14}$$

Coupling to the lowest-order surface-wave mode becomes significant above the frequency

$$f_s = \frac{c_0}{2\pi h}\sqrt{\frac{2}{\varepsilon_{r1}-1}} \arctan \varepsilon_{r1} = \frac{30.7 \text{ GHz}}{h[\text{mm}]}. \tag{1.15}$$

Radiation losses can be described by a radiation quality factor Q_r of a $\lambda/2$ microstrip resonator. An approximate value is given by

$$Q_r = \frac{3\varepsilon_{r1} Z_0 \lambda_0^2}{32\eta_0 h^2}. \tag{1.16}$$

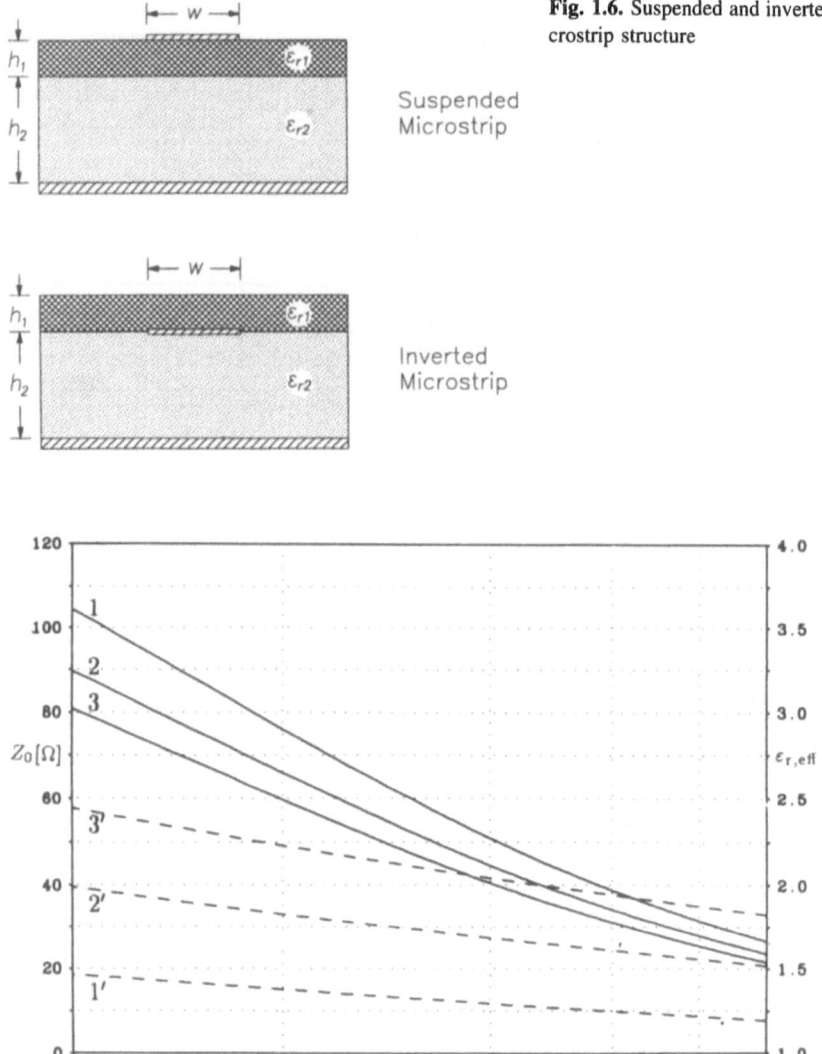

Fig. 1.6. Suspended and inverted microstrip structure

Suspended
Microstrip

Inverted
Microstrip

Fig. 1.7. Characteristic impedance (———) and effective dielectric constant (- - - - -) of suspended microstrip line on silicon: 1,1': $h1/h2 = 0.2$; 2,2': $h1/h2 = 0.6$; 3,3': $h1/h2 = 1.0$

Thus for a $Z_0 = 50\Omega$ microstrip line on a $h = 100\mu$m thick silicon substrate ($w = 80$ μm) (1.14) yields $f_{c,\mathrm{HE_1}} \doteq 200$ GHz and (1.15) yields $f_s \doteq 380$ GHz. At 100 GHz a quality factor of $Q_r \doteq 125$ is obtained.

The dielectric losses in microstrip lines may be significantly reduced using the suspended and inverted microstrip geometry (Fig. 1.6). These geometries are well suited for monolithic integration if the circuit is integrated on a thin semiconductor substrate. The substrate is then fixed on posts above a metallic mount providing

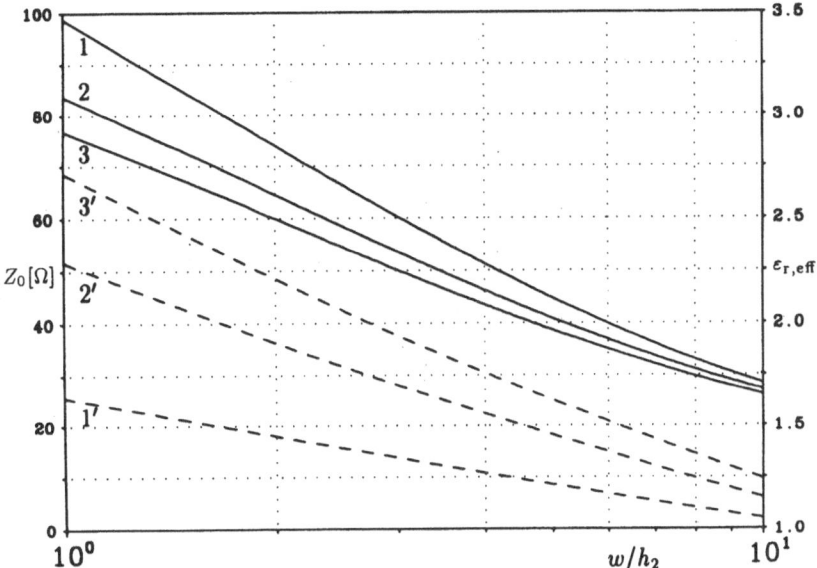

Fig. 1.8. Characteristic impedance (————) and effective dielectric constant (- - - - -) of inverted microstrip line on silicon: 1,1': $h1/h2 = 0.2$; 2,2': $h1/h2 = 0.6$; 3,3': $h1/h2 = 1.0$

the ground plane or simply glued onto a thick substrate with low permittivity and back side metallization. For $h_2 \gg h_1$ and $\varepsilon_{r2} \ll \varepsilon_{r1}$ the excitation of surface wave modes is efficiently suppressed, but high dispersion has to be accepted [1.47–49]. Another distinctive advantage of these structures is the wide range of impedance values achievable making them particularly suited for filters.

For $\varepsilon_{r2} = 1$ closed-form expressions can be given for Z_0 and $\varepsilon_{r,eff}$ (Sect. 1.5). For a suspended microstrip and for an inverted microstrip, Z_0 and $\varepsilon_{r,eff}$ are plotted in Figs. 1.7 and 1.8, respectively.

Slot Lines

Unlike the strip line and the microstrip line, the slot line is a balanced transmission line. Slot lines suffer from high dispersion, high radiation loss, and low-power handling capabilities. They find, however, interesting applications as resonators [1.38] and resonant antennas [1.39,40,50]. A hybrid mode propagates along the slot with the longitudinal component of the electric field much smaller than the transverse components. All the components of the magnetic field are of the same order of magnitude. Thus, the fundamental mode of the slot line is a quasi-TE mode. In general, due to the hybrid nature of the fundamental mode, treatment of the slot line requires a full wave analysis. *Cohn* [1.51] presented a simplified approach regarding the slot-line as a dielectric-backed diaphragm in a hollow waveguide. He gave closed-form expressions found by a functional approximation, which are

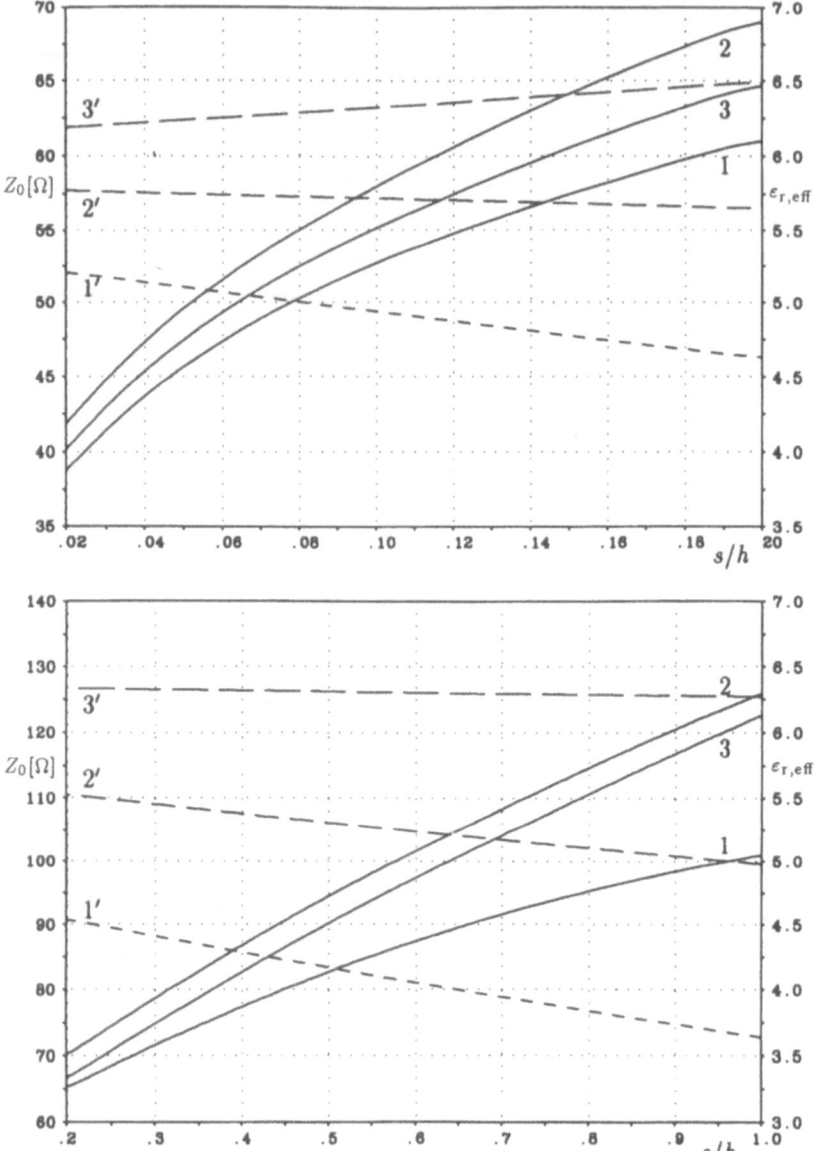

Fig. 1.9. Characteristic impedance (———) and effective dielectric constant (- - - -) of slot line on silicon: 1,1': $h/\lambda_0 = 0.015$, 2,2': $h/\lambda_0 = 0.04$, 3,3': $h/\lambda_0 = 0.075$

summarized in Sect. 1.5. Figure 1.9 shows plots of Z_0 and $\varepsilon_{r,eff}$ of a slot line on a silicon substrate. *Cohn* gave different formulas for the ranges below and above $s/h = 0.2$. This results in small steps of the curves at $s/h = 0.2$.

Dispersion characteristics for wide slot lines on low-permittivity substrates were reported in [1.52].

Coplanar Lines and Coplanar Strip Lines

A coplanar geometry implies that all the conductors are in the same plane, i.e. the top surface of the substrate. Coplanar waveguides find extensive applications in millimeter-wave integrated circuits. They extend substantially the flexibility in the circuit design and exhibit some distinctive advantages. Shunt and series connections are equally easy and no via holes are needed. The coplanar waveguide is an unbalanced transmission line while its complementary, the coplanar strip line, is a balanced transmission line. Their radiation losses and power-handling capabilities lie between the values for the microstrip line and the slot line. The total loss of the coplanar waveguide increases with increasing characteristic impedance. The coplanar strip line behaves vice versa. The fundamental mode propagating along coplanar lines at low frequencies is quasi-TEM and may be interpreted as an even coupled slot-line mode. With increasing frequency the fields and surface currents concentrate around the slots and finally the odd coupled slot-line mode may occur.

Coplanar lines have been investigated theoretically and experimentally [1.53,54] and particularly with respect to applications in hybrid and monolithic integrated circuits [1.55–58]. For $w, s, h \ll \lambda_0$ closed-form expressions for Z_0 and $\varepsilon_{r,eff}$ can be found employing a quasi-static approach (Sect. 1.5). The effective dielectric constants of the coplanar waveguide and the coplanar strip line are equal. Figures 1.10 and 1.11 show plots of Z_0 and $\varepsilon_{r,eff}$ of a coplanar waveguide and a coplanar strip line on silicon, respectively.

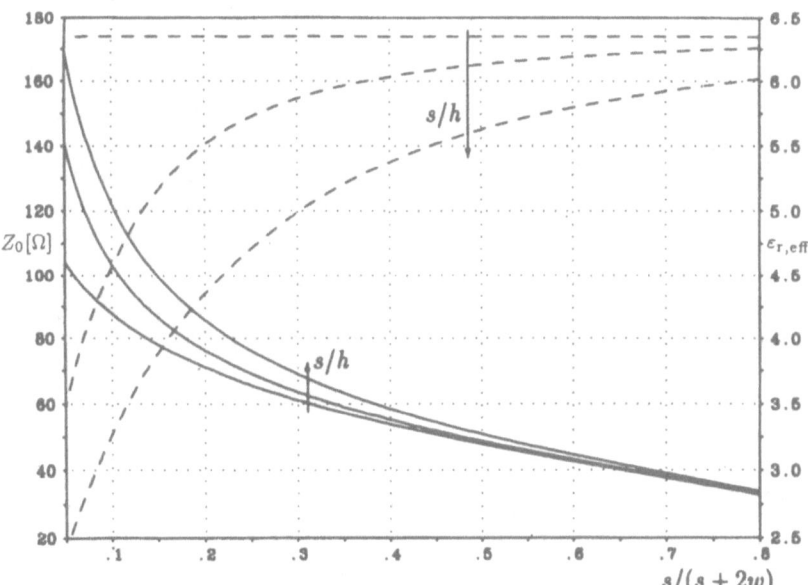

Fig. 1.10. Characteristic impedance (———) and effective dielectric constant (-----) of coplanar waveguide on silicon: $s/h = 0.01, 0.5, 1.0$

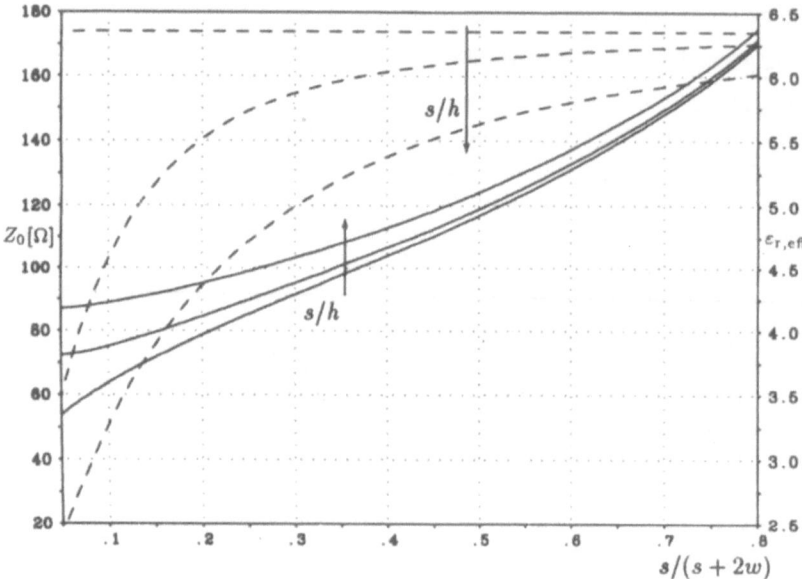

Fig. 1.11. Characteristic impedance (———) and effective dielectric constant (-----) of coplanar stripline on silicon: $s/h = 0.01, 0.5, 1.0$

Microshield Lines

The microshield line is fabricated on a SiO_2-Si_3N_4-SiO_2 membrane [1.59,60]. The membrane is grown on a silicon substrate. Under the line the substrate is removed completely by wet etching. Figure 1.12 exhibits a cross-sectional view of a microshield line. The etching process is anisotropic and yields pyramidal cavities with very steep side walls. The advantage of the microshield line are the low dielectric losses and the low radiation losses at discontinuities. The microshield line can be fabricated completely in monolithic technology. Closed-form expressions for the microshield-line characteristic impedance have been derived using a conformal mapping method [1.59] and results are plotted in Fig. 1.13.

The microshield geometries are not limited to one-dimensional designs. Using anisotropic etching of silicon T's and crosses may be fabricated.

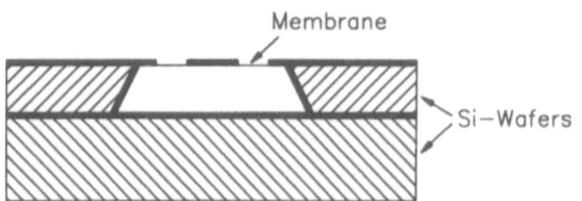

Fig. 1.12. Cross-sectional view of a microshield line

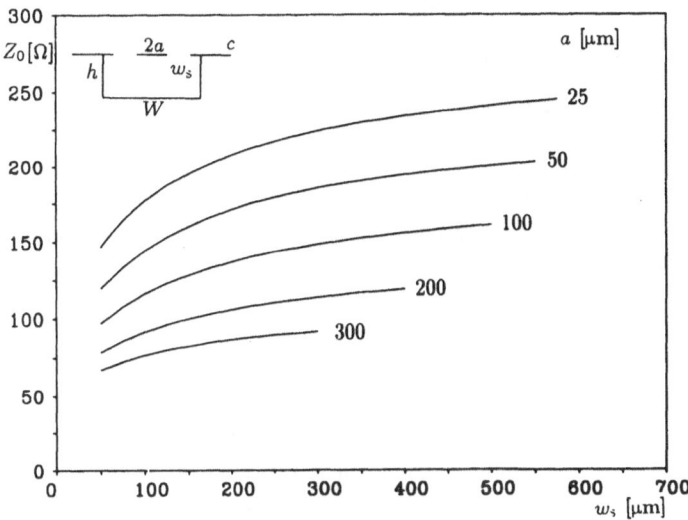

Fig. 1.13. Characteristic impedance of microshield line vs. slotwidth w_s

Full Wave Analysis of Transmission Lines

The spectral domain approach is one of the most efficient full-wave methods for the analysis of transmission line characteristics in the millimeter-wave region [1.61]. The problem may be stated in an electric field integral equation, which reduces to an algebraic equation

$$-\begin{bmatrix} \tilde{E}_x \\ \tilde{E}_z \end{bmatrix} = \begin{bmatrix} \tilde{G}^E_{xx} & \tilde{G}^E_{xz} \\ \tilde{G}^E_{yz} & \tilde{G}^E_{zz} \end{bmatrix} \cdot \begin{bmatrix} \tilde{J}_x \\ \tilde{J}_z \end{bmatrix} \tag{1.17}$$

in the spectral domain, where x and z are the transversal and the longitudinal direction, respectively. The tilde indicates the spectral representation. The components of the dyadic Greens function are given by

$$\tilde{G}^E_{xx} = \frac{k_x^2}{k_t^2}\tilde{G}^e + \frac{k_z^2}{k_t^2}\tilde{G}^h, \tag{1.18}$$

$$\tilde{G}^E_{xz} = \tilde{G}^E_{zx} = \frac{k_x k_z}{k_t^2}(\tilde{G}^e - \tilde{G}^h), \tag{1.19}$$

$$\tilde{G}^E_{zz} = \frac{k_z^2}{k_t^2}\tilde{G}^e + \frac{k_x^2}{k_t^2}\tilde{G}^h, \tag{1.20}$$

$$k_t^2 = k_x^2 + k_z^2. \tag{1.21}$$

The scalar Greens functions $\tilde{G}^e$ and $\tilde{G}^h$ may be found by the solution of the wave equation for the layered dielectric medium or by the transmission-line analogy [1.61]. Applying now Galerkin's method to (1.17) yields the homogeneous linear equation system

$$\begin{bmatrix} [\int \int \tilde{J}_{x,i}\tilde{G}_{xx}\tilde{J}_{x,j}dk_xdk_z] & [\int \int \tilde{J}_{x,i}\tilde{G}_{xz}\tilde{J}_{x,j}dk_xdk_z] \\ [\int \int \tilde{J}_{z,i}\tilde{G}_{zx}\tilde{J}_{x,j}dk_xdk_z] & [\int \int \tilde{J}_{z,i}\tilde{G}_{zz}\tilde{J}_{z,j}dk_xdk_z] \end{bmatrix} \cdot \begin{bmatrix} [A_{x,i}] \\ [A_{z,i}] \end{bmatrix} = 0,$$

$$(1.22)$$

where $\tilde{J}_{x,i}$ and $\tilde{J}_{z,i}$ represent the x-directed and z-directed basis (and weighting) functions and $A_{x,i}$, $A_{x,j}$ the corresponding amplitude coefficients, respectively. In order to find a nontrivial solution one has to solve an eigenvalue problem. Starting with an assumed value for the longitudinal wavenumber k_z the determinant of the matrix is calculated. Applying a root-seeking procedure yields the correct values of k_z (eigenvalue). From k_z the effective permittivity can be calculated. Subsequently the vectors $[A_{x,i}]$ and $[A_{z,i}]$ (eigenvectors) may be determined. With $A_{x,i}$, $\tilde{J}_{x,i}$, $A_{z,i}$, and $\tilde{J}_{z,i}$ the current distribution is given. Insertion of the current distribution in (1.17) yields the electric field in the spectral domain. By application of Parseval's theorem the carried power P may be calculated in the spectral domain. The integral over the current distribution gives the total current I_0. Then the characteristic impedance of the transmission line is given by

$$Z_0 = \frac{P}{I_0^2}. \qquad (1.23)$$

1.2.3 Planar Transmission-Line Discontinuities

Typical discontinuities in planar circuits are open circuits and short circuits, bends, step change in strip and/or slot width, and T- and cross-junctions. Since the discontinuity dimensions are usually much smaller than a wavelength, the discontinuities may be modelled by lumped-element equivalent circuits. At the discontinuities electric and magnetic energy is stored and power is scattered into radiation and surface-wave modes. Usually the power leakage may be neglected and the lumped-element equivalent circuit contains only reactive elements accounting for the energy storage. The longitudinal dimensions of open circuits, short circuits, and changes in width are very short. Thus, the equivalent circuit consists of a single shunt or series-connected reactance located at the point of the discontinuity. Discontinuities of some longitudinal extent such as junctions and bends may be represented by a π- or T-network of reactances.

For a variety of strip line and microstrip line discontinuities equivalent circuits and closed-form expressions for the reactances have been found [1.62,28]. However, very few models are available on coplanar waveguides and slot lines. In [1.63] open circuits, short circuits, and step changes in width of coplanar waveguides have been analysed by means of a 3-D finite-difference method and some plots for the values of the equivalent reactances have been presented. Open-end and short-end coplanar waveguide stubs for filter applications have been characterized theoretically and experimentally in [1.64]. The theoretical method is based on a space-domain integral equation. The stubs are modelled by an equivalent circuit containing three capacitances and three inductances. Radiation losses of short-circuited coplanar waveguides have been investigated theoretically and experimentally [1.65]. The resistance is typically $0.025 \cdot Z_0$ for a symmetric coplanar waveguide and increases with increasing frequency and with increasing asymmetry (ratio of the width of the slots on either side of the core conductor). In [1.58] the mode conversion at coplanar waveguide discontinuities has been studied and it has been found that depending on the type of the discontinuity substantial conversion can occur.

A comparison between a short-circuited slot line and an open-circuited microstrip line was given in [1.66]. The open-circuited microstrip line exhibits greater loss than the short-circuited slot line for the same substrate height and the same width of slot and strip. This is probably due to a stronger excitation of the TM_0 surface wave mode by the microstrip open-end than by the slot-line short-circuit.

1.2.4 Planar Resonators

At low frequencies up to 10 GHz resonators can be built by lumped elements. In the millimeter-wave region distributed resonators have to be used, which are most commonly based on transmission lines. The various forms of transmission-line resonators can be divided into two classes, one-port resonators and two-port resonators. Two-port resonators are frequently employed as transmission components providing filtering and signal shaping. One-port resonators find extensive applications in oscillators. A conventional one-port resonator consists of a piece of closed transmission line like a hollow waveguide or a coaxial line, which is shorted on both ends by a conducting plane. In the resulting cavity electric and magnetic energy is stored in the electric and magnetic fields, respectively. The energy storage can be described by a network of lumped elements, as shown in Fig. 1.14, where the capacitance C stands for the storage of electrical energy and the inductance L for the storage of magnetic energy. The resistance R describes the energy loss due to dielectric and ohmic losses. The equivalent circuit representation is only valid in the vicinity of the resonant frequency. Figure 1.14a depicts the equivalent circuit for a parallel resonance, which applies when the resonator length is an odd multiple of an half wavelength $\lambda/2$. The series network in Fig. 1.14b appropriates, when the resonator length is an even multiple of $\lambda/2$.

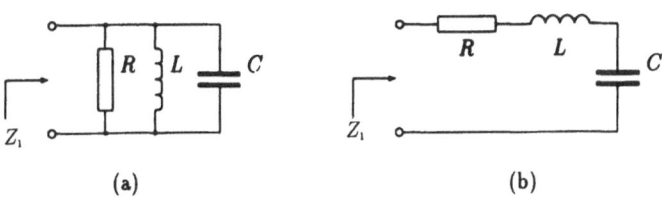

(a) (b)

Fig. 1.14. Equivalent circuit of microstrip resonator for (a) parallel and (b) series resonance

Unlike cavity resonators, the planar resonators are open structures and suffer from energy leakage into radiating and surface-wave modes. Thus, in the internal quality factor Q the radiation losses have to be taken into account additionally. In contrast to the dielectric and ohmic losses, however, the radiated energy is not dissipated, but can be used for signal transmission. Planar resonant antennas take advantage of this phenomenon and will be treated later. In principle, all the planar transmission-line types discussed in Sect. 1.2.2 can be utilized as basis structure for planar resonators. However, the microstrip structure is most extensively used due to its low radiation loss. The slot resonator finds interesting applications as a resonant antenna and will be discussed in detail in Sect. 1.3.1. Coplanar resonators have been employed in high-T_c superconducting oscillators [1.67,68]. In this text we will give some fundamental data on the basic microstrip resonating structures, the rectangular microstrip and the circular disc.

Rectangular Microstrip Resonator

Figure 1.15 shows a rectangular microstrip resonator. The resonator consists of a rectangular metallized strip of length l and width w on a grounded substrate of height h with the relative dielectric constant ε_{r1}. For $l \gg w$ the resonator can be viewed as an open-ended transmission line of length l. The transmission line will be resonant, if

$$l + 2\Delta l = n\frac{\lambda_g}{2}, \quad n = 1, 2, 3, \ldots \tag{1.24}$$

with the guide wavelength

$$\lambda_g = \frac{\lambda_0}{\sqrt{\varepsilon_{r,\text{eff}}}}. \tag{1.25}$$

The open ends of the transmission line are not truly open. Due to fringing fields the open ends behave like capacitances that can be modelled as an extension Δl at both ends of the microstrip line. This has been taken into account in (1.24). An approximate expression for Δl on silicon, based on a formula evaluated by

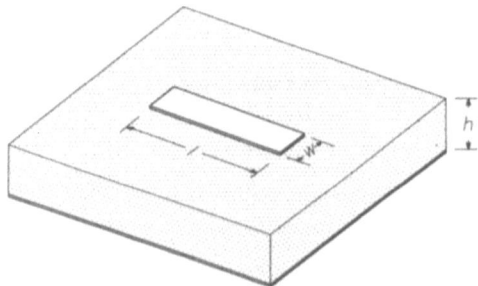

Fig. 1.15. Microstrip rectangular resonator

Hammerstad [1.44], is given by

$$\frac{\Delta l}{h} = 0.432\frac{w/h + 0.262}{w/h + 0.813}. \tag{1.26}$$

with an error less than 5% for $0.3 \leq w/h \leq 2$ and $1 \leq \varepsilon_{r1} \leq 50$. Substituting (1.25) in (1.24) yields the resonance frequency for the case of silicon substrates

$$f_0 = \frac{c_0}{2\sqrt{\varepsilon_{r,\text{eff}}}(l + 2\Delta l)} \tag{1.27}$$

with Δl from (1.26) and $\varepsilon_{r,\text{eff}}$ from (1.49).

A more accurate analysis of the rectangular resonator by means of a spectral domain approach can be found in [1.69].

Circular Microstrip Disc Resonator

The geometry of a circular microstrip disc resonator consisting of a disc of radius a on a dielectric substrate with height h and relative dielectric constant ε_{r1} is shown in Fig. 1.16. For electrically thin substrates ($h \ll \lambda_0$) the disc resonator can be modelled as a cylindrical cavity with electrical walls on top and bottom, and a magnetic wall at the circumference. The fields in the cavity correspond to those of the TM_{nm0} modes of the circular hollow waveguide resonator. The resonance frequency is given by

$$f_{0,nm0} = \frac{\xi'_{nm}c_0}{2\pi a_{\text{eff}}\sqrt{\varepsilon_{r1}}} \tag{1.28}$$

where ξ'_{nm} is the mth root of the equation $J'_n(ka) = 0$, and J'_n is the first derivative with respect to the argument of the Bessel function of the first kind and order n. The first few roots ξ'_{nm} are listed in Table 1.2. The fringing fields at the edge of the disc have to be taken into account in a manner similar to the open-ended microstrip. Thus, the calculation of the resonance frequency with (1.28) is based on an effective disc radius that is slightly larger than the physical radius a. An approximative expression for a_{eff} is given by

$$\left(\frac{a_{\text{eff}}}{a}\right)^2 = 1 + \frac{2h}{\pi a \varepsilon_{r1}}\left(\ln\frac{\pi a}{2h} + 1.7726\right). \tag{1.29}$$

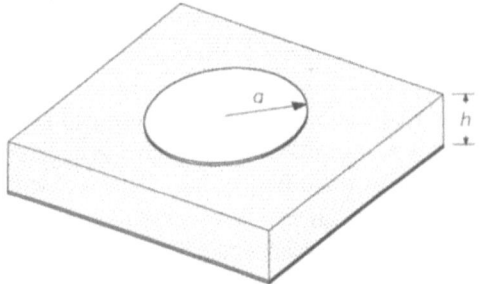

Fig. 1.16. Microstrip disc resonator

Table 1.2. Roots of the equation $J_n'(ka) = 0$

ξ_{nm}'	$m = 1$	$m = 2$	$m = 3$
$n = 0$	3.832	7.016	10.173
$n = 1$	1.84	5.33	8.54
$n = 2$	3.054	6.706	9.969

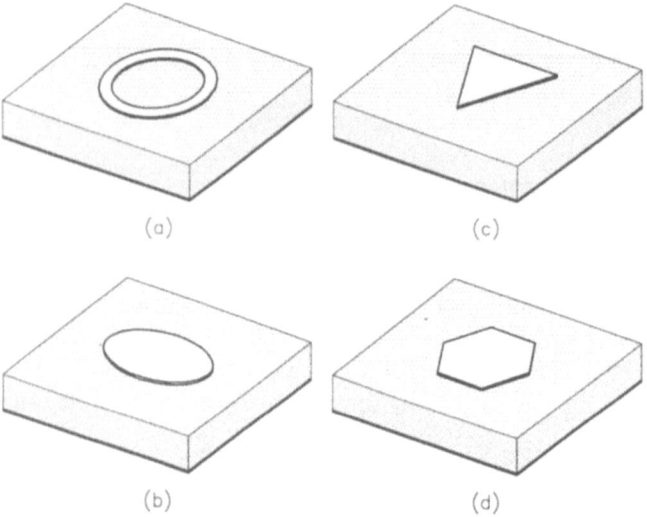

Fig. 1.17. Some geometries of microstrip resonators: (a) annular ring, (b) elliptic patch, (c) trigonal patch, (d) hexagonal patch

An accurate analysis of the circular-disc resonator by means of a spectral domain approach has been reported in [1.70].

Resonators with more complex geometrical shapes are depicted in Fig. 1.17. These structures have been extensively analysed in the past. Excellent design information is available for the circular ring [1.71–73], the elliptic disc and elliptic ring [1.74–76], the triangular microstrip [1.77–79], and the hexagonal microstrip [1.80].

1.3 Planar Millimeter-Wave Antennas on Silicon

A variety of antennas at millimeter-wave frequencies has been designed and studied in recent years. Initially conventional microwave antennas have been used just by scaling from microwave- to millimeter-wave frequencies. However, with the progress in integration of millimeter-wave circuits new demands arose on millimeter-wave antennas such as small size, light weight, ruggedness, reliability, and low cost. Compatibility to integrated active circuits requires a planar geometry of the radiating elements. The resulting planar antennas are based on a common principle, the perturbation of a planar transmission line such that it radiates in the desired manner. Further development of integration has led to integrated antennas, where solid-state devices are integrated with the radiating elements in one unit that is directly coupled to the electromagnetic field.

The antenna gain of single antenna elements is not sufficient in many applications. Thus, the beam has to be formed either by external devices such as horns or lenses or by use of antenna arrays. Antenna arrays will be discussed in Sect. 1.3.2.

1.3.1 Antenna Elements

In this subsection we will discuss some basic planar antenna structures, the microstrip patch antenna, the resonant slot antenna, and the bow-tie antenna. The microstrip patch is the most commonly used planar millimeter-wave antenna and the basic antenna element for antenna arrays. The resonant slot is very attractive for monolithic integrated antennas, since the active devices can be placed within the radiating aperture and the slot simultaneously acts as a resonator. The bow-tie antenna is an example of a non-resonant planar antenna and particularly useful in broad-band millimeter-wave receivers.

The intention of this subsection is to discuss basic design concepts of planar millimeter-wave antennas and their theoretical treatment in an exemplary manner. For additional information about millimeter-wave antennas we refer to [1.81–86].

Patch Antennas

Microstrip patch antennas have been studied extensively at microwave frequencies and are well understood by now. They are easy to fabricate, compatible with integrated microstrip circuits, and low cost. The application of these antennas, however, is limited by their narrow bandwidth. At millimeter-wave frequencies they suffer from excitation of surface-wave modes and from losses in the feed lines. Excitation of surface-wave modes results in poor radiation efficiency and in mutual coupling of antenna elements in antenna arrays.

Typical layouts of microstrip patch antennas with microstrip feed and coaxial feed are shown in Fig. 1.18. The patch is about a half electrical wavelength long and has the width w. In the simplest analytical description, the transmission-line model, the patch is regarded as a line resonator with no transversal field variations. Radiation occurs from the fringing fields at the open ends of the resonator. The transmission-line equivalent circuit of the patch antenna consists of two admittances that are connected by an ideal transmission line. The admittances account for radiation and storage of electrical energy at the open ends, thus consisting of a conductance and a capacitance in parallel. The capacitance is given by

$$C = \Delta l \frac{\sqrt{\varepsilon_{r,\text{eff}}}}{c_0 \eta_0} \tag{1.30}$$

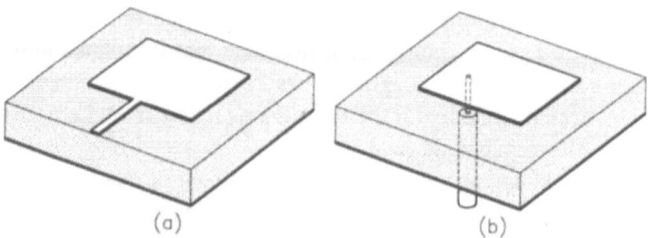

Fig. 1.18. Microstrip rectangular patch antenna: (a) microstrip feed, (b) coaxial feed

with Δl form (1.26), and $\varepsilon_{\mathrm{r,eff}}$ from (1.49). An approximate expression for the conductance is given by

$$
G = \begin{cases} \dfrac{w^2}{90\lambda_0^2}, & \text{for } w \ll \lambda_0 \\[2ex] \dfrac{w}{120\lambda_0}, & \text{for } w \gg \lambda_0. \end{cases} \tag{1.31}
$$

The input impedance of the patch antenna at resonance is $Y_{\mathrm{i}} = 2G$ for a feed point at the edge of the patch. This simple approach cannot be applied to other geometries and neglects the transversal field components.

A more accurate description is obtained by the modal-expansion technique. This approach was discussed in detail in [1.87] and will here be sketched only briefly. The patch is considered as a cavity with electrical walls at the top and the bottom, and leaky magnetic walls at the circumference. For $h \ll \lambda_0$ the modes of the cavity are TM_{mnl} with $l = 0$. The field between the patch and the ground plane is expanded in terms of cavity resonant modes

$$
E_z(x, y) = \sum_m \sum_n A_{mn} e_{mn}(x, y) \tag{1.32}
$$

where A_{mn} are the mode amplitude coefficients, and e_{mn} are the z-directed orthonormalized electric field mode vectors. The modal expansion functions satisfy the Helmholtz equation and the boundary conditions for non-radiating cavities yielding real eigenvalues k_{mn}. Expansion functions and eigenvalues for patches of various shapes have been given in [1.88]. In the case of leaky magnetic walls, the magnetic field vectors have no longer a zero tangential component on the cavity walls. Since the perturbation is very small, the electric field representation in (1.32) is still valid. The eigenvalues, however, become complex.

A coaxial line feed is modelled by a z-directed current probe at (x_0, y_0). Then, the input impedance seen by the feed line is given by

$$
Z_{\mathrm{i}} = -j\eta_0 kh \sum_m \sum_n \frac{\psi_{mn}^2(x_0, y_0)}{k^2 - k_{mn}^2} G_{mn} \tag{1.33}
$$

where $\psi_{mn}(x_0, y_0) = \sqrt{\varepsilon_{\mathrm{r1}}\varepsilon_0 h} \, e_{mn}(x_0, y_0)$ and the factor G_{mn} accounts for the width of the feed. The term $m, n = 0, 0$ with $k_{00} = 0$ (1.33) gives the static capacitance in parallel with a resistance to represent the loss in the substrate. The term $m, n = 1, 0$ represents the dominant TM_{10} mode.

How to find the complex eigenvalue k_{mn} depends on the geometrical shape of the patch. For rectangular and circular patch some approaches were discussed in [1.87].

By means of a full wave analysis patch antennas of arbitrary shape and almost arbitrary feed system can be treated. A full wave analysis requires considerable numerical effort, but in many cases it is the only way to obtain correct results. Here, we refer to the literature, where numerous publications can be found treating microstrip antennas by means of various field-theoretic approaches, such as the integral-equation method, spectral domain approach, method of lines, Transmission Line Matrix method (TLM), mode matching, and combinations of these [1.89–92].

Slot Antennas

Beyond all the advantages of microstrip antennas there is always the drawback of a small bandwidth. Slot antennas are considered to have a larger bandwidth. Thus, planar slot antennas would be promising alternatives, when large bandwidth is required. In a typical application in the millimeter–wave region the slot antenna is fed by a negative impedance amplifier, i.e. an IMPATT diode, as schematically depicted in Fig. 1.19. The slot is one electrical wavelength long and acts simultaneously as a resonator and an antenna. The feed point is in the center of the slot. In such applications the behaviour of the antenna impedance seen by the active element must be known precisely in order to find the optimum design.

Unlike microstrip antennas, there hasn't been a lot of investigations on the characteristics of planar slot antennas. In particular, no closed-form expressions for the characteristic parameters are provided in the literature. In [1.89] and [1.91] results of a rigorous full-wave analysis of half-wave and full-wave planar slot antennas are reported. A spectral-domain approach has been employed, which is based on a Green's function formalism for the magnetic fields. These Green's functions are obtained by a transmission line analogy [1.61] based on impressed

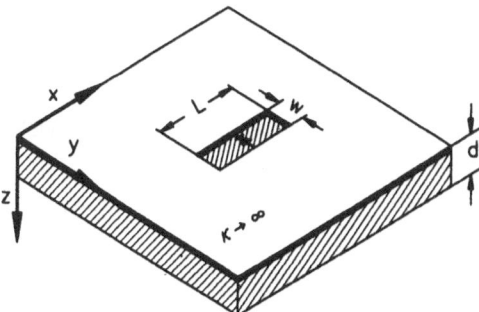

Fig. 1.19. Schematic structure of integrated slot antenna

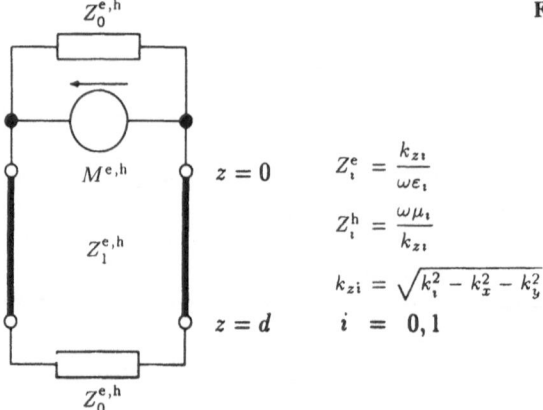

Fig. 1.20. Transmission line analogy

$$Z_i^e = \frac{k_{zi}}{\omega \varepsilon_i}$$

$$Z_i^h = \frac{\omega \mu_i}{k_{zi}}$$

$$k_{zi} = \sqrt{k_i^2 - k_x^2 - k_y^2}$$

$$i = 0, 1$$

voltage sources, as shown in Fig. 1.20. This results in

$$\begin{bmatrix} \tilde{H}_x \\ \tilde{H}_y \end{bmatrix} = \begin{bmatrix} \tilde{G}_{xx}^H & \tilde{G}_{xy}^H \\ \tilde{G}_{yx}^H & \tilde{G}_{yy}^H \end{bmatrix} \cdot \begin{bmatrix} \tilde{M}_x \\ \tilde{M}_y \end{bmatrix}, \tag{1.34}$$

where $\tilde{H}$ is the magnetic field and $\tilde{M}$ the magnetic current density in the slot. We use the tilde as a reminder that the spectral representation is used.

Since the slot width is far below a wavelength ($\omega \ll \lambda_0$), the transverse components are neglected. This simplifies (1.34) to

$$\tilde{H}_x = \tilde{G}_{xx}^H \tilde{M}_x. \tag{1.35}$$

The Green's function is given by

$$\tilde{G}_{xx}^H = \frac{k_x^2}{k_t^2} \tilde{G}^h + \frac{k_y^2}{k_t^2} \tilde{G}^e, \tag{1.36}$$

where

$$k_t = \sqrt{k_x^2 + k_y^2}, \tag{1.37}$$

$$\tilde{G}^{e,h} = \frac{1}{Z_A^{e,h}} + \frac{1}{Z_B^{e,h}} \tag{1.38}$$

with

$$Z_A^{e,h} = Z_0^{e,h}, \tag{1.39}$$

$$Z_B^{e,h} = Z_1^{e,h} \cdot \frac{Z_0^{e,h} + j Z_1^{e,h} \tan(k_{z1} d)}{Z_1^{e,h} + j Z_0^{e,h} \tan(k_{z1} d)}. \tag{1.40}$$

Applying the Galerkin's method with piecewise sinusoidal basis functions yields the linear equation system

$$\begin{bmatrix} J_1 \\ \vdots \\ J_N \end{bmatrix} = \begin{bmatrix} \int\int B_1 \tilde{G}^{xx} B_1 dk_x dk_y & \cdots & \int\int B_1 \tilde{G}^{xx} B_N dk_x dk_y \\ \vdots & \ddots & \vdots \\ \int\int B_N \tilde{G}^{xx} B_1 dk_x dk_y & \cdots & \int\int B_N \tilde{G}^{xx} B_N dk_x dk_y \end{bmatrix} \cdot \begin{bmatrix} A_1 \\ \vdots \\ A_N \end{bmatrix}, \tag{1.41}$$

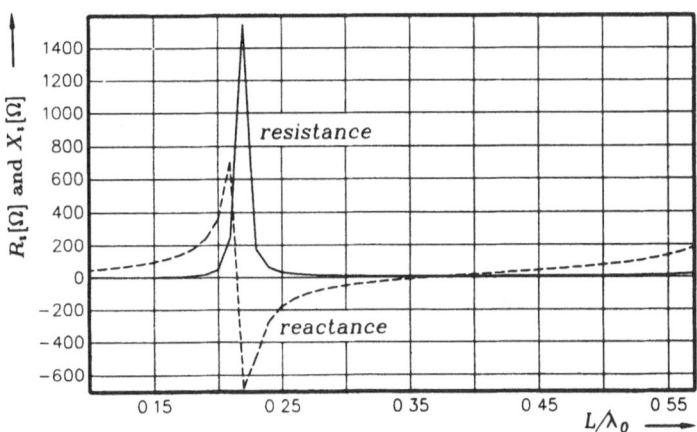

Fig. 1.21. Slot impedance $Z_1(R_1 + jX_1)$ for silicon ($w = L/14$, $d = L/11.2$)

where the excitation vector $\vec{J} = \vec{n} \times \vec{H}$ represents the impressed currents. B_i are the basis functions and A_i the corresponding weighting coefficients. The IMPATT diode was modelled by an impressed line current located at the center of the slot. Thus, the excitation vector contains one non-zero element

$$\vec{J} = [0, \ldots, 0, 1, 0, \ldots, 0]^T. \tag{1.42}$$

Since the excitation current is normalized to unity, the input impedance Z_i is given by the coefficient of the center basis function.

For two different substrate materials (silicon: $\varepsilon_r = 11.6$, PTFE (teflon): $\varepsilon_r = 2.55$) the input impedance versus various parameters has been calculated. Figure 1.21 depicts the input impedance Z_i versus L/λ_0, where λ_0 represents the free space wavelength and L the length of the slot. The slot width is $w = L/14$ and the substrate thickness is $d = L/11.2$. As can be seen the half-wave ($L/\lambda_0 \doteq 0.22$) and full-wave ($L/\lambda_0 \doteq 0.37$) resonance represents a parallel-type and series-type resonance, respectively. Since the input resistance at the parallel-type resonance is very high, a full-wave slot has to be used for IMPATT diode oscillators. The resistance of the resonator seen by the IMPATT diode terminals should be typically in the range of $2 \ldots 5\,\Omega$.

In the next two figures the behaviour of the input resistance at the series-type resonance is shown versus the substrate height d/λ_0 (Fig. 1.22) and the slot width ω/λ_0 (Fig. 1.23), respectively. A minimum of the resistance is obtained for $d/\lambda_0 = 0.037$ for silicon and $d/\lambda_0 = 0.07$ for PTFE. That means, there exists an optimum substrate thickness where the quality factor of the resonator becomes maximum. Since there is only a small variation of the resistance around the minimum, the accuracy of the substrate thickness is not very critical. This is important for reproducible design of slot antennas.

As can be seen in Fig. 1.23, decreasing the width of the slot yields an almost linear decrease of the resonance resistance. Thus, a desired resonance resistance can easily be adjusted by choosing the proper slot width.

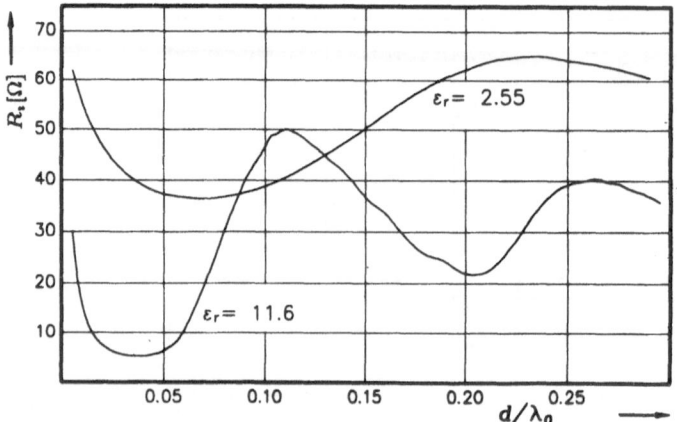

Fig. 1.22. Resonance resistance R_1 for PTFE and silicon ($w = L/14$)

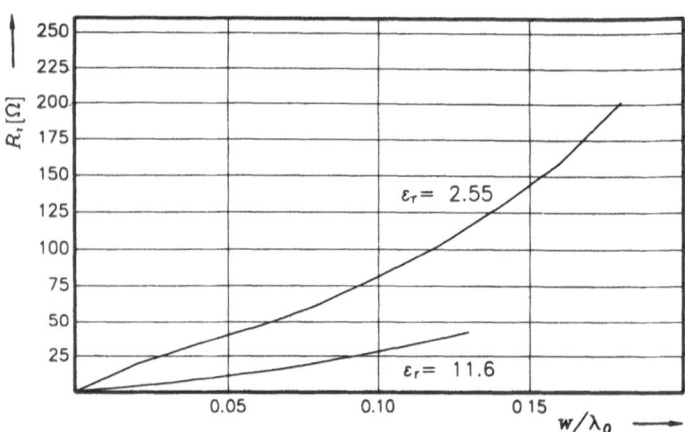

Fig. 1.23. Resonance resistance R_1 for PTFE and silicon ($d = L/11.2$)

Bow-Tie Antennas

An example of a broad-band planar millimeter-wave antenna is the bow-tie antenna. The typical layout is depicted in Fig. 1.24. Investigations of bow-tie antennas on a dielectric half space were reported in [1.93]. For a $2\Phi_0 = 60°$ bow-tie on fused quartz ($\varepsilon_r = 3.83$) the radiated power in the dielectric is 20 times larger than the power radiated into free space. This radiation may be coupled out of the dielectric with a substrate lens (Fig. 1.25). Bow-tie antennas are well suited for integrated antenna designs, where the detector is placed in the apex of the antenna.

The quasistatic impedance and the quasistatic wave number of an infinite bow-tie on a dielectric halfspace is obtained by conformal mapping as

$$Z_{qs} = \eta_0 \sqrt{\frac{2}{\varepsilon_{r1} + 1} \frac{K(k)}{K'(k)}}, \tag{1.43}$$

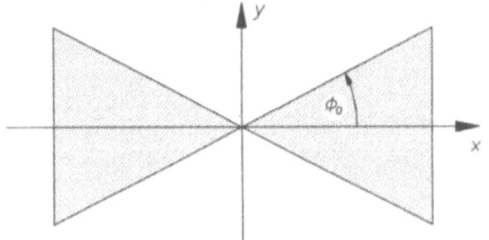

Fig. 1.24. Layout of a bow-tie antenna

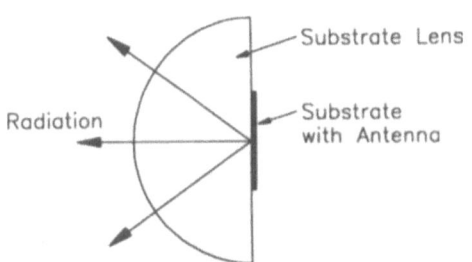

Fig. 1.25. Substrate lens

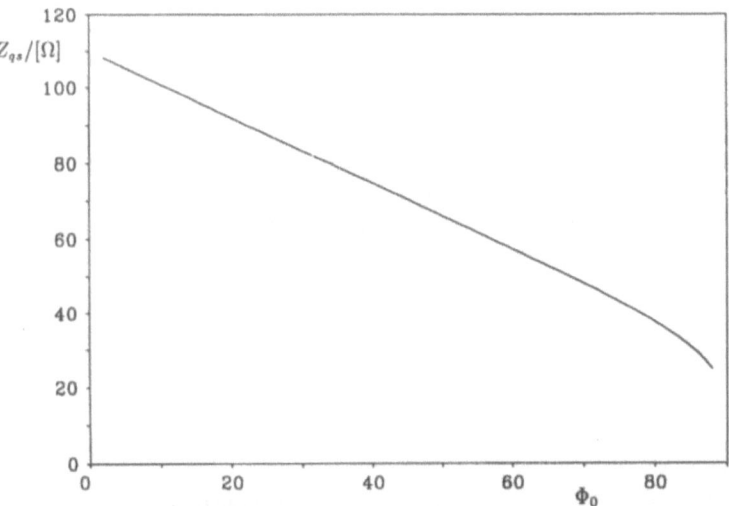

Fig. 1.26. Characteristic impedance of infinite bow-tie versus bow angle

$$k_{qs} = k_0 \sqrt{\frac{\varepsilon_{r1} + 1}{2}} \tag{1.44}$$

with the argument $k = \tan^2(45° - \Phi_0/2)$ of the elliptic integrals. An approximate expression for $K(k)/K'(k)$ gives (1.45). Figure 1.26 shows Z_{qs} versus the bow-half-angle Φ_0. The expressions (1.43 and 44) agree well with microwave measurements of long bow-tie antennas with $2a \gg \lambda_0$. Input impedance and radiation

Fig. 1.27. Layout of a log periodic antenna

pattern of short bow-tie antennas have been calculated by means of the moment method [1.93]. For wide bows the dominant mode is a wave propagating along the axis of the bow at the dielectric wave number $k_d = \sqrt{\varepsilon_{r1}}k_0$. As the bow narrows, the dominant current becomes an edge current with the quasistatic wave number k_{qs}. The current of the bow-tie antenna may be modelled as a traveling-wave current. Due to the strong radiation from this current only little reflection of this current from the ends of the antenna occurs. Thus, the bow-tie antenna can be modelled as a long wire-antenna fed at its center. This makes clear that the radiation pattern contains two major lobes on each side of the antenna plane. In some applications (particularly, in antenna arrays) this behaviour may present problems.

This drawback of bow-tie antennas can be overcome by a planar log periodic antenna design based on a bow-tie antenna, as shown in Fig. 1.27 [1.93,94]. The radiation pattern of this antenna has its maximum normal to the antenna plane and has comparable E- and H-plane beam widths.

Quasi-Planar Antennas

In quasi-planar circuits the substrate is structured vertically in a manner that is completely compatible to monolithic integration. An example is the microshield-line, which has been discussed above. This concept has been used by *Rebeiz* for quasi-planar antennas [1.84]. A cross-sectional view of an integrated horn antenna array is depicted in Fig. 1.28. On the backside of a silicon wafer a silicon-nitride membrane is deposited and subsequently the antenna structures are fabricated. Then, the material below the antenna structures is removed by an anisotropic etching process, resulting in pyramidal holes, which are subsequently coated with gold. A second wafer with pyramidal, gold coated holes is placed behind the first one so that the holes in both of the wafers form horn antennas.

The aperture of the integrated horn antenna ranges between $1.0\lambda_0^2$ and $1.6\lambda_0^2$. The efficiency of the antenna is defined as the power delivered to the load related to the power delivered by a plane wave into the aperture of the antenna. For an aperture of $1.0\lambda_0^2$ an efficiency of 80% has been obtained at 93 GHz [1.95]. On

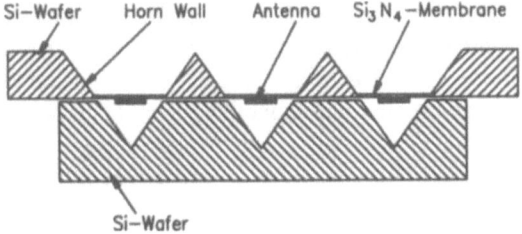

Si–Wafer   Horn Wall   Antenna   Si$_3$N$_4$–Membrane

Si–Wafer

the basis of these antennas integrated horn antenna arrays have been realized for 92 GHz, 240 GHz [1.96] and 802 GHz [1.97].

1.3.2 Antenna Arrays

A common requirement on planar antennas is a high directivity which implies the use of planar antenna arrays. Provided that no mutual coupling between the antenna elements occurs, the radiation pattern of the array can easily be calculated by superposition of the far-field components excited by the antenna elements. At microwave frequencies the mutual coupling is so weak that the radiation pattern of even very large arrays can precisely predicted. A comprehensive review of microstrip arrays can be found in [1.83].

The design of such arrays in the millimeter-wave region involves new problems, since the substrates become electrically thick. Surface waves trapped in the substrate are likely to increase the mutual coupling between the antenna elements. Furthermore, one has to deal with the losses in the feed lines limiting the size of antenna arrays.

Much work has been done to overcome these problems. Mutual coupling between microstrip array elements has been analyzed analytically [1.98,99]. It has been shown that the excitation of surface waves can be suppressed by a substrate/superstrate configuration [1.47,100]. The feed problems can be overcome by two-layer substrates [1.101], where the feed lines are printed on the lower layer which is electrically thin. So the energy guided by the feed lines is tightly bound and only little radiation from the feedlines will occur. The feed system and the patch antennas can be isolated by a ground plane between two dielectric layers. The patches are printed on the top surface of the upper layer and the feed system is printed on the bottom surface of the lower layer. The patches are excited by near-field coupling via slots in the ground plane. This concept is called aperture-coupled microstrip antenna [1.102–107].

A way out of the feed problems of large antenna arrays is the use of integrated antennas as array elements. As transmitters planar MESFET grid oscillators operating at X-band and Ku-band have been fabricated [1.108]. A 16-element grid has produced 335 mW of power at 11.6 GHz. This design was scaled to produce a 36-element grid oscillator with output power of 235 mW at 17 GHz. The planar grid is placed in a Fabry-Perot cavity providing quasi-optical power combining. The radiating array elements are the source and drain leads of the MESFETs. As a receiver a 100-element planar Schottky diode grid mixer has been fabricated [1.109]. At 10 GHz an improvement in the dynamic range of somewhat below 20 dB over an equivalent single-diode mixer has been obtained. The extension of these concepts into the millimeter-wave region and monolithic integration seems to be staightforward.

1.4 Planar Millimeter-Wave Circuits Containing Active and Nonlinear Elements

For the planar millimeter-wave oscillators IMPATT diodes were used as the active elements. IMPATT diodes yield high cw output power at frequencies up to above 100 GHz. In order to obtain a high output power and a narrow spectrum IMPATT diodes must be connected with a low-impedance series resonant circuit. Since the obtainable quality factor of planar resonators typically is below 100 care must be taken in order to obtain a low-impedance planar resonator design. Best results were achieved using a planar circular-disk resonator with the IMPATT diode in the center and a through contact beyond the IMPATT diode to the circuit ground plane [1.8,11,12,14,15,110].

Based upon optimized circuit layouts cw power-output values as high as 20 mW at 93 GHz [1.12] and 200 mW at 73 GHz [1.14,15] were achieved. Figure 1.29 illustrates the layout of the 200 mW oscillator. The circuit dimensions are 6×4.5 mm^2. The substrate thickness is 110 μm. The through metallization beyond the IMPATT diode has been realized using a hybrid technology. The planar circuit has a hole in the center of the disk resonator. Within this hole the IMPATT diode is mounted on a cylindric heat-sink continuation reaching into the hole. Figure 1.30 shows the cross section through the planar resonator with the inserted IMPATT diode. A double-drift IMPATT diode is used in this circuit. At the oscillation frequency of 73 GHz in continuous-wave operation 200 mW output power at an efficiency of 4.5% was achieved.

The dependence of the output power, the efficiency on the dc bias current and the measured power spectrum of the planar 200 mW oscillator are shown in

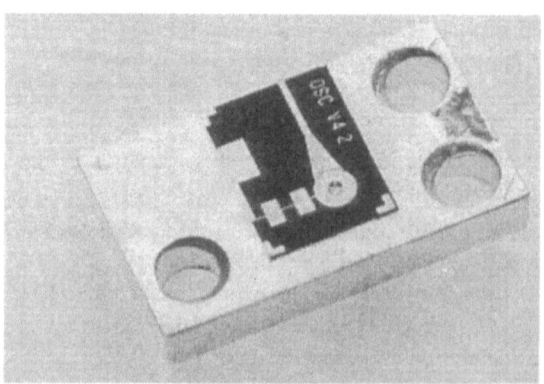

Fig. 1.29. The planar 200 mW IMPATT oscillator

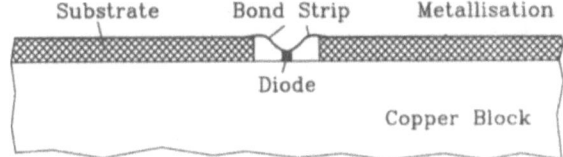

Fig. 1.30. Cross section through the planar resonator with inserted IMPATT diode

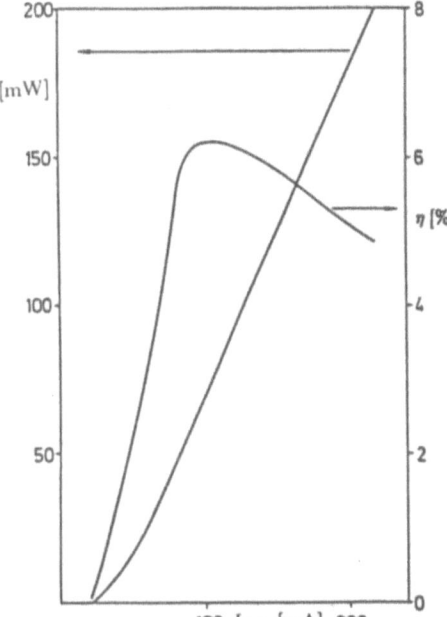

Fig. 1.31. Bias dependence of the cw output power P and the efficiency η

Figs. 1.31 and 1.32, respectively. The substrate thickness of only 110 μm minimizes the resonator impedance. Due to the low impedance of IMPATT diodes and the limited quality factor of planar resonators the impedance level of the resonator should be chosen as low as possible in order to obtain maximum output power. As discussed in detail in [1.111] a lower resonator impedance suppresses the low-frequency bias oscillations up to a higher output level.

In future developments the via hole in the center of the planar IMPATT oscillator may be made by a selective etching process from the ground plane to the epitaxial diode structure. The conical via hole may be filled with gold by a galvanic process. The conical geometry of an etched via hole will be advantageous for the oscillator design, since it reduces the series inductance between IMPATT diode and disk resonator.

Many applications of millimeter-wave oscillators require electronic frequency tuning. Figure 1.33 shows a varactor diode tunable planar IMPATT oscillator [1.19,20]. The oscillator output power is 18 mW at 80.2 GHz. A tuning range of 425 MHz has been achieved.

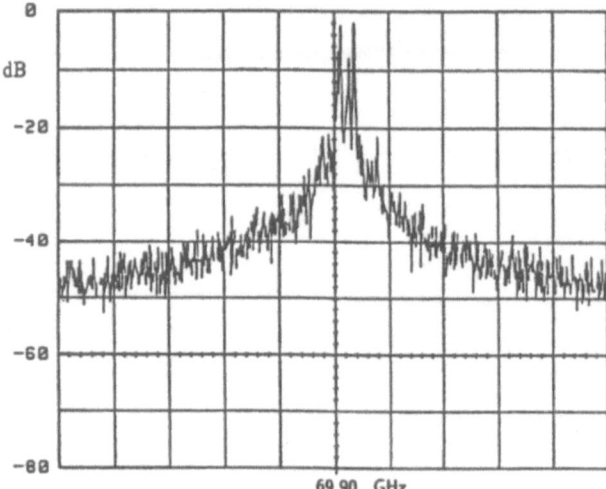

Fig. 1.32. Power spectrum of the 200 mW oscillator

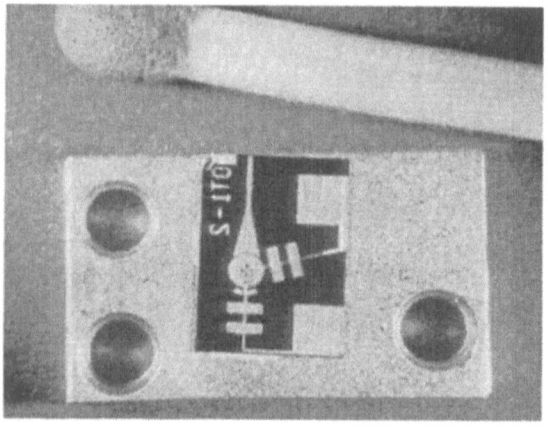

Fig. 1.33. Electronically frequency tunable planar IMPATT oscillator

Using coplanar IMPATT diodes, as shown in Fig. 1.34, no via holes are required. Several coplanar oscillator designs have already been discussed [1.13,20, 38,110]. Figure 1.35 exhibits a coplanar IMPATT oscillator. The coplanar IMPATT diode is connected to the center of a slot line [1.38]. The oscillator was fabricated on a 100 μm substrate in order to minimize the surface wave losses. The slot length is 1.12 mm and the slot width is 80 μm. The IMPATT diode diameter is 20 μm. A radiated output power of 7 mW was measured at 109 GHz.

A 93 GHz SIMMWIC receiver with a planar 36 element antenna and a Schottky diode has been realized [1.12,15]. Figure 1.36 shows the receiver. A coplanar Schottky diode is connected to the antenna via a microstrip line. The Schottky diode is terminated by a λ/4 line. The dc bias lines include low-pass filters. The receiver has been integrated monolithically on a highly insulating

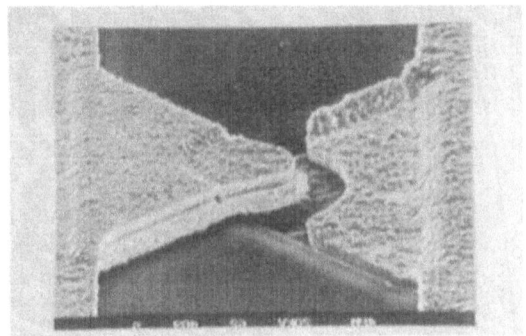

Fig. 1.34. Coplanar IMPATT diode

Fig. 1.35. Coplanar IMPATT oscillator

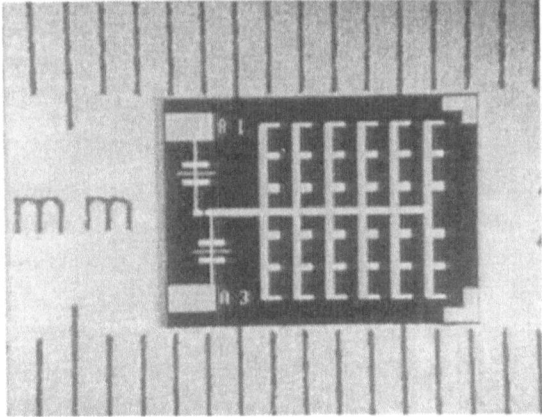

Fig. 1.36. SIMMWIC receiver

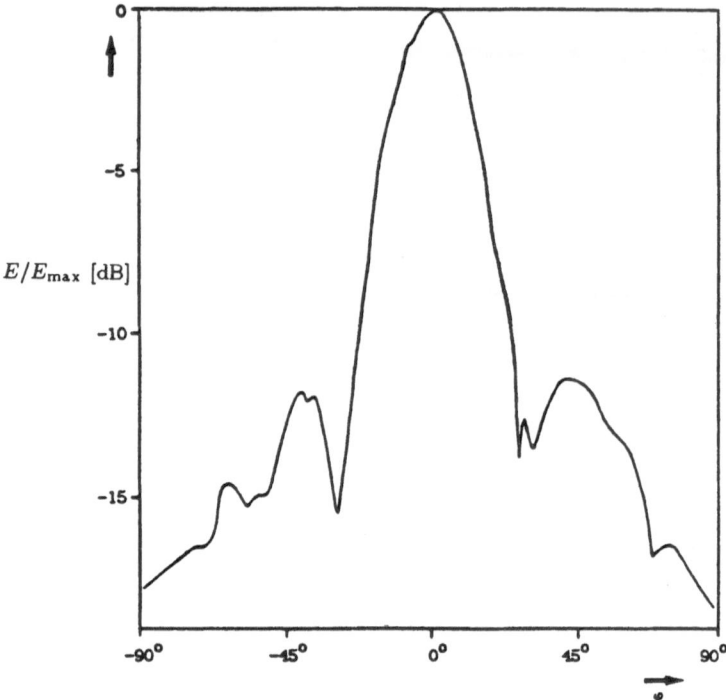

Fig. 1.37. Characteristics of the receiver antenna

silicon substrate. The antenna dimensions are 5.4×5.6 mm^2. The Schottky diode exhibits 4Ω series resistance and a junction capacity of 0.1 pF. Figure 1.37 shows the measured antenna characteristics. The measured receiver sensitivity is $65 \ \mu V \cdot \mu W^{-1} \ cm^2$.

1.5 Appendix: Closed-Form Expressions for Transmission-Line Characteristics

Some of the following expressions have been obtained by conformal mapping. These equations contain elliptic functions $K(k)$ and $K'(k)$. K represents a complete elliptic function of the first kind and K' its complementary function. A very accurate approximation (error less than 10 ppm) for K/K' is given by

$$\frac{K(k)}{K'(k)} = \begin{cases} \left[\dfrac{1}{\pi} \ln \left(2 \dfrac{1+\sqrt{k'}}{1-\sqrt{k'}} \right) \right]^{-1} & \text{for } 0 \leq k \leq \dfrac{1}{\sqrt{2}} \\[3mm] \dfrac{1}{\pi} \ln \left(2 \dfrac{1+\sqrt{k}}{1-\sqrt{k}} \right) & \text{for } \dfrac{1}{\sqrt{2}} \leq k \leq 1 \end{cases} \tag{1.45}$$

with $K'(k) = K(k')$ and $k' = \sqrt{1-k^2}$.

Strip Lines

For an infinitesimally thin center conductor the characteristic impedance

$$Z_0 = \frac{\eta_0}{4\sqrt{\varepsilon_{r1}}} \frac{K'(k)}{K(k)} \quad \text{with} \quad k = \tanh\left(\frac{\pi w}{2h}\right) \tag{1.46}$$

is obtained, where $\eta_0 = \sqrt{\mu_0/\varepsilon_0}$ is the characteristic impedance of free space.

Microstrip Lines

The following expressions are valid within a large range ($0.01 \le w/h \le 100$ and $1 \le \varepsilon_{r1} \le 128$). The characteristic impedance is represented as

$$Z_0 = \frac{Z_0(\varepsilon_{r1} = 1)}{\sqrt{\varepsilon_{r,\text{eff}}}} \tag{1.47}$$

where the effective dielectric constant $\varepsilon_{r,\text{eff}}$ takes into account the ratio of the field in the air to the field in the substrate. For $\varepsilon_{r1} = 1$ the characteristic impedance

$$Z_0(\varepsilon_{r1} = 1) = \frac{\eta_0}{2\pi} \ln\left[\frac{F_1 h}{w} + \sqrt{1 + \left(\frac{2h}{w}\right)^2}\right] \tag{1.48}$$

with

$$F_1 = 6 + (2\pi - 6) \exp\left[-\left(30.666\frac{h}{w}\right)^{0.7528}\right]$$

is obtained by a functional approximation to exact values obtained by conformal mapping. The effective dielectric constant

$$\varepsilon_{r,\text{eff}} = \frac{\varepsilon_{r1} + 1}{2} + \frac{\varepsilon_{r1} - 1}{2} \cdot \left(1 + 10\frac{h}{w}\right)^{-ab} \tag{1.49}$$

$$a = 1 + \frac{1}{49} \ln \frac{\left(\frac{w}{h}\right)^4 + \left(\frac{w}{52h}\right)^2}{\left(\frac{w}{h}\right)^4 + 0.432} + \frac{1}{18.7} \ln\left[1 + \left(\frac{w}{18.1h}\right)^3\right]$$

$$b = 0.564 \left(\frac{\varepsilon_{r1} - 0.9}{\varepsilon_{r1} + 3}\right)^{0.053}$$

has been found by functional approximation to numerical results based on the static Green's function method. The following equations characterize the dispersive behavior of microstrip lines. They are valid with an error below 3% for the parameter range $2 \le \varepsilon_{r1} \le 16$, $0.06 \le w/h \le 16$ and $0.003 \le h/\lambda_0 \le 0.3$.

$$\varepsilon_{r,\text{eff}}(f) = \left(\frac{\sqrt{\varepsilon_{r1}} - \sqrt{\varepsilon_{r,\text{eff}}}}{1 + 4F^{-1.5}} + \sqrt{\varepsilon_{r,\text{eff}}}\right)^2 \tag{1.50}$$

where

$$F = \frac{4h\sqrt{\varepsilon_{rl} - 1}}{\lambda_0} \left\{ 0.5 + \left[1 + 2\log\left(1 + \frac{w}{h}\right) \right]^2 \right\} \tag{1.51}$$

and $\varepsilon_{r,eff}$ from (1.49). The frequency dependence of the characteristic impedance is given by

$$Z_0(f) = Z_0 \frac{\varepsilon_{r,eff}(f) - 1}{\varepsilon_{r,eff} - 1} \sqrt{\frac{\varepsilon_{r,eff}}{\varepsilon_{r,eff}(f)}}, \tag{1.52}$$

where Z_0 represents the quasistatic value of the characteristic impedance from (1.47) with (1.48).

Suspended and Inverted Microstrip Lines

For $\varepsilon_{rl} = 1$ the structures are identical with the classical microstrip and, thus, $Z_0(\varepsilon_{rl} = 1)$ can be taken from (1.48) substituting $h \equiv h_1 + h_2$ for suspended microstrip and $h \equiv h_2$ for inverted microstrip. For suspended microstrip $\varepsilon_{r,eff}$ is given by

$$\varepsilon_{r,eff} = \left[1 + \frac{h_1}{h_2}\left(a_1 - b_1 \ln\frac{w}{h_2}\right)\left(\frac{1}{\sqrt{\varepsilon_{rl}}} - 1\right) \right]^{-2},$$

$$a_1 = \left(0.8621 - 0.1251 \ln\frac{h_1}{h_2}\right)^4, \tag{1.53}$$

$$b_1 = \left(0.4986 - 0.1397 \ln\frac{h_1}{h_2}\right)^4,$$

and for inverted microstrip by

$$\varepsilon_{r,eff} = \left[1 + \frac{h_1}{h_2}\left(a_2 - b_2 \ln\frac{w}{h_2}\right)\left(\sqrt{\varepsilon_{rl}} - 1\right) \right]^2,$$

$$a_2 = \left(0.5173 - 0.1515 \ln\frac{h_1}{h_2}\right)^2, \tag{1.54}$$

$$b_2 = \left(0.3092 - 0.1047 \ln\frac{h_1}{h_2}\right)^2.$$

For semiconductor substrates the accuracy of (1.53) and (1.54) is within $\pm 2\%$ for $1 < w/h_2 \leq 8$ and $0.2 \leq h_1/h_2 \leq 1$.

Slot Lines

Cohn [1.51] gave the following closed-form expressions found by functional approximation. They are, however, valid only in a relatively small parameter range. Fortunately, typical parameters of millimeter-wave monolithic integrated circuits are covered by the parameter range

$$9.7 \le \varepsilon_{r1} \le 20,$$
$$0.02 \le s/h \le 1.0, \tag{1.55}$$
$$0.01 \le h/\lambda_0 \le h/\lambda_{g,TE_0},$$

where $h/\lambda_{g,TE_0}$ is the cutoff value for the TE_0 trapped surface wave mode of the single side metallized dielectric slab and can calculated from (1.7). For a silicon substrate we obtain $h/\lambda_{g,TE_0} = 0.0768$. The following expressions have an accuracy of about 2% within the given parameter range.

For $0 \le s/h \le 0.2$:

$$\frac{1}{\sqrt{\varepsilon_{r,eff}}} = 0.923 - 0.195 \ln \varepsilon_{r1} + 0.2\frac{s}{h}$$
$$- \left(0.126\frac{s}{h} + 0.02\right) \ln\left(100\frac{h}{\lambda_0}\right), \tag{1.56}$$

$$Z_0 = 72.62 - 15.283 \ln \varepsilon_{r1} + 50\frac{h}{s}\left(\frac{s}{h} - 0.02\right)\left(\frac{s}{h} - 0.1\right)$$
$$+ \ln\left(100\frac{s}{h}\right)(19.23 - 3.3693 \ln \varepsilon_{r1})$$
$$- \left[0.139 \ln \varepsilon_{r1} - 0.11 + \frac{s}{h}(0.465 \ln \varepsilon_{r1} + 1.44)\right]$$
$$\times \left(11.4 - 2.636 \ln \varepsilon_{r1} - 100\frac{h}{\lambda_0}\right)^2. \tag{1.57}$$

For $0.2 \le s/h \le 1.0$:

$$\frac{1}{\sqrt{\varepsilon_{r,eff}}} = 0.987 - 0.21 \ln \varepsilon_{r1} + \frac{s}{h}(0.111 - 0.0022\varepsilon_{r1})$$
$$- \left(0.053 + 0.041\frac{s}{h} - 0.0014\varepsilon_{r1}\right) \ln\left(100\frac{h}{\lambda_0}\right), \tag{1.58}$$

$$Z_0 = 113.19 - 23.257 \ln \varepsilon_{r1} + 1.25\frac{s}{h}(114.59 - 22.531\varepsilon_{r1})$$
$$+ 20\left(\frac{s}{h} - 0.02\right)\left(1 - \frac{s}{h}\right)$$
$$- \left[0.1 \ln \varepsilon_{r1} + 0.15 + \frac{s}{h}(0.899 \ln \varepsilon_{r1} - 0.79)\right]$$
$$\times \left[10.25 - 2.171 \ln \varepsilon_{r1} + \frac{s}{h}(2.1 - 0.617 \ln \varepsilon_{r1}) - 100\frac{h}{\lambda_0}\right]^2. \tag{1.59}$$

Coplanar Lines and Coplanar Strip Lines

For $w, s, h \ll \lambda_0$ closed form expressions for Z_0 and $\varepsilon_{r,eff}$ can be found employing the quasi-static approach. The error is less than 1.5% compared to results of a full wave analysis.

The effective dielectric constants of the coplanar waveguide and the coplanar stripline are equal and given by

$$\varepsilon_{r,eff} = 1 + \frac{\varepsilon_{r1} - 1}{2} \frac{K(k')}{K(k)} \frac{K(k_1)}{K(k'_1)}, \tag{1.60}$$

$$k = \sqrt{1 - k'^2} = \frac{s}{s + 2w}, \tag{1.61}$$

$$k_1 = \sqrt{1 - k_1'^2} = \frac{\sinh \dfrac{\pi s}{4h}}{\sinh \dfrac{\pi (s + 2w)}{4h}}. \tag{1.62}$$

An approximate expression for $K(k)/K(k') = K(k)/K'(k)$ can taken from (1.45). The characteristic impedance of the coplanar waveguide

$$Z_0 = \frac{\eta_0}{4\sqrt{\varepsilon_{r,eff}}} \frac{K(k')}{K(k)} \tag{1.63}$$

and of the coplanar stripline

$$Z_0 = \frac{\eta_0}{4\sqrt{\varepsilon_{r,eff}}} \frac{K(k)}{K(k')}. \tag{1.64}$$

is obtained by conformal mapping.

Microshield Lines

By use of the conformal mapping method the characteristic impedance of microshield lines may be represented as parallel capacitors C_a and C_w [1.59], where

$$C_a = 2\varepsilon_0 \frac{K(k)}{K(k')} \tag{1.65}$$

with

$$k = \frac{a}{b} \sqrt{\frac{1 - b^2/c^2}{1 - a^2/c^2}}, \tag{1.66}$$

$$k' = \sqrt{1 - k^2} \tag{1.67}$$

characterizes the half space on the top of the structure. The influence of the cavity below of the transmission line is determined by

$$C_w = 2\varepsilon_0 \frac{K(\zeta)}{K(\zeta')}, \tag{1.68}$$

where

$$\zeta = \frac{sn(a/\beta)}{sn(b/\beta)}, \tag{1.69}$$

$$\zeta' = \sqrt{1 - \zeta^2}, \tag{1.70}$$

$$\beta = \frac{W}{2K(\gamma)}, \tag{1.71}$$

$$\gamma = \left(\frac{e^{\pi W/2h} - 2}{e^{\pi W/2h} + 2}\right)^2 \tag{1.72}$$

and $sn(\theta)$ is the elliptic Jacobian function. In Fig. 1.12 the parameters a, b and W are depicted, where $b = a + w_s$. With (1.65,68) the characteristic impedance

$$Z_0 = \frac{1}{c_0(C_a + C_w)}. \tag{1.73}$$

of a microshield line is obtained.

References

Section 1.0

1.1   J. Magarshack: Civil applications with mm-wave MMICs in Europe. GAAS '92 Europ. Galluim Arsenide and Related Compounds Applications Symp. (ESTEC, Noordwijk, Netherlands 1992)

1.2   H. Dämbkes, L.-P. Schmidt: MMIC technology in Europe for millimeterwave applications. GAAS '92 Europ. Galluim Arsenide and Related Compounds Applications Symp. (ESTEC, Noordwijk, Netherlands 1992)

1.3   D. Pavlidis: Microwave/millimeter-wave monolithic integrated circuits. Rept, Dept. Electrical Engineering and Computer Science, University of Michigan (1992)

1.4   A. Rosen, M. Caulton, P. Stabile, A.M. Gombar, W.M. Janton, C.P. Wu, J.F. Corboy, C.W. Magee: Silicon as a millimeter-wave monolithically integrated substrate. RCA Rev. **42**, 633–660 (1981)

1.5   P. Russer: Silicon monolithic millimeterwave integrated circuits. Proc. 21st Europ. Microwave Conf. (Stuttgart, 1991) pp. 55–71

1.6   J.-F. Luy: Silicon monolithic millimeter-wave integrated circuits. Proc. IEE-H **139**, 209–215 (1992)

1.7   K.M. Strohm, J. Büchler, P. Russer, E. Kasper: Silicon high resistivity substrate millimeter-wave technology. IEEE Microwave and Millimeter-Wave Monolithic Circuits Symp. (Baltimore, MD) Digest. pp. 93–97

1.8   J. Büchler, E. Kasper, P. Russer, K.M. Strohm: Silicon high-resistivity-substrate millimeter-wave technology. IEEE Trans. MTT – **34**, 1516–1521 (1986)

1.9   J. Büchler, E. Kasper, P. Russer, K.M. Strohm: Silicon high-resistivity-substrate millimeter-wave technology. IEEE Trans. ED – **33**, 2047–2052 (1986)

1.10 J.-F. Luy, A. Casel, W. Behr, E. Kasper: A 90-GHz double-drift IMPATT diode made with SI MBE. IEEE Trans. ED – **34**, 1084–1089 (1987)

1.11   J. Büchler, E. Kasper, P. Russer, K.M. Strohm: Planar millimeter-wave circuits on silicon substrate in microwave applications. *Proc. 8th Int'l Congr. Laser* 87, ed. by H. Groll, W. Waidelich (Springer, Berlin, Heidelberg 1987) pp. 108–113

1.12   J. Büchler, E. Kasper, J.F. Luy, P. Russer, K.M. Strohm: Silicon millimeter-wave circuits for receivers and transmitters. IEEE Microwave and Millimeter-Wave Monolithic Circuits Symp. (New York 1988) Digest, pp. 67–70

1.13   J.-F. Luy, K.M. Strohm, J. Büchler: Monolithically integrated coplanar 75-GHz silicon IMPATT oscillator, Microwave and Optical Technology Lett. **1**, 117–119 (1988)

1.14   J. Büchler, E. Kasper, J.F. Luy, P. Russer, K.M. Strohm: 70 GHz integrated silicon oscillator. Electron. Lett. **24**, 977–978 (1988)

1.15   J. Büchler, E. Kasper, J.F. Luy, P. Russer, K.M. Strohm: Planar W-band receiver and oscillator. 18th Europ. Microwave Conf. (Stockholm 1988) Digest, pp. 364–369

1.16   K.M. Strohm, J.F. Luy, E. Kasper, J. Büchler, P. Russer: Silicon technology for monolithic integrated millimeter wave circuits. Mikrowellen & HF Magazin **14**, 750–760 (1988)

1.17   J.-F. Luy, U. König, K.M. Strohm, J. Büchler: Silicon monolithic millimeter wave integrated oscillators. IEE Colloq. on Microwave and Millimeter Wave Monolithic Integrated Circuits, (London 1988), Digest No.: 1988/117, pp. 10/1–10/4

1.18   K.M. Strohm, J. Büchler, E. Kasper, J.F. Luy, P. Russer: Millimeter wave transmitter and receiver circuits on high resistivity silicon. IEE Colloq. Microwave and Millimeter Wave Monolithic Integrated Circuits (London 1988) Digest No: 1988/117, pp. 11/1–11/4

1.19   J. Büchler, J.-F. Luy, K.M. Strohm: Varactor-tuned planar W-band oscillator, Proc. 1989 IEEE MTT-S Int'l Microwave Symp. (Long Beach, CA 1989) Digest, pp. 1205–1206

1.20   J.-F. Luy, J. Büchler: 90 GHz SIMMWIC (Silicon Monolithic Microwave Integrated Circuits). MIOP '89 Conf. (Sindelfingen 1989)

1.21   P. Russer: Millimeterwave integrated circuits, 9th Natl Meeting on Electromagnetics (Assisi, Italy 1992)

1.22   K.M. Strohm, J.-F. Luy, J. Büchler, F. Schäffler, A. Schaub: Planar 100 GHz silicon detector circuits. 21th Europ. Solid State Device Res. Conf. ESSDERC 91 (Lausanne 1991)

1.23   H. Meinel: System design, applications and development trends in the mm-wave range, 18th Europ. Microwave Conf. (Stockholm 1988) Digest, pp. 1203–1217

1.24   J. Schroth: Millimeterwellen-Sensoren. MIOP 89 Conf. (Sindelfingen 1989) Digest, pp. IB4

Section 1.1

1.25   T.M. Hyltin: Microstrip transmission on semiconductor dielectrics. IEEE Trans. MTT- **13**, 777–781 (1965)

1.26   H. Storck: Streifenleitungen auf Halbleitermaterialien, Dissertation, Rheinisch-Westfälische Technische Hochschule Aachen (1971)

Section 1.2

1.27   I.J. Bahl: Transmission lines, in *Handbook of Microwave and Optical Components*, ed. by K. Chang, (Wiley, New York, 1989) Vol. 1, Chap. 1

1.28   I.J. Bahl: Transmission lines and lumped elements, in *Microwave Solid State Circuit Design*, ed. by I. Bahl, P. Bhartia Wiley, New York, 1988) Chap. 2

1.29   F.K. Schwering, A.W. Glisson, M.A. Morgan: Antennas I: Fundamentals and numerical methods, in *Handbook of Microwave and Optical Components*, ed. by K. Chang, Wiley, New York, 1989) Vol. 1, Chap. 10

1.30   W.R.J. Hoefer: Time domain electromagnetic simulation for microwave CAD applications. IEEE Trans. MTT- **40**, 1517–1527 (1992)

1.31   T. Itoh, W. Menzel: A full-wave analysis method for open microstrip structures. IEEE Trans. AP- **29**, 63–67 (1981)

1.32   E.K. Miller: A selective survey of computational electromagnetics. IEEE Trans. AP- **36**, 1281–1303 (1988)

1.33   A. Wexler: Computation of electromagnetic fields. IEEE Trans. MTT- **17**, 416–439 (1969)

1.34   R.F. Harrington: *Time-Harmonic Electromagnetic Fields* (McGraw-Hill, New York 1961)

1.35   N.G. Alexopoulos, P.B. Katehi, D.B. Rutledge: Substrate optimization for integrated circuits. IEEE Trans. MTT- **31**, 550–557 (1983)

1.36   A. Ishimaru: *Electromagnetic Wave Propagation, Radiation, and Scattering* (Prentice-Hall, London 1991) Chap. 3, pp. 31–75

1.37   D.M. Pozar: Rigorous closed-form expressions for the surface wave loss of printed antennas, Electronic Lett., **26**, pp. 954–956 (1990)

1.38   J. Buechler, K.M. Strohm, J.-F. Luy, T. Goeller, P. Russer: Coplanar monolithic silicon IMPATT transmitter. 21th Europ. Microwave Conf. (Stuttgart, 1991)

1.39   Y. Yoshimura: A microstripline slot antenna. IEEE Trans. MTT- **20**, 760–762 (1972)

1.40   J. Zmuidzinas, N.G. LeDuc: Quasi-optical slot antenna sis mixers. IEEE Trans. MTT- **40**, 1797–1804 (1992)

1.41   D.F. Filipovic, W.Y. Ali-Ahmad, G.M. Rebeiz: Millimeter-wave double-dipole antennas for high-gain integrated reflector illumination. IEEE Trans. MTT- **40**, 962–967 (1992)

1.42   S.B. Cohn: Characteristic impedance of shielded strip transmission line. IRE Trans. MTT- **2**, 52–55 (1954)

1.43   M. Abramowitz, I.A. Stegun: *Handbook of Mathematical Functions* (Dover New York, 1972) p. 591

1.44   E.O. Hammerstad: Equations for microstrip circuit design. Proc. 5th Europ. Microwave Conf., (Kent 1975) pp. 268–272

1.45   E.O. Hammerstad: Accurate models for microstrip computer-aided design. IEEE MTT-S Int'l Microwave Symp. (1986) Digest, pp. 407–409

1.46   R.K. Hoffmann: *Handbook of Microwave Integrated Circuits* (Artech House, Norwood 1987)

1.47   N.G. Alexopoulos, D.R. Jackson: Fundamental superstrate (cover) effects on printed circuit antennas. IEEE Trans. AP- **32**, 807–816 (1984)

1.48   N.G. Alexopoulos, D.R. Jackson, P.B. Katehi: Criteria for nearly omnidirectional radiation pattern for printed antennas. IEEE Trans. AP- **33**, 195–205 (1985)

1.49   D.R. Jackson, N.G. Alexopoulos: Analysis of planar strip geometries in a substrate-superstrate configuration. IEEE Trans. AP- **34**, 1430–1438 (1986)

1.50   K. Itoh, N. Aizawa, N. Goto: Circularly polarized printed array antennas composed of strips and slots. Electr. Lett. **15**, 811–812 (1979)

1.51   S.B. Cohn: Slotline on a dielectric substrate. IEEE Trans. MTT- **17**, 768–778 (1969)

1.52   R. Janaswami, D.H. Schaubert: Dispersion characteristics for wide slotlines on low-permittivity substrates. IEEE Trans. MTT- **33**, 723–726 (1985)

1.53   V. Fouad Hanna, D. Thebault: Theoretical and experimental investigation of asymmetric coplanar waveguides. IEEE Trans. MTT- **32**, 1649–1651 (1984)

1.54   A. Gopinath: Losses in coplanar waveguides. IEEE Trans. MTT- **30**, 1101–1104 (1982)

1.55   G. Ghione, C. Naldi: Analytical formulas for coplanar lines in hybrid and monolithic MIC's. IEEE Electron. Lett. **20**, 179–181 (1984)

1.56   G. Ghione, C. Naldi: Coplanar waveguides for MMIC applications: Effect of upper shielding, conductor backing, finite extent ground planes, and line-to-line coupling, IEEE Trans. MTT- **35**, 260–267 (1987)

1.57   R.W. Jackson: Considerations of the use of coplanar waveguide for millimeter-wave integrated circuits. IEEE Trans. MTT- **34**, 1450–1465 (1986)

1.58   R.W. Jackson: Mode conversion due to discontinuities in modified coplanar grounded waveguide. IEEE MTT-S Microwave Symp. (New York 1988) Digest, pp. 203–206

1.59   N.I. Dib, L.P.B. Katehi: Impedance calculation for the microshield line. IEEE Microwave and Guided Wave Lett. **2**, pp. 406–408 (1992)

1.60   L.P.B. Katehi: Novel transmission lines for the submillimeter-wave region. Proc. IEEE **80**, pp. 1771–1787 (1992)

1.61   T. Itoh: Spectral domain approach for dispersion characteristics of generalized printed transmission lines. IEEE Trans. MTT- **28**, 733–736 (1980)

1.62   K.C. Gupta: Transmission-line discontinuities, in *Handbook of Microwave and Optical Components*, ed. by K. Chang, (Wiley, New York 1989) Vol. 1, Chap. 2

1.63   M. Naghed, I. Wolff: Equivalent capacitances of coplanar waveguide discontinuities and interdigitated capacitors using three-dimensional finite difference method. IEEE Trans. MTT- **38**, 1808–1815 (1990)

1.64   N.I. Dib, L.P.B. Katehi, G.E. Ponchak, R.N. Simons: Theoretical and experimental characterization coplanar waveguide discontinuities for filter applications. IEEE Trans. MTT- **39**, 873–881 (1991)

1.65   M. Drissi, F. Hanna, J. Citerne: Analysis of radiating end effects of symmetric and asymmetric coplanar waveguide using integral equations technique. IEEE MTT-S Int. Microwave Symp. (Long Beach 1989) Digest, pp. 791–794

1.66   J. McLean, H. Ling, T. Itoh: Spectral domain analysis of electrically wide short-circuit dicontinuities in slotline. IEEE AP-S Int'l Symp. (San Jose 1989) Digest, pp. 1242–1245

1.67   R. Klieber, R. Ramisch, R. Weigel, M. Schwab, R. Dill, A.A. Valenzuela, P. Russer: High-temperature superconducting resonator-stabilized coplanar hybrid-integrated oszillator at 6.5 GHz. IEEE IEDM Int'l Electron Devices Meeting (Washington 1991) Digest, pp. 923–926

1.68   R. Klieber, R. Ramisch, A.A. Valenzuela, R. Weigel, P. Russer: A coplanar transmission line high-T_c superconductive oscillator at 6.5 GHz on a single substrate. IEEE Microwave Guided Wave Lett. **2**, pp. 22–24 (1992)

1.69   T. Itoh: Analysis of microstrip resonators. IEEE Trans. MTT – **22**, 946–942 (1974)

1.70   T. Itoh, R. Mittra: Analysis of microstrip disk resonators. Arch. Elekt. Übertragung. **27**, pp. 456–458 (1973)

1.71   I. Wolff, N. Knoppik: Microstrip ring resonator and dispersion measurement on microstrip line. Electron. Lett. **7**, pp. 779–781 (1971)

1.72   A.K. Sharma, B. Bhat: Spectral domain analysis of microstrip ring resonators. Arch. Elekt. Übertragung. **33**, 130–132, Mar. 1979

1.73   Y.S. Wu, F.J. Rosenbaum: Mode chart for microstrip ring resonators. IEEE Trans. MTT- **21**, 487–489 (1973)

1.74   J.G. Kretzschmar: Theoretical results for the elliptic microstrip resonator. IEEE Trans. MTT- **20**, 342–343 (1972)

1.75   A.K. Sharma, B. Bhat: Spectral domain analysis of elliptic microstrip disk resonators. IEEE Trans. MTT- **28**, 573–576 (1980)

1.76   A.K. Sharma: Spectral domain analysis of an elliptic microstrip ring resonator. IEEE Trans. MTT- **32**, 212–218 (1984)

1.77   J. Helszajn, D.S. James: Planar triangular resonators with magnetic walls. IEEE Trans. MTT- **27**, 95–100 (1978)

1.78   J. Helszajn, D.S. James, W.T. Nisbet: Circulators using planar triangular resonators. IEEE Trans. MTT- **28**, 188–193 (1979)

1.79   A.K. Sharma, B. Bhat: Analysis of triangular microstrip resonators. IEEE Trans. MTT- **30**, 2029–2031 (1982)

1.80   A.K. Sharma, W.J.R. Hoefer: Spectral domain analysis of a hexagonal microstrip resonator. IEEE Trans. MTT- **30**, 825–828 (1982)

Section 1.3

1.81   P. Bhartia, I.J. Bahl: *Millimeter Wave Engineering and Applications* (Wiley, New York 1984) Chap 9, pp. 477–616

1.82   R.J. Mailloux, F.K. Schwering, A. Oliner, J.W. Mink: Antennas III: Array, millimeterwave, and integrated antennas *Handbook of Microwave and Optical Components*, ed. by K. Chang, (Wiley, New York, 1989) Vol. 1, Chap. 12

1.83   R.J. Mailloux, J.F. McIlvenna, N.P. Kernweis: Microstrip array technology. IEEE Trans. AP-**29**, 25–37 (1981)

1.84   G.M. Rebeiz: Millimeter-wave and terahertz integrated circuit antennas. Proc. IEEE **80**, 1748–1770 (1992)

1.85   D.B. Rutledge, D.P. Neikirk, D.P. Kassiligam: Integrated-circuit antennas, in *Infrared and Millimeter Waves* **10**, 1–90 (Academic, New York 1983)

1.86   F.K. Schwering: Millimeter wave antennas. Proc. IEEE **80**, 92–102 (1992)

1.87   K.R. Carver: Microstrip antenna technology. IEEE Trans. AP- **29**, 2–24 (1981)

1.88   Y.T. Lo, S.M. Wright, M. Davidovitz: Antennas IV: Microstrip antennas, in *Handbook of Microwave and Optical Components*, ed. by K. Chang, (Wiley, New York 1989) Vol. 1, Chap 10

1.89   E.M. Biebl, J. Müller, H. Ostner: Analysis of planar millimeter wave slot antennas using a spectral domain approach. IEEE MTT-S Int'l Microwave Symp. (Albuquerque 1992) Digest, pp. 381–384

1.90   J.R. Mosig, F.E. Gardiol: Dielectric losses, ohmic losses, and surface waves effects in microstrip antennas. Proc. URSI Int'l Symp. on Electromagnetic Theory (Spain 1983) pp. 524–428

1.91   H. Ostner, T. Ostertag, E.M. Biebl: Calculation of the impedance of planar slot antennas. IEEE Asia-Pacific Microwave Conf. (Adelaide (1992)) Digest, pp. 137–140

1.92   K. Wu, M. Yu, R. Vahldieck: Rigorous analysis of 3-d planar circuit discontinuities using the space-spectral domain approach (SSDA). IEEE Trans. MTT- **40**, 1475–1483 (1992)

1.93   R.C. Compton, R.C. McPhedran, Z. Popovic, G.M. Rebeiz, P.T. Tong, D.B. Rutledge: Bow-tie antennas on dielectric half-space: Theory and experiment. IEEE Trans. AP- **35**, 622–631 (1987)

1.94   P.H. Siegel: A planar log-periodic mixtenna for millimeter and submillimeter wavelengths. IEEE MTT-S Int'l Microwave Symp. (1986) Digest, pp. 649–652

1.95   Y. Guo, K. Lee, P.A. Stimson, K.A. Potter, D.B. Rutledge: Aperture efficiency of integrated horn antennas. Microwave and Opt. Techn. Lett. **4**, pp. 6–9 (1991)

1.96   G.M. Rebeiz, D.P. Kasilingam, P.A. Stimson, Y. Guo, D.B. Rutledge: Monolithic milli-meter-wave two-dimensional horn imaging arrays. IEEE Trans. AP- **38**, 1473–1482 (1986)

1.97   W.Y. Ali-Achmad, G.M. Rebeiz, H. Davee, G. Chin: 802 GHz integrated horn antennas imaging array. Int'l J. Infrared Millimeter Waves **12**, pp. 481–486 (1991)

1.98   P.B. Katehi: A generalized method for the evaluation of mutual coupling in microstrip arrays. IEEE Trans. AP- **35**, 125–133 (1987)

1.99   P.B. Katehi: Mutual coupling between microstrip dipoles in multielement arrays. IEEE Trans. AP- **37**, 275–280 (1989)

1.100 D.R. Jackson, N.G. Alexopoulos: Gain enhancement method for printed circuit antennas. IEEE Trans. AP- **33**, 976–987 (1985)

1.101 P.B. Katehi, N.G. Alexopoulos: On the modeling of electromagnetically coupled microstrip antennas: The printed strip dipole. IEEE Trans. AP- **32**, 1179–1186 (1984)

1.102 A.C. Buck, D.M. Pozar: Aperture-coupled microstrip antenna with perpendicular feed. Electr. Lett. **22**, pp. 125–126 (1986)

1.103 X.Z. Gao, K. Chang: Network modelling of an aperture coupling between microstrip line and patch antenna for active array application. IEEE Trans. MTT- **36**, 505–513 (1988)

1.104 G. Gronau, I. Wolf: Aperture-coupling of a rectangular microstrip resonator. Electr. Lett. **22**, pp. 554–556 (1986)

1.105 D.M. Pozar: Microstrip antenna aperture-coupled to a microstripline. Electr. Lett. **21**, pp. 49–50 (1985)

1.106 D.M. Pozar: A reciprocity method of analysis for printed slot and slot-coupled microstrip antennas. IEEE Trans. AP- **34**, 1439–1446 (1986)

1.107   P.L. Sullivan, D.H. Schaubert: Analysis of an aperture-coupled microstrip antenna. IEEE
        Trans. AP- **34**, 977–984 (1986)
1.108   R.M. Weikle, M. Kim, J.B. Hacker, M.P. DeLisio, D.B. Rutledge: Planar MESFET grid
        oszillator using gate feedback. IEEE Trans. MTT- **40**, 1997–2003 (1992)
1.109   J.B. Hacker, R.M. Weikle, M. Kim, M.P. DeLisio, D. Rutledge: A 100-element planar schot-
        tky diode grid mixer. IEEE Trans. MTT- **40**, 557–562 (1992)

Section 1.4

1.110   J.-F. Luy, K.M. Strohm, J. Büchler: Silicon monolithic millimeter wave IMPATT oscillators.
        18th Europ. Microwave Conf. (Stockholm 1988) Digest, pp. 382–387
1.111   C.A. Brackett: The elimination of tuning-induced burnout and bias-circuit oscillations in
        IMPATT oscillators. Bell Syst. Techn. J. 271–306 (1973)

2 Transit-Time Devices

J.-F. LUY

Daimler-Benz Research Center, Wilhelm Runge Str. 11, 89081 Ulm, Germany

It was shown already in 1934 [2.1] that transit-time effects may lead to a frequency-dependent negative resistance which may be used for the generation of oscillations. Later these ideas were implemented in semiconductor devices [2.2]. This concept is the basis for IMPATT (IMPact Avalanche Transit-Time), BARITT (BARier Injection Transit-Time), TUNNETT (TUNNEl injection Transit-Time) diodes and related versions.

In IMPATT diodes the generation of carriers is performed by avalanche multiplication, in BARITT diodes minority carriers are injected via a potential well and in TUNNETT diodes a tunneling mechanism is used to supply the carriers in the drift region. In double-drift devices electrons and holes are used whereas in single-drift devices only electrons or holes contribute to the existence of a negative resistance. More sophisticated structures are used in low-high-low-doping profiles: doping spikes confine and separate injection and drift regions.

In this chapter the properties, technology and the performance of transit-time devices are discussed. The drift mechanism common to all devices under consideration is described. Then, the different injection mechanisms are treated. Common to all transit-time devices are problems related to skin effect losses, thermal properties and technology aspects. The performance and the possible applications are strongly dependent on the device type: IMPATT diodes provide the largest output power and efficiency – BARITT and TUNNETT diodes are low-power and low-noise sources.

2.1 Principles of Transit-Time-Induced Negative Resistance

The transport of carriers through a drift region is of basic importance for high-frequency transit-time devices. The injection of carriers in the drift region is possible by mechanisms as impact ionisation, thermionic emission or tunnel injection, which shall be treated separately. The configuration of an idealized transit-time device is depicted in Fig. 2.1.

The model is used to study only carrier transport by drift within $0 \leq x \leq d$. The positive polarity of the applied voltage is connected to the injector and the negative polarity to the collector. A general analysis of this structure is possible with the following assumptions [2.3,4]:

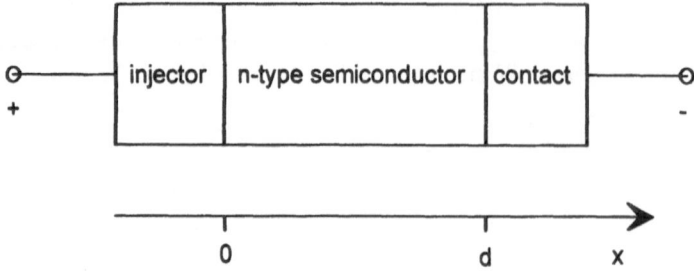

Fig. 2.1. Model structure of a general transit-time device with an n-doped drift region

- only one kind of carriers is considered: holes in this case,
- the analysis may be performed one-dimensional,
- diffusion may be neglected and the transit time is small compared to the carrier lifetime,
- there is no generation or recombination of carriers in the drift region.

Carrier transport can then be described by

$$\frac{\partial E}{\partial x} = \frac{e}{\varepsilon_0 \varepsilon_r}(N_D + p) \quad \text{Poisson equation,}$$

$$\frac{\partial p}{\partial t} = -\frac{1}{e}\frac{\partial J}{\partial x} \qquad \text{continuity equation,} \tag{2.1}$$

$$J = evp \qquad\qquad \text{particle current density equation.}$$

The time-dependent quantities J, p, v, E are assumed to follow $J = \Re\{J_0 + J_1 e^{j\omega t}\}$, $p = \Re\{p_0 + p_1 e^{j\omega t}\}\ldots$, which represents a small signal assumption. If only ac contributions are considered and higher-order terms are neglected, we obtain

$$\frac{\partial E_1}{\partial x} = \frac{e}{\varepsilon_0 \varepsilon_r} p_1,$$

$$j\omega p_1 = -\frac{1}{e}\frac{\partial J_1}{\partial x}, \tag{2.2}$$

$$J_1 = e(v_0 p_1 + v_1 p_0).$$

Assuming constant carrier mobility leads to the second-order differential equation

$$\frac{d^2 E_1}{dx^2} = -\left(\frac{p_0 \mu e}{v_0 \varepsilon} + \frac{j\omega}{v_0}\right)\frac{dE_1}{dx} \tag{2.3}$$

from which the general solution is derived

$$E_1(x) = ae^{kx} + b,$$

$$k = -\left(\frac{p_0 \mu e}{v_0 \varepsilon} + j\frac{\omega}{v_0}\right) \tag{2.4}$$

with the complex propagation constant k of the travelling wave. For the determination of a and b we need boundary conditions: The total current density is the sum of displacement current and particle current: $J_t = J_1 + j\omega\varepsilon_0\varepsilon_r E_1(x)$. By comparison with the differential equation and the general solution, we obtain $b = J_t/j\omega\varepsilon_0\varepsilon_r$ and $a = -J_1(0)/j\omega\varepsilon_0\varepsilon_r$. The solution for the ac field is then

$$E_1(x) = \frac{J_t}{j\omega\varepsilon_0\varepsilon_r}\left(-\frac{J_1(0)}{J_t}e^{kx} + 1\right). \tag{2.5}$$

The impedance of the semiconductor structure is calculated by integration over the drift region:

$$Z_D = \frac{\int_0^d E_1(x)}{J_t A} = -\frac{j}{\omega C}\left[1 + \frac{J_1(0)}{J_t}\frac{1}{kd}\left(1 - e^{kd}\right)\right] \text{ with } C = \frac{\varepsilon_0\varepsilon_r A}{d}, \tag{2.6}$$

A being the device area. Some features of transit time devices shall be discussed using this expression.

First, it is obvious that the impedance level is inversely proportional to the capacitance and device area, respectively. Second, we recognize the importance of the complex injection ratio which has the general form

$$\frac{J_1(0)}{J_t} = A(\omega)e^{-j\phi(\omega)} \tag{2.7}$$

with an amplitude $A(\omega)$ and an angle $\phi(\omega)$ denominating the phase shift between the injection current density $J_1(0)$ and the total current density J_t. This expression describes the injection mechanism and causes the differences in the impedances of IMPATT, BARITT or TUNNETT diodes.

To study the importance of the injection ratio it is convenient to investigate the real and imaginary parts of the drift region impedance:

$$\Re\{Z_D\} = \frac{A}{\theta\omega C}[\cos\phi - \cos(\theta + \phi)],$$

$$\text{Im}\{Z_D\} = \frac{-1}{\theta\omega C}\{\theta + A[\sin\phi - \sin(\phi + \theta)]\} \tag{2.8}$$

where $\theta = \omega d/v_s$ is the transit angle, and the propagation constant has been restricted to the case where the applied field is high enough to punch through the region, and carriers are drifting at the saturated velocity v_s. In this case, $k = -j\omega/v_s$.

In the calculations plotted in Figs. 2.2 and 3, for simplification, $A/\omega C = 1$ has been assumed. It can be seen, that the real part may become negative only for values of the injection phase angle $\phi \geq 0$. The optimum transit angle for a maximum real part depends on the injection phase angle and decreases with increasing phase angle – simultaneously the amount of the negative resistance increases. For example, an injection phase angle $\phi = \pi/2$ in BARITT diodes requires a transit angle near $\theta = 3\pi/2$ to achieve the largest negative resistance. With $\phi = \pi$ in IMPATT diodes, a transit angle of $\theta = 3\pi/4$ leads to the largest negative resistance.

From the previous discussion, it is apparent that a mechanism in which the carriers are injected at a large phase angle without a decrease in amplitude may be

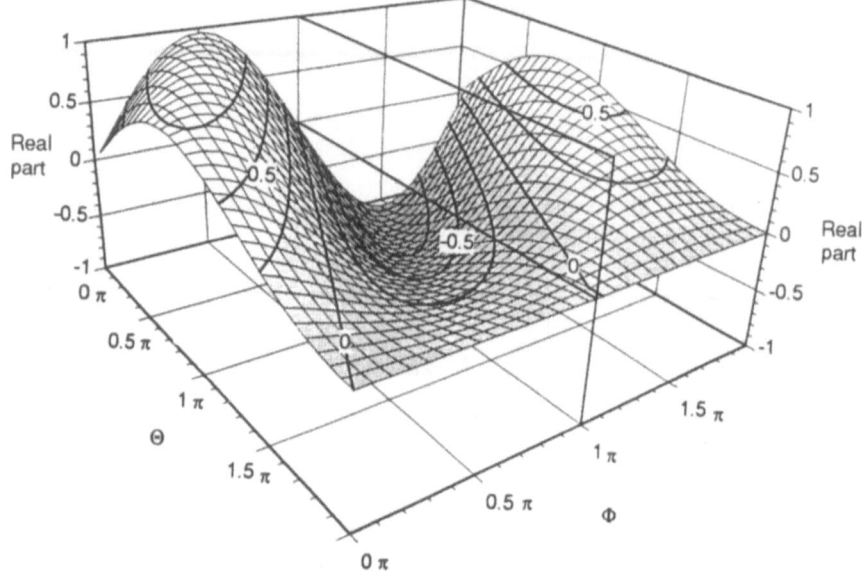

Fig. 2.2. 3D representation of the real part of the drift-region impedance as a function of the transit angle and the injection phase angle

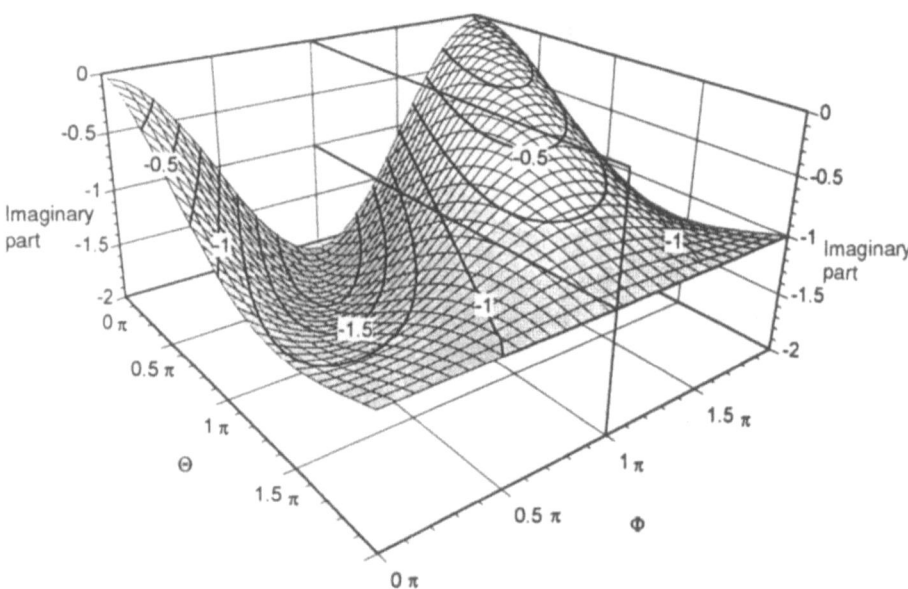

Fig. 2.3. 3D representation of the imaginary part of a drift-region impedance as a function of the transit angle and the injection phase angle

favourable. Furthermore, the injection mechanism should ensure that the carriers are injected with the saturated velocity. Another approach would be to let $\phi = 0, \pi, \ldots$ and to make $A(\omega) \leq 0$ – which occurs in the case of an IMPATT diode, as will be shown in the next section.

2.2 Injection Mechanisms

The transit-time diodes can be distinguished by considering the injection mechanisms: Carrier injection from a high-field region where impact ionization occurs (IMPATT diode), thermionic emission across a potential barrier (BARITT diode), tunnel injection (TUNNETT and QWITT diode) or diodes with intermixed drift and injection mechanisms (MISAWA type).

2.2.1 Impact Ionisation – IMPATT Diode

Read recognized the suitability of the impact-ionization and avalanche-multiplication mechanisms for the injection of hot carriers, and proposed the IMPATT diode [2.5]. One of the key parameters for this device is the ionization rate, i.e., the number α of electron-hole pairs which are generated by a carrier per unit distance. For the device analysis and design it is necessary to know the dependence on the "driving force" – the energy or the electric field in a local model. The ionization rates may be calculated from the band structure using Monte Carlo simulations [2.7]. There is a good convergence between these calculations and the measurements [2.6,8]. An analytical representation is obtained by the semiempirical equation [2.9]

$$\alpha_{n,p}(E) = \frac{eE}{W_{i_{n,p}}} \exp\left(-\frac{F_{i_{n,p}} F_{r_{n,p}}}{E^2}\right) \tag{2.9}$$

with the high-field ionization threshold energies $W_{i_n} = 3.6$ eV for electrons and $W_{i_p} = 5.0$ eV for holes, the threshold fields compensating for Coulomb scattering $F_{i_n} = 1.954 \times 10^6$ eV/cm, $F_{i_p} = 3.091 \times 10^6$ eV/cm and for optical-phonon scattering $F_{r_n} = 1.069 \times 10^5$ eV/cm, $F_{r_p} = 1.110 \times 10^5$ eV/cm for electrons and holes, respectively. These values are obtained by a fitting to the measurements of *Grant* [2.8]. The temperature dependence of the ionization rates can be introduced using elements of the "Baraff theory" [2.10]: The mean-free path for optical-phonon scattering is

$$\lambda_{n,p} = \lambda_{0_{n,p}} \tanh\left(\frac{W_p}{2kT}\right) \tag{2.10}$$

with the optical-phonon energy $W_p = 63$ meV for electrons and holes. If we assume a homogeneous field between two scattering events, the threshold field due to optical-phonon scattering can be determined from [2.11]

$$F_{r_{n,p}} = \frac{W_p}{\lambda_{n,p}} = \frac{W_p}{\lambda_{0_{n,p}} \tanh(W_p/2kT)} \tag{2.11}$$

with $\lambda_{0_n} = 7.6$ nm and $\lambda_{0_p} = 5.5$ nm. Figure 2.4 exhibits the ionization rates at two different temperatures as a function of the electric field.

The injection current in an IMPATT diode is described following *Read's* equation for avalanche multiplication [2.5]

$$\tau_i \frac{\partial J}{\partial t} = (\alpha l_a - 1)J \tag{2.12}$$

with $\tau_i = (1/3)(l_a/v)$ which follows from a higher-order solution of the *Read* equation [2.12]. For an approximate large-signal solution of the *Read* equation, the electric field is assumed to be spatially constant but cosinusoidal time varying [2.13]. This may be described by a Taylor series for the ionization rate with the modulation index m

$$\alpha(E(t)) = \alpha_0(E) + \alpha'(E)mE \cos \omega t \tag{2.13}$$

where second and higher orders are neglected. The solution is then

$$J(t) = e^{z \sin \omega t} \left(\frac{J_0}{I_0(z)}\right) = J_0\left(1 + 2\frac{I_1(z)}{I_0(z)} \sin \omega t\right) \tag{2.14}$$

if $z = \alpha' l_a mE/\omega\tau_i = \omega_a^2 \varepsilon mE/J_0\omega$ where the avalanche frequency $\omega_a = \sqrt{\alpha' l_a J_0/\varepsilon\tau_i}$ is introduced. The total current is obtained by switching to the

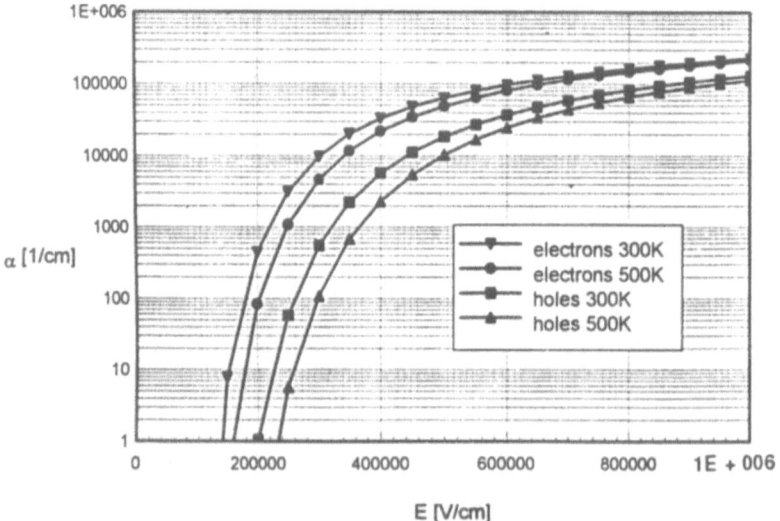

Fig. 2.4. Ionization rates for electrons and holes as a function of the electric field

complex notation and adding the displacement current $\omega \varepsilon m E$ to the frequency dependent component of $J(t)$:

$$J_t = J_0 z \left(\frac{\omega^2}{\omega_a^2} - \frac{2I_1(z)}{z I_0(z)} \right). \tag{2.15}$$

The injection ratio is now

$$\frac{J_1(0)}{J_t} = \frac{1}{1 - \dfrac{z I_0(z)}{2 I_1(z)} \dfrac{\omega^2}{\omega_a^2}} = A(\omega) e^{-j\phi} \tag{2.16}$$

which yields $\phi = \pi$ for the IMPATT-diode operation regime! The final expressions for the real and imaginary parts are then

$$\Re\{Z_D\} = A(\omega) \cdot \frac{1}{\omega C_d} \cdot \frac{1 - \cos\theta}{\theta}$$

$$\mathrm{Im}\{Z_D\} = -\frac{1}{\omega C_d} \left[1 - A(\omega) \frac{\sin\theta}{\theta} + \frac{l_a}{l_d} \left(\frac{1}{1 - \dfrac{\omega_a^2}{\omega^2} \dfrac{2I_1(z)}{z I_0(z)}} \right) \right]. \tag{2.17}$$

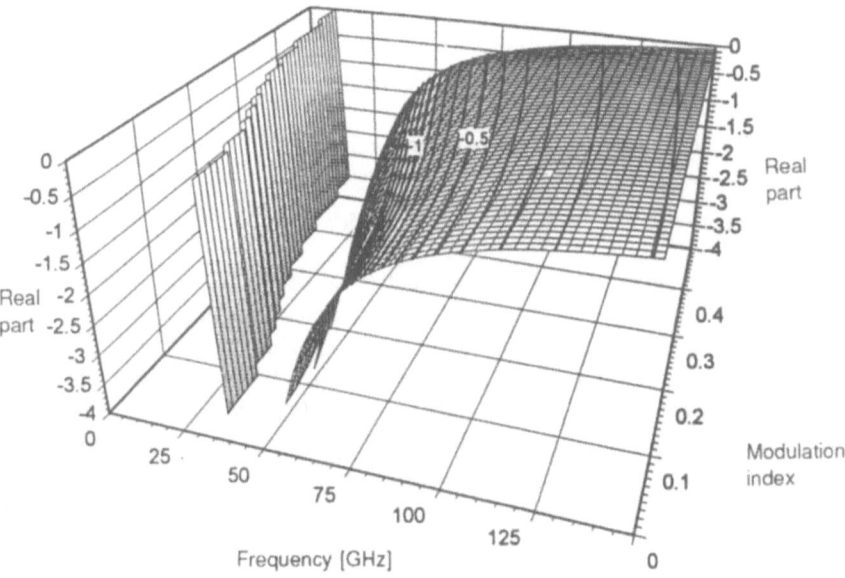

Fig. 2.5. Real part of the impedance of an IMPATT diode as a function of the frequency and the modulation index. Contour lines every 0.1 Ω. Parameters for the calculation: Drift velocity 0.65×10^7 cm/s; drift region 0.25 μm, $E = 5 \times 10^5$ V/cm. Current density 10 kA/cm^2. Real part normalized to an area of 7×10^{-6} cm^2

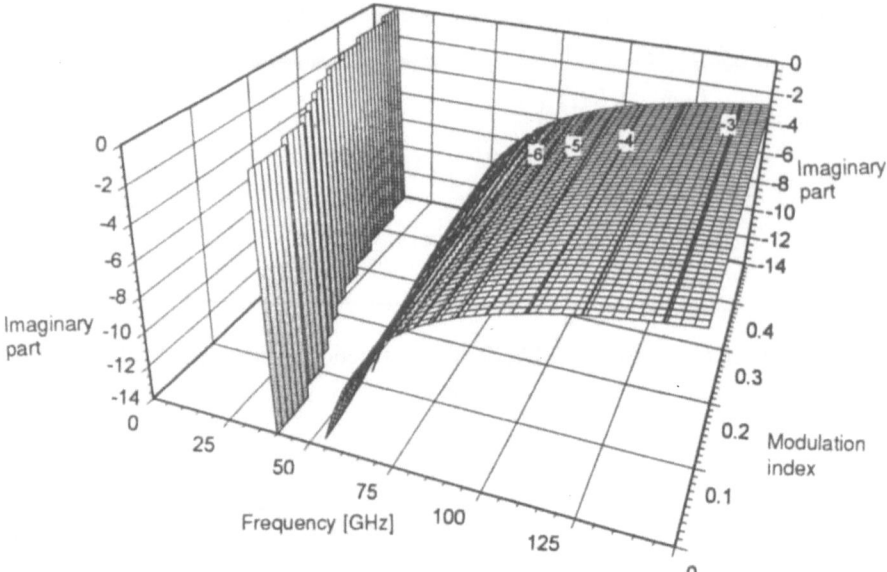

Fig. 2.6. Imaginary part of an IMPATT diode as a function of the frequency and the modulation index (Contour lines every 0.5 Ω, and same parameters as in Fig. 2.5)

Figure 2.5 shows the effect of increasing the modulation index – the amount of the real part decreases. The avalanche resonance is also dependent on the modulation index and is shifted towards lower frequencies with increased modulation.

The imaginary part is always capacitive above the avalanche resonance frequency. The dependence on the modulation index is weaker than for the real part.

2.2.2 Thermionic Emission – BARITT Diode

The simplest version of a BARITT device consists of a $p^+ - n - p^+$, $M - n - p^+$ or an inverse structure. One $p - n$, $(M - n)$ junction is in reverse direction [2.14].

If the applied voltage is increased, at a certain value the reverse-biased depletion region will reach through to the forward-biased depletion region. The corresponding voltage is called reach through voltage U_{RT}. Holes are injected across the potential barrier at the forward-biased $p - n$ junction. The holes injected at x_0 drift through the n-region and are collected at $x = d$.

This thermionic emission leads to an injection-current density which is in-phase with the driving RF-voltage: $J_1(0) = \sigma E_1$ following Ohm's law with the ac component E_1 of the electric field and the injection conductivity σ. The total-current density is obtained adding the displacement current: $J_t = \sigma E_1 + j\omega\varepsilon E_1$ with the dielectric permittivity $\varepsilon = \varepsilon_0\varepsilon_r$.

Comparing with the expression for the injection ratio we obtain

$$A(\omega) = \frac{\sigma}{\sqrt{\sigma^2 + \omega^2\varepsilon^2}} \tag{2.18}$$

p-region   n-region   p-region

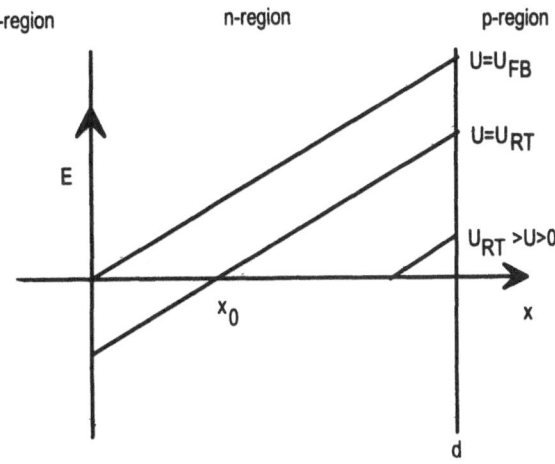

Fig. 2.7. Electric field in a BARITT diode at different bias conditions

and

$$\phi(\omega) = \arctan(-\omega\varepsilon/\sigma).$$ (2.19)

For the real part of the diode impedance one ends up with

$$\Re\{Z_D\} = \frac{\sigma^2}{\theta\omega C(\sigma^2 + \omega^2\varepsilon^2)} \left(1 - \cos\theta + \frac{\omega\varepsilon}{\sigma}\sin\theta\right)$$ (2.20)

and for the imaginary part with

$$\mathrm{Im}\{Z_D\} = -\frac{1}{\theta\omega C}\left[\theta + \frac{\sigma^2}{\sigma^2 + \omega^2\varepsilon^2}\left(-\frac{\omega\varepsilon}{\sigma} + \frac{\omega\varepsilon}{\sigma}\cos\theta - \sin\theta\right)\right]$$ (2.21)

where $C = \varepsilon A/d$.

Still unknown is the injection conductivity σ which can be calculated by using the Schottky approximation: The flat band voltage follows from Poisson's equation [2.15]

$$U_{FB} = \frac{e}{2\varepsilon}N_D d^2.$$ (2.22)

The operation regime of a BARITT diode is between the reach through voltage and the flat band voltage. The reach through voltage is $U_{RT} = U_{FB} - \sqrt{4U_{FB}U_D}$, where U_D is the diffusion voltage at the junction and is 0.9 V for a Pt-Schottky contact on n-type silicon. The potential barrier caused by the applied voltage is $U_{pot} = (U_{FB} - U)^2/4U_{FB}$ and the injected hole current is $J_p = J_{FB}e^{-(U_{pot}/U_t)}$ with $U_t = kT/e$. The prefactor J_{FB} is dependent on the kind of the forward-biased junction: $M - n$ or $p - n$. The conductivity is then

$$\sigma = \frac{dJ}{dE} = J_p\frac{U_{FB} - U}{2U_{FB}U_t}d$$ (2.23)

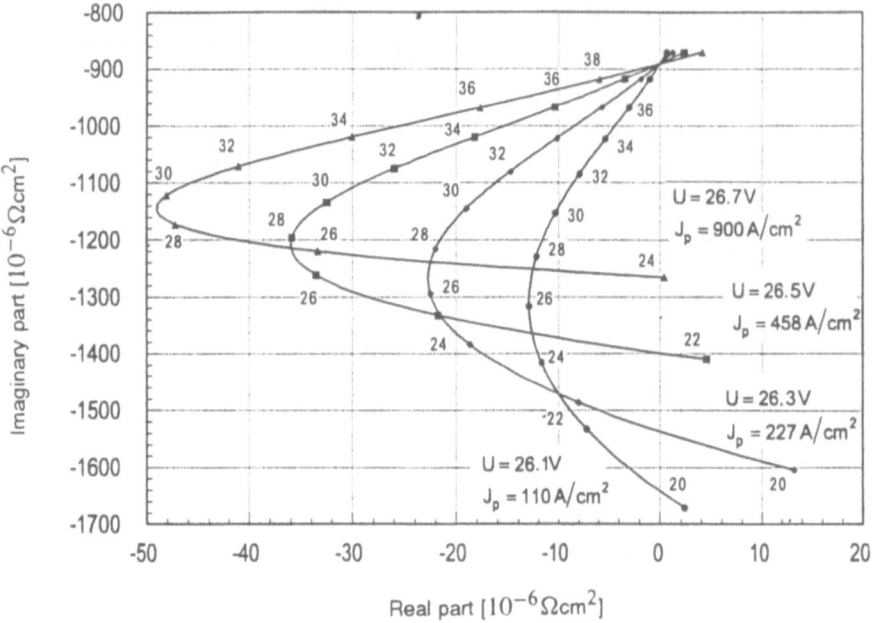

Fig. 2.8. Impedance of a BARITT diode, $d = 2.3$ μm, $N = 8 \times 10^{15}$ cm^{-3}, $f_{opt} = 30$ GHz

or using (2.22)

$$\sigma = J_p \frac{\varepsilon(U_{FB} - U)}{N_D dkT}. \tag{2.24}$$

The impedance of a BARITT diode with an n-layer thickness of 2.3 μm and a doping level of 8×10^{15} cm^{-3} is shown in Fig. 2.8 for the X-Y plane. The maximum amount of the real part is achieved at an operation frequency of 30 GHz at a current density of 900 A/cm^2. At a typical diode diameter of 60 μm for Ku-band devices this corresponds to a current of 25 mA and a real part of -1.7Ω.

2.2.3 Tunnel Injection

Tunnel injection has already been considered by *Read*, however, as a loss mechanism disturbing the favourable phase relations of the IMPATT diode [2.5]. *Nishizawa* and *Watanabe* proposed the TUNNEl injection Transit Time (TUNNETT) diode [2.16]. If the thickness of the transit time diode is reduced in order to increase the oscillation frequency, the avalanche phenomenon tends to vanish and then the tunnel injection becomes dominant in the reverse biased p-n junction.

The injection ratio of a pure TUNNETT diode is similar to that of a BARITT diode [2.4] with $\sigma = \sigma_t = 1/\rho_g$ the conductivity due to tunneling which can be derived from the resistance due to tunneling [2.17].

Theoretical and experimental investigations of the interband tunnel generation rate in silicon have shown that electric fields in excess of 1 MV/cm are necessary to provide a significant amount of tunneling-generated carriers [2.18,19]. Up to now, no convincing demonstration of a silicon-based TUNNETT diode has been shown.

Tunnel injection may also be performed by tunneling through potential barriers. This was introduced as the QWITT (Quantum Well) diode concept [2.20]. The injection region is represented by a resonant-tunneling diode (Chap. 8). If the device is biased so that during the RF cycle at $\omega t = \pi/2$ the quantum well is at resonance, carriers will be injected in the drift region. The operation mode is then similar to BARITT or TUNNETT diodes. The QWITT diode can also be biased so that current injection can peak at $3\pi/2$ and a high conversion efficiency can be expected. An experimental realization of a Si/SiGe QWITT structure showed the expected asymmetry and resonant states in the I-V characteristic but due to the difficulties with room-temperature operation of Si/SiGe resonant tunneling elements no RF-results have been given [2.21].

2.2.4 The Misawa Mode

Misawa considered the possibility of using a structure with superimposed impact-ionization and drift mechanisms [2.22]. This operation mode can be used to explain a *pin* structure.

In a strongly simplified representation (Fig. 2.9) the occurence of a negative resistance is explained as follows: At a time step "a" a small excess of electrons

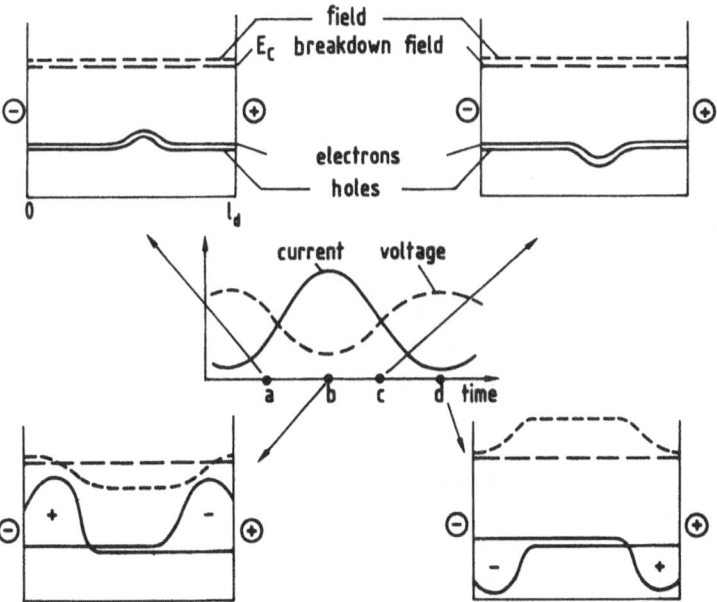

Fig. 2.9. Dynamic behaviour of a pin structure [2.23]

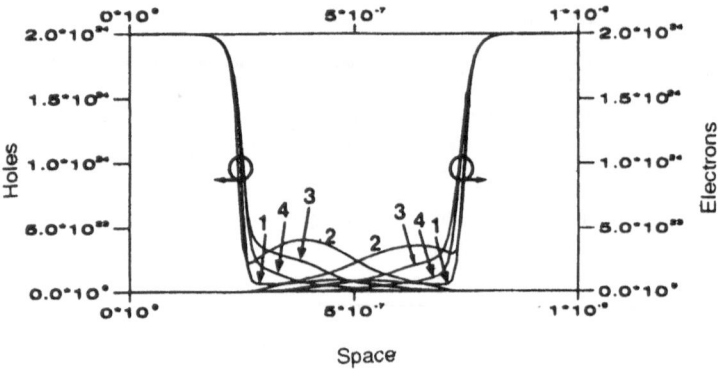

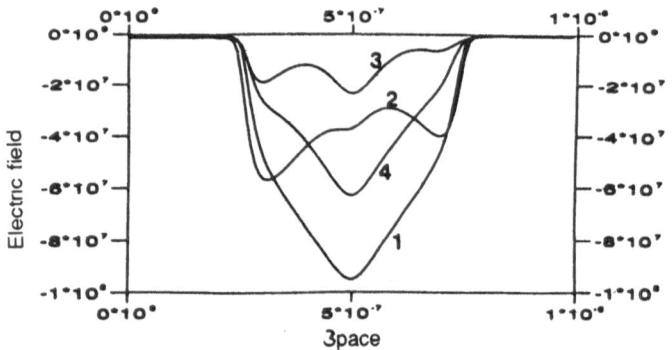

Fig. 2.10. Spatial evolution of the free-carrier densities and the electric field, at various times during a 100 GHz cycle, in a symmetrical flat-doping profile Si DDR IMPATT diode [2.26]. Space coordinates in [m], electric field in [V/m], hole and electron densities in [m^{-3}]. $L_n = L_p = 0.23$ μm, $N_a = N_D = 2 \times 10^{17}$ cm^{-3}, $A = 10^{-4}$ cm^2, $T = 500$ K, $U_0 = 20$ V, $J_0 = 200$ kA/cm^2, $1 \rightarrow t = T_0/4$, $2 \rightarrow T_0/2$, $3 \rightarrow T_0/4$, $4 \rightarrow t = T$

and holes is assumed in the center of the diode. These carriers will drift to the electrodes with increasing concentration due to avalanche multiplication ("b"). The large amount of carriers causes a decrease of the electric field in the center of the diode and consequently a minima during the voltage cycle. At the same time the current is maximum due to the maximum of the carrier density. Due to the decreased field in the center of the diode there is now a lack of carriers ("c") which will drift in the opposite direction. At time step "d" the electric field is increased in the center of the diode due to the carriers which have reached the electrodes. This means voltage is maximum and current is minimum. To close the cycle, the field maximum in the center of the diode will lead to the situation of time step "a". The occurence of a negative resistance is obvious from the phase shift of π between voltage and current.

However, operating pin diodes at moderate current densities leads only to a phase shift of slightly more than $\pi/2$, as can be seen from numerical simulations.

A phase shift of π is only obtained close to the avalanche resonance due to a displacement current.

It is theoretically predicted that operation at high current densities near the avalanche resonance frequency might lead to an improved device performance [2.24]. The required high current densities in pin diodes lead to a breakdown of the electric field in the centre of the diode. This can be avoided by the introduction of a p- and n-doped layer which simply leads to the flat-profile double-drift diode [2.25]. A self-consistent drift-diffusion model identifies the avalanche resonance mode in p^+-p-n-n^+ structures at current densities of about 200 kA/cm^2 which precludes cw operation except for very-small-area diodes [2.26].

The small-signal avalanche frequency (220 GHz) of the structure simulated in Fig. 2.10 is well above the operation frequency (100 GHz), the large-signal avalanche frequency is still very close to the operation frequency [2.25]. The evolutions clearly indicate that the diode is no longer working in a *Read*-like mode of operation since the avalanche phenomenon takes place in the entire active zone.

The simulation yields an optimum design of a 94 GHz IMPATT structure for peak output power in excess of 50 W under low duty cycle: $L_n = L_p = 0.15$ μm, $N_A = N_D = 3 \times 10^{17}$ cm^{-3}, $A = 1.4 \times 10^{-4}$ cm^2. This operation mode is predicted to be useful for frequencies up to 200 GHz where still output powers of several watts should be possible [2.27].

2.3 Numerical Large-Signal Simulations

The analytical models which have already been discussed are all based on the assumption that there exists an injection region and a drift region, which can be treated independently. This is however a crude approximation in several cases: in mm-wave flat-profile IMPATTs a continuous transition between impact ionization and drift mechanism exists [2.28].

For the description of mixed generation and drift mechanisms the semiconductor equations have to be treated:

$$\frac{\partial n}{\partial t} = \frac{1}{e}\frac{\partial J_n}{\partial x} + G,$$
$$\frac{\partial p}{\partial t} = -\frac{1}{e}\frac{\partial J_p}{\partial x} + G,$$
$$G = \alpha_n n v_n + \alpha_p p v_p,$$
$$J_n = -e v_n n + e D_n \frac{\partial n}{\partial x},$$
$$J_p = e v_p p - e D_p \frac{\partial p}{\partial x},$$
$$\varepsilon \frac{\partial E}{\partial x} = -e(N_A - N_D + p - n).$$

$$(2.25)$$

This set of equations is known as representing the drift-diffusion model and can be found in a hierarchy of semiconductor modelling approaches which is explained using Fig. 2.11.

The quantum transport theory is still far beyond the scope of device simulation. The quantum-mechanical nature of electron transport in semiconductors may be described using a "quasi-free-particle" approximation. The electrons are treated as particles with effective masses taken from a band-structure calculation. Screening processes and acceleration due to an electric field yield band structure data different from the case of cold electrons. With known band-structure data and collision terms the Boltzmann-equation can be solved with, e.g., Monte-Carlo methods. These simulations may be used for basic investigations of carrier-transport phenomena as impact ionization but are still too expensive for device simulations. Therefore, further simplifications are appropriate. Instead of treating single carriers or an exact distribution function an approximate distribution function, e.g. a Maxwell distribution, is utilized. The equations for the unknown material parameters may then be obtained from the transport equation in the phase space. Further, it is assumed that the distribution function is in equilibrium with the electric field. Then the drift diffusion model is obtained.

Nearly all large-signal transit-time simulation schemes are based on a grid of points in space and time, and the evaluation of the electric field and carriers at these grid points [2.30]. Using appropriate boundary conditions the device terminal

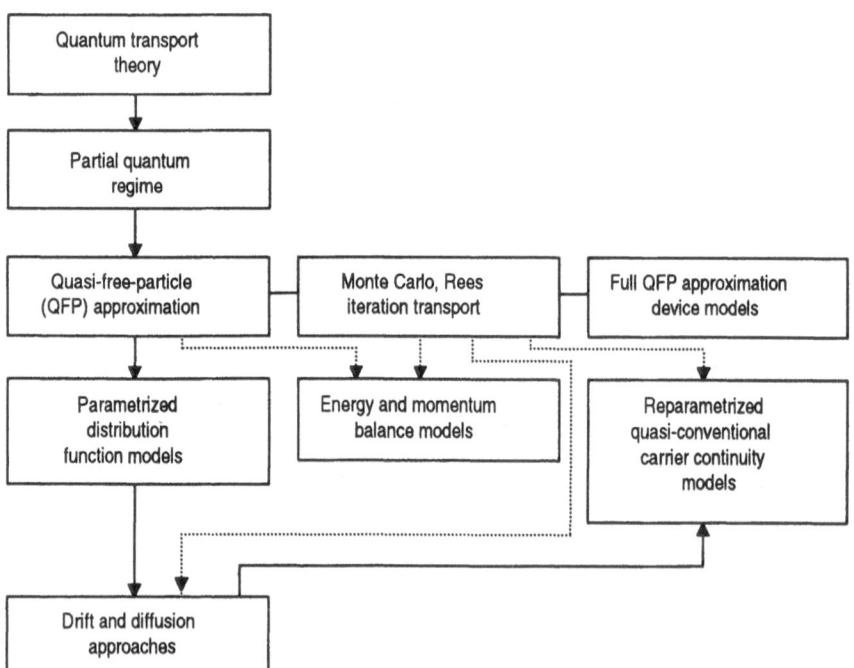

Fig. 2.11. Modells for the carrier transport in semiconductors [2.29]

voltage and current can be calculated at each time level. If a steady state is obtained, Fourier analysis leads to impedance, output power and efficiency.

In detail, the discretization of the quantities and the solution algorithm is much more difficult than what one would think from this general description. One of the first successful algorithms was described by *Scharfetter* and *Gummel* [2.31], an implicit method which relies on the idea of treating the current-density equations as differential equations in p and n with J_n, J_p, μ_p, μ_n and E assumed constant between the mesh points. Problems with pseudo-diffusion and CPU time are attributed to this method. In explicit schemes a double-interleaved spatial mesh is used, with carrier densities on one mesh and fields and currents on the other [2.32,33]. A FORTRAN source code was given by *Thoren* [2.34].

Recently, the effects of momentum and energy relaxation have been considered [2.35,36]. It has been observed that, when the energy relaxation is taken into account, an additional injection delay is introduced. This delay tends to improve device performance due to a smaller transit time. However, as the energy relaxation time is approaching $\tau_e = 0.1$ ps at high energies in silicon [2.37], an approximate phase delay of $\varphi_e = \arctan(\omega\tau_e) = 3.5°$ at 100 GHz will not influence the device performance remarkably. The momentum relaxation time is still below the energy relaxation time. The models are used for the design and analysis of different structures. For the 50 GHz double low-high-low structure in Fig. 2.12 an output power of 1.1 W at a current density of 12 kA/cm^2 with a device area of 2×10^{-5} cm^2 was calculated [2.38].

Using a drift-diffusion model different doping profile structures have been compared for 94 GHz operation accounting for thermal limitations [2.39]. The

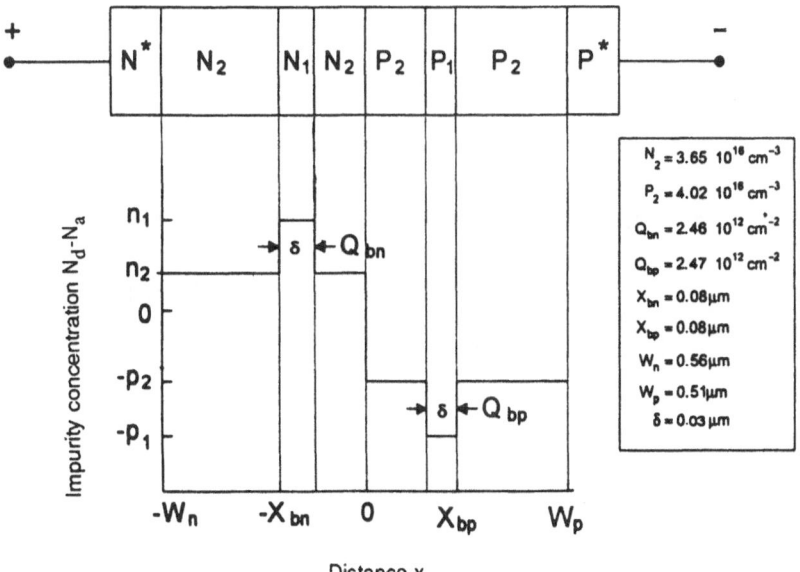

Fig. 2.12. Double low-high-low 50 GHz IMPATT structure

comparison is difficult due to the necessary assumptions on the loading circuit and can be divided as follows:

– If a high-Q waveguide cavity is used and the lowest possible thermal resistance is achieved, flat-profile double-drift diodes are expected to yield the highest output power. This is due to the fact that the output power increases with the RF modulation. The highest RF power levels are obtained at low terminal negative-resistance levels.

– If more lossy circuits are used, such as planar ones, then the higher negative resistance level of double low-high-low or quasi-*Read* diodes makes them the best choice.

2.4 Skin Effect

One of the limitations of the performance of a transit-time diode at millimeter-wave frequencies arises from an increase in its series resistance at high frequencies due to the ac current flow primarily at the surface ("skin effect"). At microwave frequencies this phenomenon has been found too small to account for an increase in series resistance [2.40]. The importance of the skin effect in the substrate at millimeter-wave frequencies becomes evident from a simple model [2.41], where a hollow shell is assumed with a uniform current flow in a thickness

$$\delta_p = \frac{1}{\sqrt{\omega \mu_0 \sigma}} \qquad (2.26)$$

which is the penetration depth in cylinder symmetric conductors. For example, at 100 GHz a penetration depth of 3.5 μm is expected for a substrate resistivity of $1/\sigma = 1$ mΩcm. This is less than a typical device diameter, and one conclusion leads to the recommendation of extreme substrate thinning. However, this simple approximation overestimates the effect of current displacement at high frequencies, and device dimensions in the region of the skin depth. A calculation of the current distribution in cylindric semiconductor material solving Maxwell's equations is used to evaluate an effective value of the skin effect resistance [2.42]

$$R_{\text{eff}} = \frac{L}{\sigma \pi r^2} \left[1 + \left(\frac{r}{\delta_p} \right)^4 \Re \left\{ \sum_{s=1}^{\infty} \frac{1}{j_{1,s}^2 k_s^2} \left(1 - \frac{2r}{k_s L} \tanh \frac{k_s L}{2r} \right) \right\} \right] \qquad (2.27)$$

where L denotes the substrate thickness r the diode radius, and $j_{1,s}$ are the zeros of the Bessel function J_1, $k_s = j_{1,s}^2 + j(r/\delta_p)^2$.

An application of this formula to W-band diodes reveals that the substrate thickness of the considered high-conductivity silicon material should be below 9 μm if $R_{\text{eff}} < 0.25$ Ω for a diode with 30 μm diameter. An extension of this calculations to consider the skin effect in contact layers leads to an optimum substrate thickness for minimum skin effect resistance [2.43].

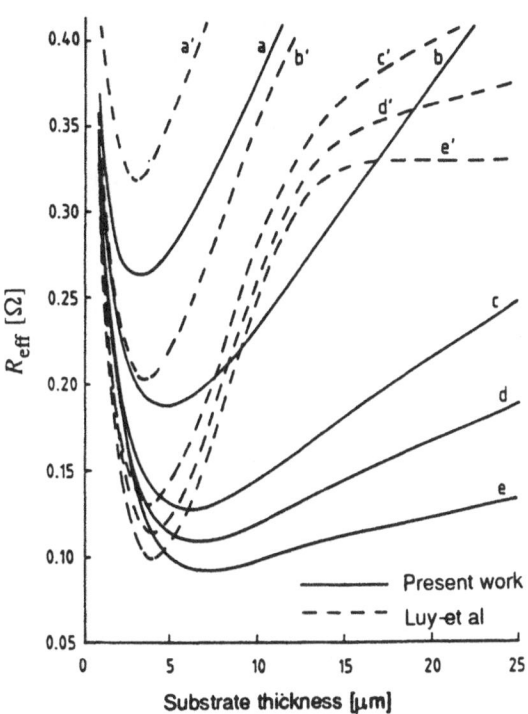

Fig. 2.13. Skin-effect resistance of a semiconductor-gold composite for different diode radii as parameter [2.43]. Uniform excitation and collection. $a = 15$ μm; $b = 25$ μm; $c = 50$ μm; $d = 75$ μm; $e = 150$ μm. Solid lines correspond to the calculation considering a thick gold contact layer [2.43], dashed lines according to (2.27)

It should be noticed that the occurence of an optimum substrate thickness different from zero is the result of the assumption of a relatively thick gold contact layer (5 μm) leading to a current displacement in the contact region and consequently a δ-function ring-type collection at the substrate-contact interface. In this case, non-uniformly doped contact regions might yield lower skin-effect resistances [2.44].

2.5 Thermal Properties

Transit-time devices are operated at elevated junction temperatures due to the limited conversion efficiency. The Mean Time Between Failure (MTBF) is influenced by the temperature distribution as well as the RF performance. Therefore, it is important to know the temperatures in the devices under operation conditions.

These devices are mounted on heat-sinks. From a thermal point of view diamond-type IIa is the best choice for the heat-sink material. In MMICs, however, the heat has to flow through a semiconductor substrate before a heat-sink is reached.

The conduction of heat in a solid with a temperature-dependent thermal conductivity is generally governed by [2.45]:

$$\vec{j} = -\kappa \nabla T \tag{2.28}$$

with the flux of heat $\mathbf{j}$, the thermal conductivity κ, and the temperature T. The continuity equation may be derived assuming that the heat content of a volume element with a density ρ and a specific heat c can be described by

$$Q = c\rho T. \tag{2.29}$$

The transient change of heat content has to be provided by the change of the heat flux with respect to the space coordinates which yields

$$\dot{Q} + \mathrm{div}\vec{j} = 0 \tag{2.30}$$

if no heat is generated or wiped out within the medium. In the case of steady temperature in which T does not vary with time and for the cylinder symmetric case we get

$$-\frac{1}{r}\frac{\partial}{\partial r}\left(\kappa r\frac{\partial T}{\partial r}\right) + \frac{\partial}{\partial z}\left(\kappa\frac{\partial T}{\partial r}\right) = 0 \tag{2.31}$$

or

$$\frac{1}{r}\left[\frac{\partial\kappa}{\partial T}r\left(\frac{\partial T}{\partial r}\right)^2 + \kappa\left(r\frac{\partial^2 T}{\partial r^2} + \frac{\partial T}{\partial r}\right)\right] + \left[\frac{\partial\kappa}{\partial T}\left(\frac{\partial T}{\partial z}\right)^2 + \kappa\frac{\partial^2 T}{\partial z^2}\right] = 0. \tag{2.32}$$

which shows the nonlinearity of the differential equation. Following *Carslaw* and *Jaeger* [2.45] a suitable transformation is

$$\theta = \int_{T_0}^{T} \kappa(T)dT \tag{2.33}$$

which yields

$$\vec{j} = -\nabla\theta \tag{2.34}$$

for the heat flux and for the heat flux potential

$$\Delta\theta = 0. \tag{2.35}$$

This is a Laplace equation with θ being the only temperature-dependent variable.

Using the model in Fig. 2.14 and appropriate boundary conditions an analytical solution can be found [2.46].

$$\theta_1(r, 0 \le z \le H) = jr_1\left[\left(1 - \frac{z-H}{r_2}\right)\frac{r_1}{r_2}\right.$$

$$\left. + \sum_{n=1}^{\infty} J_0\left(j_{1,n}\frac{r}{r_2}\right)e^{-j_{1,n}\frac{z}{r_2}}\pi\frac{J_1\left(j_{1,n}\frac{r_1}{r_2}\right)}{j_{1,n}}\right]. \tag{2.36}$$

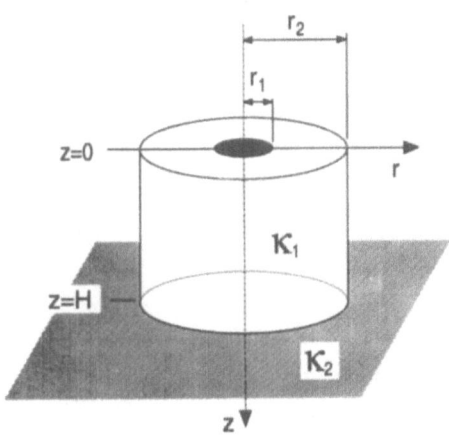

Fig. 2.14. Model for the calculation of the temperature stress in transit time devices: Focal plane with radius r_1 on a cylinder with radius r_2 and temperature dependent thermal conductivity $\kappa_1(T)$ on a semi-infinite space with thermal conductivity κ_2. The temperature is assumed to be constant at $z \to \infty$

The temperature dependence of the thermal conductivity of silicon can be expressed as [2.47]:

$$\kappa_1(T) = \frac{\kappa_{10}}{T/T_0 - 1} \tag{2.37}$$

with $K_{10} = 4\text{W/(cm·K)}$ and $T_0 = 80$ K. The agreement of this representation with measured values from [2.48] is rather good (Fig. 2.15).

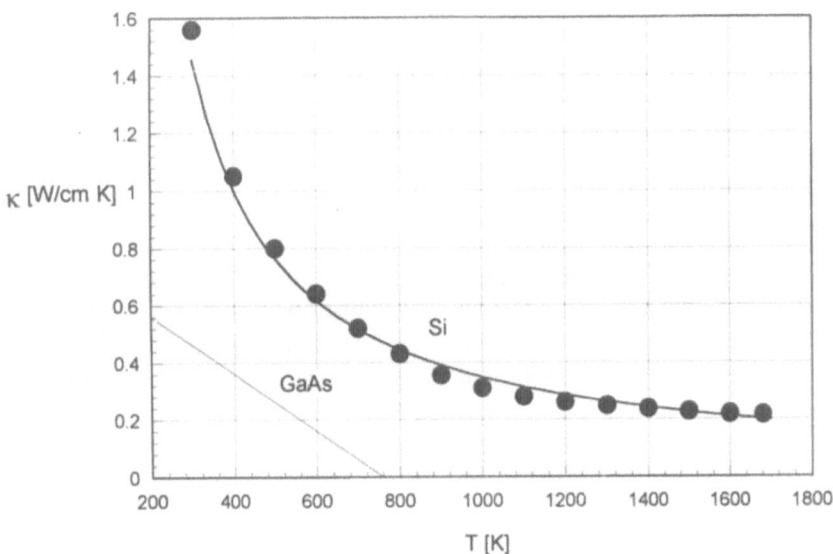

Fig. 2.15. Thermal conductivity of silicon as a function of the temperature

The back transformation using (2.33) yields

$$\theta_1 - \theta_0 = \int\limits_{T(r,H)}^{T(r,z)} \kappa_1(T)dT = T_0 K_{10} \ln \left| \frac{T(r,z)}{T_0} - 1 \right|$$

$$- T_0 K_{10} \ln \left| \frac{T(r,H)}{T_0} - 1 \right| \tag{2.38}$$

with

$$T(r,H) = \frac{\theta_0(r)}{\kappa_2} + 300 \text{ K} \tag{2.39}$$

It is important to consider a reference temperature in order to avoid the complicated splitting of (2.36). We obtain

$$T(r,z) = [T(r,H) - T_0]e^{\frac{\theta_1 - \theta_0}{K_{10}T_0}} + T_0. \tag{2.40}$$

With the definition of the thermal resistance

$$R_{th} = \frac{\Delta T}{I}, \tag{2.41}$$

the temperature difference ΔT, and the heat-generation rate $I = jr_1^2\pi$, the thermal resistance can be calculated. Because only the temperature difference is of interest, the reference temperature (e.g., 300 K) again has to be subtracted and we obtain

$$R_{th}(\kappa_1(T)) = \frac{\left(\frac{\theta_0(r)}{\kappa_2} + 300 \text{ K} - T_0\right) e^{\frac{\theta_1 - \theta_0}{K_{10}T_0}} + T_0 - 300 \text{ K}}{I}. \tag{2.42}$$

The thermal resistance is dependent on the supplied heat flux density: j is appearing in θ_1 and θ_0, and therefore the thermal resistance has an exponential j-dependence. This is explained by the decreasing thermal conductivity with increasing heat-flux density – the thermal resistance is dependent on the input power!

2.5.1 Integrated Transit-Time Devices

A monolithically integrated coplanar SIMMWIC diode in mesa configuration (Chap. 5) is modeled as a focal plane on a semiconductor substrate. Figure 2.16 shows the calculated thermal resistances for silicon substrate material. It can be seen that in small devices the self-heating effect is more pronounced, i.e., for small input powers the thermal resistance is approximately proportional to the inverse junction diameter whereas at large input powers $R_{th} \propto r_1^{-2}$ is approached.

2.5.2 Diamond Heat Sinks

The thermal conductivity of type-IIa diamond can be described by [2.49]

$$\kappa_1(T) = K_{10}T^{-\beta} \tag{2.43}$$

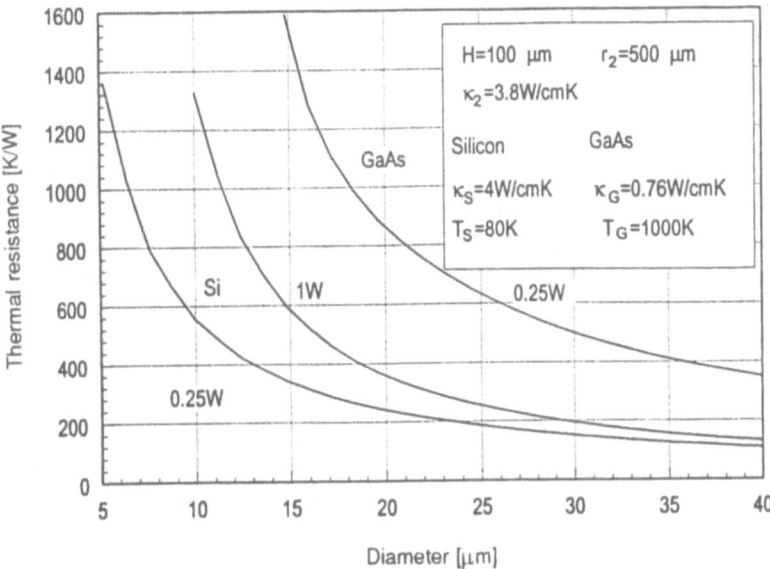

Fig. 2.16. Calculated thermal resistances versus device diameter for different input powers and silicon and GaAs

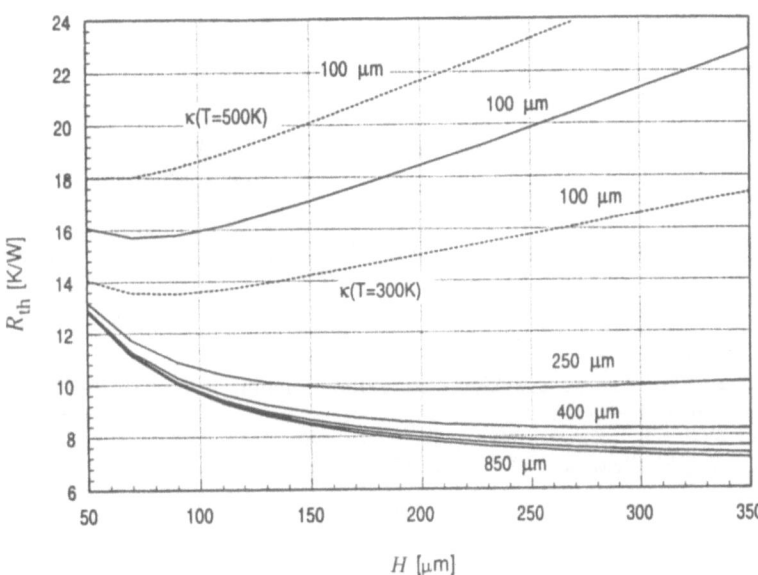

Fig. 2.17. Thermal resistance of a focal plane on a diamond heat sink as a function of the height of the diamond. (Parameter: radius of the diamond)

which yields after transformation

$$T(r, z) = \left(-\frac{(\beta - 1)(\theta_1 - \theta_0) - [T(r, H)]^{1-\beta} K_{10}}{K_{10}} \right)^{\frac{1}{1-\beta}}. \tag{2.44}$$

The reference temperature is assumed to be 300 K. The "conductivity" is $K_{10} = 27530$ and $\beta = 1.26$, the dimension of K_{10} is W/cmK.

The thermal conductivity of diamond is about one order of magnitude larger than that of silicon (diamond: $\kappa_1(300\ \text{K}) \approx 20$ W/cm·K and $\kappa_1(500\ \text{K}) \approx 10$ W/cm·K). Because of this high thermal conductivity diamond is used as a heat sink material for transit time devices [2.50]. In Fig. 2.17 it can be seen that the thermal resistance R_{th} reveals a minima at small cylinder (diamond heat sink) radii with respect to thickness. With increasing radius r_2 the minima becomes weaker and for radii $\geq 300\ \mu\text{m}$ is no longer visible. This minima is dependent on the expression for the temperature dependence of the thermal conductivity: Only for a constant thermal conductivity $\kappa(500\ \text{K})$ the rule $H = r_2/3$ seems appropriate. At a higher conductivity $\kappa(300\ \text{K})$ the minima occurs for $H > r_2/2$.

If $r_2 > 400\ \mu\text{m}$ the calculated R_{th} values are very close together – a significant improvement in thermal resistance by further increasing the diamond dimensions is not possible.

2.5.3 Ring Diodes

The solid circular-mesa diode is not thermally optimized. A reduction of the thermal resistance can be expected using ring structures [2.51]. The Laplace equation for steady-state and one-dimensional heat flow is solved using a Fourier integral approach. The thermal resistance of a ring diode can be expressed as

$$R_{\text{th ring}} = R_{\text{th mesa}} \frac{1}{\sqrt{a^2 + 2aR}} \left(\int_R^{R+a} dr \frac{r}{r+R} \frac{2}{\pi} K(m) \right) \tag{2.45}$$

with $K(m) = \int_0^{\frac{\pi}{2}} \frac{d\alpha}{\sqrt{1 - m \sin^2 \alpha}}$ (Complete elliptic integral of the first kind)

and $m = \frac{4rR}{(r+R)^2}$, $R_{\text{th mesa}} = \frac{1}{\pi \kappa \sqrt{a^2 + 2aR}}$.

Figure 2.18 displays the ratio of the thermal resistance of a solid circular diode to a ring diode of equal area as a function of the ratio of the inner radius R to the cylinder thickness a.

The results from [2.51] are represented by the dashed line: *Gibbons* and *Misawa* assumed a space-dependent heat flux density which is responsible for the derivations. Obviously, a ring structure causes a significant reduction of the thermal resistance. If the inner radius exceeds the cylinder thickness by more than a factor five the thermal resistance is reduced by a factor of two.

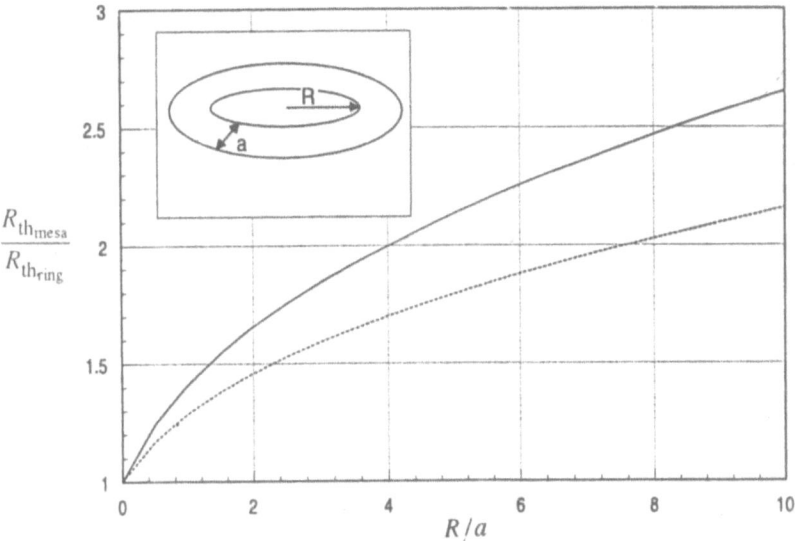

Fig. 2.18. Improvement of the thermal resistance by the use of a ring structure instead of a solid circular structure of the same area. The inset shows a scheme of the model. (Dashed line according to [2.51], solid line according to (2.45))

2.5.4 Transient Thermal Resistance

For pulsed operation the transient heat-conductivity equation has to be solved. The solution for the transient thermal resistance of a focal plane with radius r on a heat sink with constant thermal conductivity κ and thermal diffusivity α is [2.52,53]:

$$
R_{\text{th HS}} = \frac{2}{\pi r \kappa} + \frac{\sqrt{d\alpha t_1}}{\pi r^2 \kappa} \left[\frac{2}{\sqrt{\pi d}} \left(1 - e^{\frac{r^2}{4\alpha t_1}} \right) - \frac{r}{\sqrt{\alpha t_2}} + I \right.
$$

$$
\left. + \frac{r}{\sqrt{d\alpha t_1}} \, \text{erfc} \left(\frac{r}{2\sqrt{\alpha t_1}} \right) \right]
$$

with

$$
I = \frac{2}{\pi d} \int\limits_0^\infty \frac{e^{-dx^2} \left(e^{-(1-d)x^2} - e^{-x^2} \right) \left[1 - \cos(rx^2 \sqrt{\alpha t_2}) \right]}{x^2(1 - e^{-x^2})} dx
$$

and $d = t_1/t_2$ is the duty factor. Figure 2.19 shows the effect of different heat-sink materials and different diode radii on the thermal resistance. Even in pulsed operation an improvement in thermal resistance is obtained using diamond heat-sinks which may be important at very high input powers [2.54].

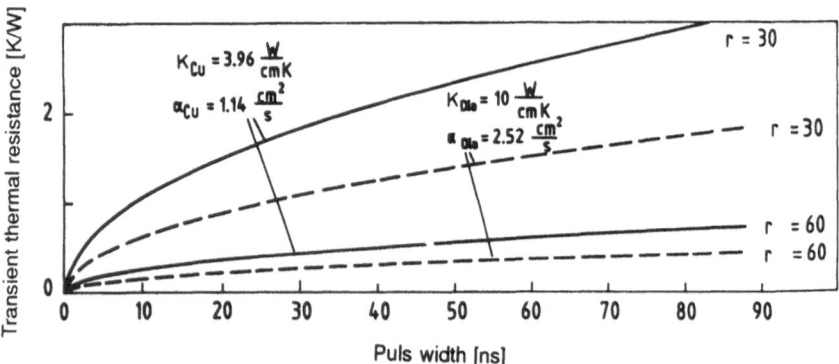

Fig. 2.19. Transient thermal resistance as a function of the pulse width for copper (solid lines) and diamond heat sink (dashed lines). Parameter: diode radius in micrometers

2.5.5 Measurement of the Thermal Resistance

The decreasing ionization rate with increasing temperature results in a temperature dependence of the breakdown voltage β

$$\beta = \frac{1}{U_b} \frac{dU_b}{dT}. \tag{2.46}$$

This temperature coefficient is dependent on the doping profile via the electric field and the temperature [2.55]. Due to this dependence it is necessary to measure β in a temperature chamber at least for every new diode lot.

The measurement of the thermal resistance is based on the fact that the temperature dependence of U_b results in a contribution to the low-frequency small-signal ac resistance of avalanche diodes [2.56]. This contribution, called the thermal-ohmic resistance $R_{t\Omega}$, has to be separated from the contributions to the total diode resistance. This separation is relatively easy because $R_{t\Omega}$ is strongly frequency dependent. *Haitz* et al. defined a cut-off frequency

$$f_c = \frac{\kappa}{4\pi c^* \rho L^2} \tag{2.47}$$

with the specific heat $c^* = 0.73$ J/gK, the density $\rho = 2.3$ g/cm^3, length of the heat-flow section L, and the thermal conductivity $\kappa = 0.8$ W/cmK

For $f > f_c$ the thermal-ohmic resistance $R_{t\Omega}$ vanishes. $R_{t\Omega}$ is simply obtained as the difference between a low-frequency value ($f < f_c$) and a high-frequency value ($f > f_c$) of the diode resistance. The thermal resistance is then calculated from

$$R_{th} = \frac{R_{t\Omega}}{\beta U_b^2}. \tag{2.48}$$

A similar method has been adopted to the measurement of thermal resistances of BARITT diodes [2.57].

2.6 Design Constraints

The design of transit-time devices is strongly affected by the device area. *Thermal limitations* are of especial importance for mm-wave IMPATT diodes. The input power

$$P_{\text{in}} = UJA \qquad (2.49)$$

(voltage: U, current density: J, area: A) is limited by the maximum-allowed junction temperature rise

$$\Delta T \geq P_{\text{in}} R_{\text{th}}. \qquad (2.50)$$

If we use, for simplicity,

$$R_{\text{th}} = \frac{1}{\kappa \sqrt{\pi A}} \qquad (2.51)$$

we obtain the thermal current density limit

$$J_{\text{th}} \leq \frac{\kappa \Delta T}{U} \sqrt{\frac{\pi}{A}}. \qquad (2.52)$$

If ionization has to be avoided in the drift region a *space-charge limit* can be derived from Poisson's equation $Q/\varepsilon \leq \Delta E$ and with $Q = 2\pi J/\omega$ we obtain

$$J_{\text{sc}} \leq \frac{1}{2\pi} \varepsilon E \omega. \qquad (2.53)$$

For IMPATT diodes the *avalanche resonance limit* provides a further constraint on the maximum current density (Sect. 2.2). The current density must be below

$$J_{\text{av}} \leq \frac{\varepsilon \omega^2}{3 v_s \alpha} \qquad (2.54)$$

– to ensure READ type operation. These limitations may be represented in an $A-J$ plane [2.58].

For oscillator operation the two-terminal device has to be matched to a resonator circuit considering all losses. The equivalent circuit for an oscillator is shown in Fig. 2.20 [2.59].

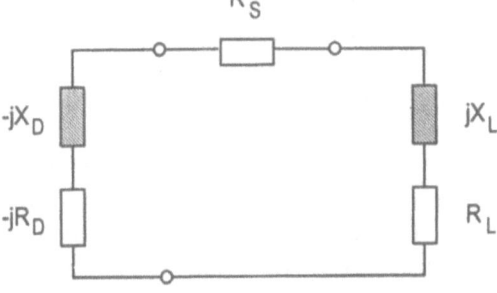

Fig. 2.20. Oscillator equivalent circuit (Real and imaginary part of diode impedance R_D, X_D; load impedance: R_L, X_L; losses are summarized in R_s)

The power delivered to the load is

$$P_{\rm L} = P_{\rm D} V \tag{2.55}$$

with

$$V = 1 - \frac{R_{\rm s}}{|R_{\rm D}|} \tag{2.56}$$

describing the matching losses. The power generated by the diode is

$$P_{\rm D} = \frac{1}{2} J_{\rm t}^2 A^2 R_{\rm D} = \eta_{\rm i} U J A. \tag{2.57}$$

The oscillator output power may now be calculated with appropriate matching assumptions. It is evident that the intrinsic efficiency $\eta_{\rm i}$ is independent of the area ($R_{\rm D} \infty A^{-1}$!). A measured $\eta(A)$ dependence [2.11] is therefore due to losses $R_{\rm s}$ which are nearly independent of the diode area.

2.7 Technology

The fabrication of mm-wave transit-time devices begins with the growth of the required epitaxial layers. For monolithic circuits the substrate may be patterned (Chap. 7). The choice of appropriate contact materials is of importance for the performance and for the lifetime. Special techniques are necessary for the handling of discrete mm-wave devices. Finally, a packaging technique is needed.

2.7.1 Material Growth

The active layers of single-drift, double-drift flat profiles and even quasi-*Read* double-drift diodes may be grown by standard chemical-vapour-deposition techniques up to the mm-wave range. The experimental realization of double low-high-low structures has been up to now exclusively done with the method of silicon Molecular Beam Epitaxy (MBE). Due to typical growth temperatures in the range of 550°C to 750°C this method provides the realization of submicrometer doping profiles with nearly arbitrary shape [2.60].

In order to obtain the required doping profile with good homogenity and with a low defect density a suited combination of cleaning procedures before MBE, an appropriate growth temperature and doping technique have to be chosen.

The measurement of the doping profile of a submicrometer double-drift diode is possible using Secondary Ion Mass Spectroscopy (SIMS). This method determines the chemically activated doping atoms. The Spreading Resistance Probe (SRP) method yields the electrically activated doping atoms – however, due to carrier spilling effects only one side of the p-n junction is resolved [2.61]. An access to the other side is possible by an additional C-V measurement. The doping level

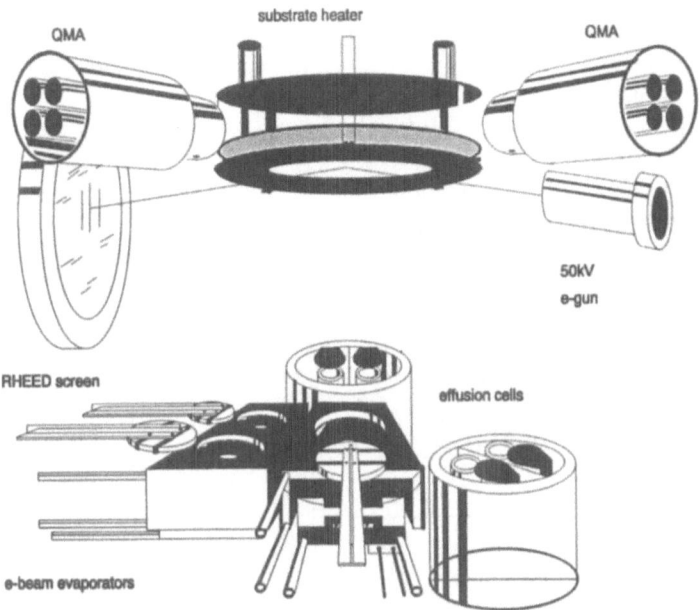

Fig. 2.21. Scheme of a silicon molecular beam epitaxy machine used for the growth of IMPATT diode layers [2.60]

	Cleaning procedure	
Temperature	HF	"RCA"
550°C	– low defect density – doping by DSI/pre build up – very abrupt transitions	high defect density
750°C	very high defect density	– very low defect density – doping by DSI/coevaporation – abrupt transitions

Fig. 2.22. Cleaning procedures, growth temperatures and doping techniques for the growth of a n doped layer by MBE

determined by the SRP method is used as input in the analysis of the C-V results:

$$N_p(V) = \left(\frac{e\varepsilon A^2}{2} \frac{d(1/C^2)}{dV} - \frac{1}{N_n} \right)^{-1} \qquad (2.58)$$

and

$$l_p(V) = \frac{\varepsilon A/C}{1 + N_p/N_n} \qquad (2.59)$$

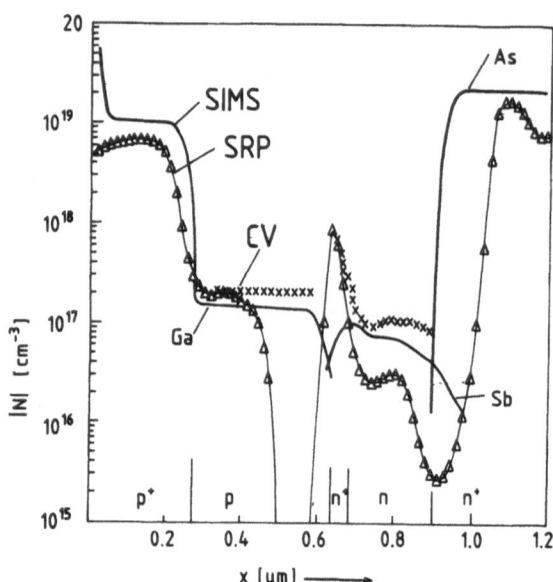

Fig. 2.23. Doping profile of a quasi-*Read* double-drift structure measured by the SIMS, SRP and CV methods [2.11]

where the n-layer doping N_n is obtained from the SRP measurement in this example. Figure 2.23 exhibits a comparison of the different doping-profile assessment methods using a quasi *Read* structure.

2.7.2 Contacts

For contacting IMPATT diodes ohmic contacts must be created which have low specific contact resistances and which must be able to withstand the thermal stress due to the high operation temperatures.

With the Ti/Au system specific contact resistances of $4 \cdot 10^{-6} \Omega cm^2$ are obtainable on (100) n-Si ($N_D = 5 \cdot 10^{19}$ cm^{-3}). This system is stable in air up to 410°C. Beyond this temperature island growth and layer decomposition begins. On heating gold diffuses through Ti to the Si/Ti interface, accumulating there up to one atomic layer. When using medium-doped substrate material ($N_D = 2 \cdot 10^{18}$ cm^{-3}) no ohmic contact is obtained [2.62]. An improvement of the contact reliability is achieved using Au/Pt/Ti with Pt as a diffusion barrier [2.63].

Silicides have been intensively investigated for their potential as interconnect materials for VLSI structures owing to their low resistivity compared to commonly used poly-Si. Nickel disilicide (NiSi$_2$) can be used advantageously due to a low lattice mismatch to silicon (0.4%) and a low Schottky barrier. The specific contact resistivities of NiSi$_2$ are below the values of the Ti/Au metallization. NiSi$_2$ formed on (100)-oriented n-type Si-MBE layers yield specific contact resistivities down to $6 \cdot 10^{-6} \Omega cm^2$ for a n-layer doping of $1.6 \cdot 10^{19}$ cm^{-3} [2.64].

For Mnp$^+$-BARITT diodes a low hole barrier is required to maximize the hole injection current (Sect. 2.2). Pt is commonly used due to its lowest hole barrier compared to other metals [2.65]

2.7.3 Handling Techniques

Transit-time diodes may be fabricated as single-mesa devices with extremely thinned substrate on a thin gold foil for low thermal resistance. The handling of these diodes is rather difficult and time consuming. For a reproducible production a method for fabricating transit-time devices with beam leads is appropriate [2.66,67]. A scheme for the fabrication of two terminal devices with integrated gold ribbons, which allows mounting on copper heat sinks as well as on diamond heat sinks,

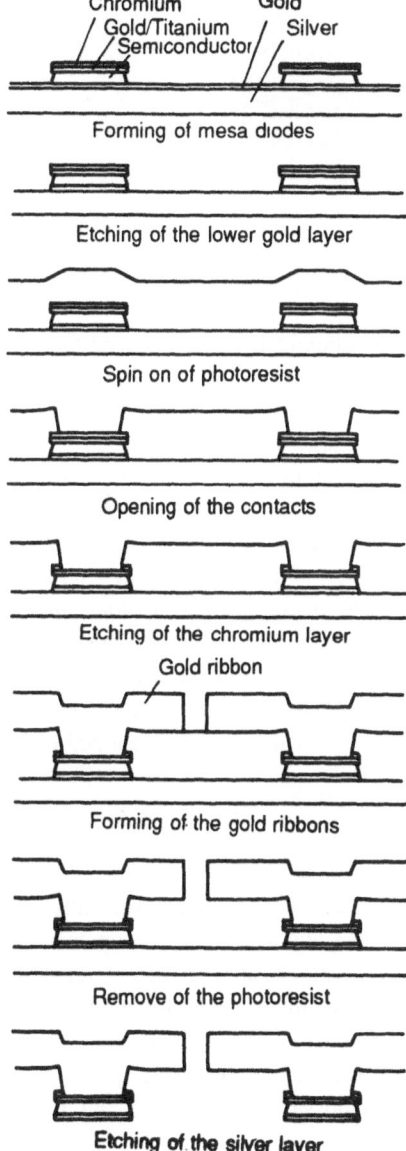

Fig. 2.24. Fabrication sequence of beam lead diodes [2.68]

Fig. 2.25. SEM of a beam lead IMPATT diode with cross ribbon [2.68]

was described [2.68]. The individual diodes are chemically separated and have an extremely thin metallization system on the heat-sink side. Figure 2.24 shows the process.

The fabrication sequence starts with epitaxy, thinning of the substrate and metallization of both ohmic contacts. Only two further metallization layers are necessary: a thick silver layer electroplated on the substrate side and thin chromium layer on the epitaxy side. A photolithography process defines the mesa diodes. The metallization and the silicon is etched. To separate the diodes the lower gold layer is etched, while the chromium layer on top of the mesas protects the upper metallization. Then a new photoresist with a thickness of 6 μm covers the mesas. The upper metallization of the mesas is opened and the chromium layer is etched. The gold ribbons are formed by evaporating and electroplating on top of the photoresist. The photoresist is removed and the silver layer is etched. Now the separated beam lead diodes are ready for insertion in a package.

2.7.4 Packaging

Discrete transit-time diodes are mounted on heat sinks. Copper heat sinks are sufficient for low-power applications or BARITT diodes which are operated at moderate input power levels. High-power IMPATT diodes are mounted on diamond heat sinks. The heat sinks are gold plated to facililtate thermo compression bonding. At mm-wave frequencies quartz rings are commonly used for the housing of the diodes. To obtain a low inductance a multi-ribbon contact technique is applied. The housings are hermetically sealed with a cap. A possible fabrication technique of the required components is described below.

Diamond Heat Sinks

A copper pin serves as a carrier for a circular type-IIa diamond. A hole with a diameter of 0.7 mm and a depth of 0.23 mm is drilled into the copper pin. The circular diamond ($\phi = 750$ μm, thickness 250 μm) is pressed into the hole. The diamond is gold plated.

Quartz Rings

Quartz rings may be fabricated from commercially available capillars. The capillars are sewed into dies of 500 μm height, then fixed with a wax and polished to a height of 300 μm. Titanium and gold is then evaporated. The rings are dissoluted from the carrier and fixed again with the plated side on the carrier. The rings are polished to the final thickness and the metallization is evaporated. Finally the rings are electroplated in a moved carrier to avoid metallization of the cylinder walls.

Caps

A 60 μm thick Cu/Be material is suitable for the fabrication of the caps. Both sides are electroplated with 5 μm thick Au. The circular caps are defined by two-side photolithography. The gold is etched and the tin is etched in a $CuCl_2$ solution until separate caps are obtained.

Thermocompression Bonding

To facilitate a partially automatic mounting all components may be thermocompression bonded. It is necessary to use a thermocompression bonder with heated-chuck and heated-bond tip. For the bonding of the quartz rings a chuck and tip temperature of 280°C has proven useful. The bonding time is 60 s. Bonding parameters of the diode depend on the diode diameter to achieve the lowest possible thermal resistance. For a 30 μm diode a chuck temperature of 250°C is used. The tip temperature is critical for BARITT diodes. Correspondingly longer bonding times are required than for IMPATT diodes where a tip temperature of 250°C is applied. Typical pressure is 10 g for 30 s for the latter device. Bonding of the ribbons on the quartz ring or on additional intermediate gold rings for improved shearing strength is as uncritical as the bonding of the cap and can be done using similar parameters as for the bonding of the quartz ring on the heatsink.

2.8 Performance

In this section the performance of discrete transit-time devices is reported. The performance of SIMMWIC circuits employing IMPATT diodes may be found in Chap. 5.

Transit-time devices are commonly characterized by their output power and efficiency which can be obtained with an optimal matched-resonator configuration. The noise level is of the same importance for applications. In the self-oscillating mixer mode the Minimum Detectable Signal (MDS) at a certain output power level is of interest (Chap. 6).

2.8.1 Power

For the measurement of high power levels at millimeter-waves attenuators are necessary to protect the power sensors. Power measurement via directional couplers should be avoided due to their, in most cases unknown behaviour at harmonics. Further care has to be taken for the calibration of the sensors. Currently thermistors seem to be most accurate over a specified waveguide band [2.69].

Keeping this in mind as well as the different allowed junction temperatures the available experimental results may be compared.

Silicon IMPATT diodes are expected to work up to frequencies of 500 GHz [2.70]. Experimentally, silicon IMPATT diodes have been operated at frequencies up to 300 GHz [2.71]. Figure 2.26 provides a survey of the best results achieved since 1976.

These results have been obtained using waveguide resonators which provide the highest flexibility in matching the diodes. Figure 2.27 shows the cross section of a coaxially-coupled reduced-height waveguide resonator.

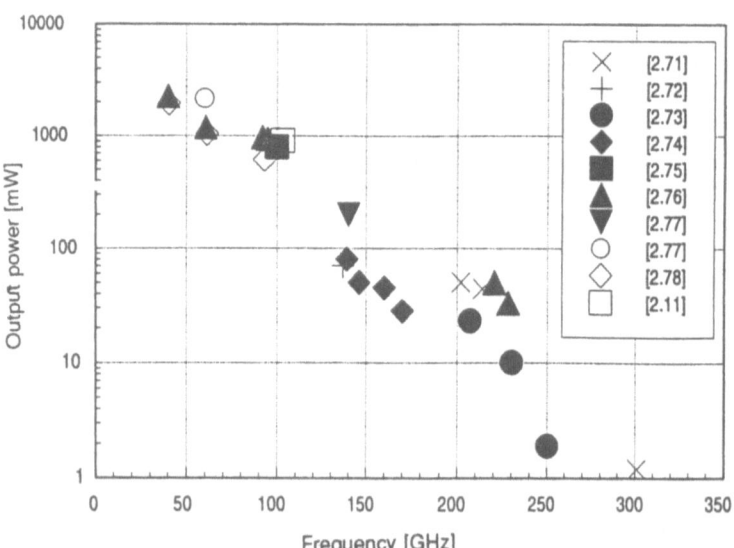

Fig. 2.26. CW output power as a function of the frequency for silicon IMPATT diodes obtained from different groups. Crossed symbols: Single-drift diodes. Closed symbols: Double-drift diodes. Open symbols: *Read*-type diodes

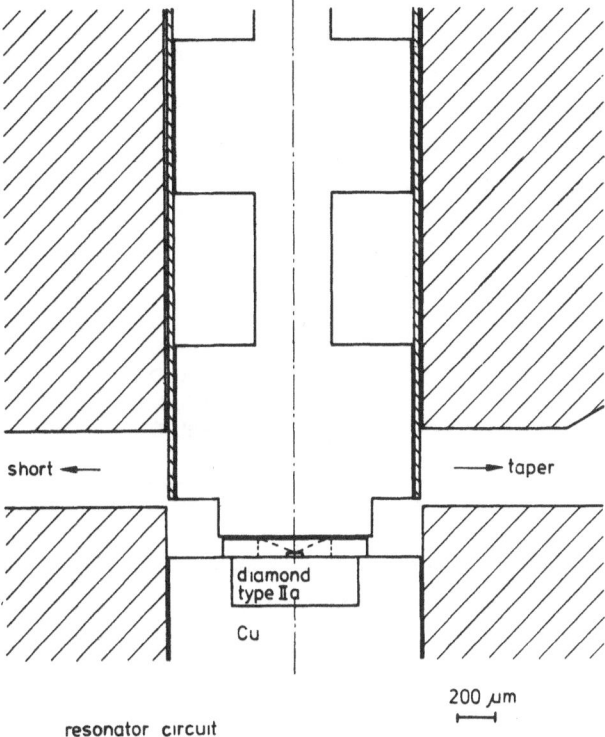

short ←   → taper

diamond
type IIa

Cu

200 μm

resonator circuit

Fig. 2.27. W-band test resonator

Due to the more unfavourable phase relations the output powers which can be obtained from BARITT diodes are much lower than those from IMPATT diodes. Nevertheless, BARITT diodes provide an interesting alternative to other active devices especially at Ka-band frequencies. In this frequency range BARITT diodes can deliver up to 6 mW of cw output power [2.80], are easy to fabricate and may be operated as sensitive self-oscillating mixers. At higher frequencies matching becomes more difficult but still 1 mW of output power has been achieved at 60 GHz [2.81].

2.8.2 Efficiency

The conversion efficiency is of special importance as well for mobile systems as for systems with preheated diodes. Experimentally a dependence of the measured external efficiency on the diode type and the diode area is observed: The amount of the real part of the transit time diode increases with decreasing area which reduces the matching losses. Additionally the optimum current densities cannot be reached with large-area diodes due to thermal limitations. At very small diode areas other mechanisms seem to prohibit reaching the theoretical expected intrinsic efficiency (Chap. 5).

Table 1. Measured conversion efficiencies and operation data as available from different references

Efficiency (%)	Bias data	Frequency	Diameter/ Capacitance	Profile	Reference
1.3	410 mA	202 GHz	23 μm	SD	2.71
3.2	170 mA	137 GHz	18 μm	SD	2.72
1		220 GHz	$\cong$ 15 μm	DD	2.76
5.8	470 mA	94 GHz	$\cong$ 40 μm	DD	,,
4.1	960 mA	94 GHz	$\cong$ 50 μm	DD	,,
6.7	44 kA/cm^2	94 GHz	1.2 pF	DD	2.82
10	570 mA	94 GHz	$\cong$ 45 μm	DD	2.75
11	18 kA/cm^2	100 GHz	0.8 pF	QRDD	2.11
8.8	0.65 A	60 GHz	1.8 pF	QRDD	2.78
12		V-Band	< 1.8 pF	QRDD	,,
13.5	8 kA/cm^2	67 GHz	45 μm/1.4 pF	DLHL	2.69
5.7	560 mA	93 GHz	1.6 pF	QRDD	2.79
13.6		61.1 GHz	1.35 pF	QRDD	,,

Double Low-High-Low (DLHL) and Quasi *Read* Double Drift (QRDD) structures show higher conversion efficiencies than Double Drift (DD) flat profile diodes which are better than Single Drift (SD) devices. Table 1 gives a survey on some of the best published results.

2.8.3 Noise

The noise behaviour is of great importance for applications. In IMPATT diodes the FM noise is significantly larger than the AM noise [2.83]. Consequently most work deals with the FM noise. Of interest for many systems is the FM single sideband noise to carrier ratio of an oscillator which is given by

$$\frac{N}{C_{\mathrm{FM_{SSB}}}} = 10 \log \left(\frac{\Delta f_{\mathrm{eff}}^2}{2 f_m^2} \right) \tag{2.60}$$

with the offset frequency f_m and the effective frequency deviation Δf_{eff}. For offset frequencies not too close to the carrier the noise to carrier ratio can be directly determined from a spectrum analyzer display. This implies that $N/C_{\mathrm{FM_{SSB}}}$ and Δf_{eff} are dependent on the resonator cavity which is expressed as the Q-value. In principle, by using a high-Q cavity the N/C can be forced down to low values. This is however counteracted by a relatively low-Q resonator needed for IMPATT diode operation near to a series resonance. To compare different diodes relative to their noise behaviour the noise measure M [2.84,92] is more suited because it is independent of the load:

$$M = 10 \log \left[\left(\frac{Q_{\mathrm{ex}} \Delta f_{\mathrm{eff}}}{f_0} \cdot \right)^2 \frac{P_{\mathrm{RF}}}{k T_0 B} \right] \tag{2.61}$$

where Q_{ex} is the external Q-value of the resonator, f_0 the oscillation frequency, P_{RF} the oscillator output power, $k = 1.38 \cdot 10^{-23}$ J/K, $T_0 = 300$ K, and B the measurement bandwidth. The Q-value of the oscillator under test is advantageously determined by a self-infection locking method [2.85]. A small part of the generated power is back injected into the oscillator (P_i). By a tunable short the phase of this signal can be continuously varied and the full synchronization range Δf_s can be observed on a spectrum analyzer display. The Q-value is then obtained from [2.84,86]

$$Q_{ex} = \frac{f_0}{2\Delta f_s} \sqrt{\frac{P_i}{P_{RF}}}. \qquad (2.62)$$

The effective frequency deviation can then be calculated from the N/C ratio measured with a spectrum analyzer. For more precise or near carrier measurements a direct measurement system using a quasi-optical resonator as FM-AM converter may be used [2.87]. The delay line/mixer frequency discriminator method is suited for noise measurements close-in to a drifting carrier (Fig. 2.28) [2.88]. The conversion is a two-part process, first converting the frequency fluctuations into phase fluctuations, which takes place in the delay line, and then converting the phase fluctuations to voltage fluctuations, which is done by the double-balanced microwave mixer. The millimeter-wave signal from the DUT is down-converted in the 2–8 GHz range by a harmonic-mixer using a phase-locked DRO as a local oscillator.

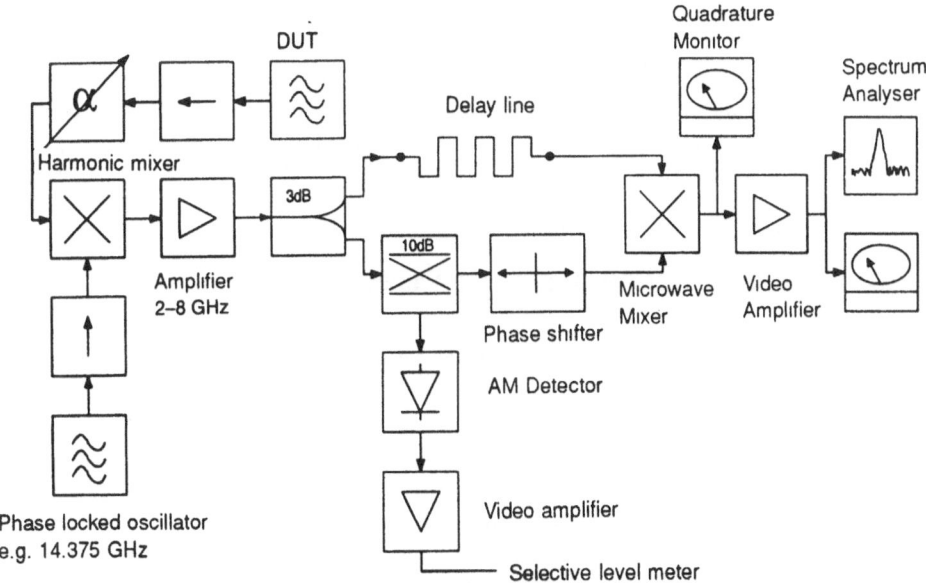

Fig. 2.28. Millimeter-wave noise measurement set-up [2.88]

A theoretical estimation of the small-signal noise measure of a transit-time device within a bandwidth B is obtained from [2.28]

$$M_0 = \frac{\overline{u_n^2}/B}{4kT_0|R_D|} \tag{2.63}$$

and the open-circuit voltage $\overline{u_n^2}$ is obtained for IMPATT diodes from [2.89]

$$\overline{u_n^2} = \frac{2eI_0}{(\omega\tau_i)^2} \frac{\left(\dfrac{1-\cos\theta}{\theta}\right)^2 + \left(\dfrac{l_a}{l_d} + \dfrac{\sin\theta}{\theta}\right)^2}{[\omega C_D(1-\omega_a^2/\omega^2)]^2} B \tag{2.64}$$

and for BARITT diodes from [2.90]

$$\overline{u_n^2} = \frac{8\mu_p kT I_0 B}{(lC_0)^2 l\nu_s}\left(1 - \frac{\sin\theta}{\theta}\right). \tag{2.65}$$

With the resulting expression for the IMPATT noise measure

$$M_0 = \frac{eI_0\left[\left(\dfrac{1-\cos\theta}{\theta}\right)^2 + \left(\dfrac{l_a}{l_d} + \dfrac{\sin\theta}{\theta}\right)^2\right]\theta}{2kT_0(\omega_a\tau_i)^2\omega C_D(1-\omega_a^2/\omega^2)(1-\cos\theta)} \tag{2.66}$$

a reasonable correlation with experimentally determined noise measures is obtained if values for the intrinsic response time τ_i, are taken from low-frequency noise measurements [2.85].

In silicon IMPATT diodes surprisingly low noise measures have been obtained at mm-wave frequencies: down to 27 dB at 75% of the maximum output power [2.91,92]. Several reasons are discussed for this observation: At the high-peak field in mm-wave IMPATTs nearly equal ionization rates of holes and electrons are obtained and increased in-phase currents require a reduced carrier multiplication. Additionally in double-drift devices the presence of both carrier types leads to lower multiplication factors and hence less noise than for single-drift devices. Further on at high frequencies, the minimum current during one period is relatively large. Therefore the current modulation is smaller and less noisy than at lower frequencies.

2.9 New Transit-Time Device Concepts

The search for new transit-time devices can be divided into two parts: the development of structures which can provide higher output powers and the research for devices with improved noise performance. To increase the output power an Integrated Series IMPATT Structure (ISIS) stacked on a single semiconductor chip has been proposed [2.93]. Higher output power is expected due to a higher operation voltage and an easier impedance matching. It is however questionable whether the thermal problems can be solved. These problems may be circumvented if the realization of a silicon/silicide structure is successful [2.94]. The device consists of a series of stacked double-drift layers separated by cobalt disilicide layers. $CoSi_2$

possesses excellent thermal stability, has a small lattice misfit (1.2%) to silicon, and the growth of Si on top of $CoSi_2$ should be possible by MBE.

A promising concept to improve the BARITT diode's performance was introduced by *Luryi* and *Kazarinov* [2.95]. The idea is to use a planar-doped triangular barrier structure for an application as an injector in a drift region. Output powers up to 10 mW at 100 GHz are expected with very good noise performance. Neglecting diffusion the maximum oscillation frequency is estimated to 1 THz. Recent experimental investigations showed that the structure needs a careful design optimization. The drift layer may not be intrinsic due to the "Kirk effect" (see also Chap. 4) caused by the drifting carriers. A moderate n-doping seems to be necessary to prevent avalanche breakdown at the end of the drift region. Further more the length of the injection region and the acceptor level in the planar doped layer is critical. The length of the injector region affects the barrier conductance and the electric field in the drift layer is extremely dependent on the planar doped layer.

It would be attractive if a double-drift BARITT diode could be realized. By this means a mechanism is required which supplies holes and electrons for injection via thermionic emission into the drift zones. It is proposed to use interband tunneling in a planar doped device as a carrier supply mechanism (Fig. 2.29) [2.96]. With increasing reverse bias the electric field E_t between the inner δ doping planes increases, giving rise to electron-hole generation by band-to-band tunneling. These carriers relax into potential wells that are formed by the δn_1 and δp_1 dopant planes. After this fast process the collected carriers are thermionic emitted over the δp_2 and δn_2 barriers as in planar doped BARITT devices [2.95]. With respect to the

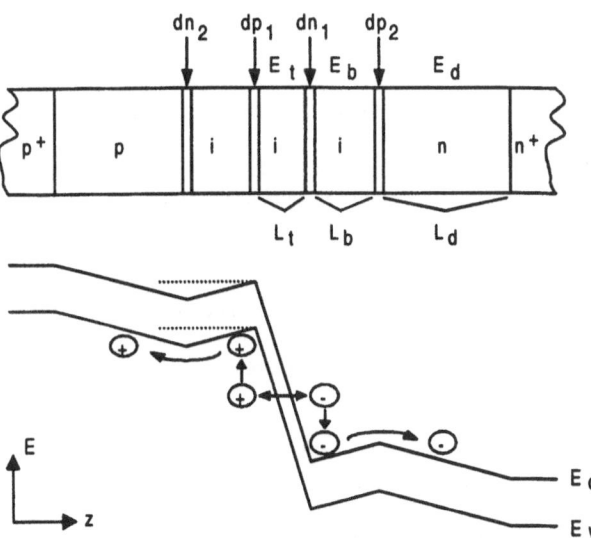

Fig. 2.29. Layer setup and schematical band diagram of a Tunneling Assisted Barrier Injection Transit-Time (TABARITT) diode. Electron-hole pairs are generated by band-to-band tunneling between the dn_1 and dp_1 dopant planes

carrier supply mechanism the device is called Tunneling Assisted Barrier Injection Transit-Time diode (TABARITT).

Another species of device concepts may be located in the above mentioned proposals: Heterostructure transit-time diodes [2.97]. The aim is to reduce the noise level of IMPATT diodes maintaining good power performance. An experimental realization of a Si/SiGe MITATT type operation yielded output powers of several mW at 100 GHz but no improvement of the noise measure [2.98]

References

Section 2.0

2.1   J. Müller: Elektronenschwingungen im Hochvakuum. Hochfrequenztechnik u. Elektroakustik **41**, 156–167 (1933)
2.2   W. Shockley: Negative resistance arising from transit time in semiconductor diodes. Bell Syst. Tech. J. **33**, 799–826 (1954)

Section 2.1

2.3   H. Johnson: A unified small-signal theory of uniform-carrier-velocity semiconductor transit-time diodes. IEEE Trans -ED **19**, 1156–1166 (1972)
2.4   C. Yeh: A unified treatment of the impedance of transit-time devices. IEEE Trans. Education **28**, 117–124 (1985)

Section 2.2

2.5   W. T. Read: A proposed high-frequency, negative resistance diode. Bell Syst. Tech. J. **37**, 401–446 (1958)
2.6   W. Maes, K. de Meyer, R. van Overstraeten: Impact ionization in silicon: A review and update. Solid State Electron. **33**, 705–718 (1990)
2.7   R. Thoma, H.J. Peifer, W.L. Engl, W. Quade, R. Brunetti, C. Jacoboni: A generalized impact-ionization model for high-energy electron transport in Si with Monte Carlo simulation. J. Appl. Phys. **69**, 2300–2311 (1991)
2.8   W. Grant: Electron and hole ionization rates in epitaxial silicon at high electric fields. Solid State Electron. **16**, 1189–1203 (1973)
2.9   K.K. Thornber: Applications of scaling to problems in high-field electronic transport. J. Appl. Phys. **52**, 279–290 (1981)
2.10 G.A. Baraff: Distribution functions and ionization rates for hot electrons in semiconductors. Phys. Rev. **128**, 2507–2517 (1962)
2.11 J.-F. Luy, E. Kasper, W. Behr: Semiconductor structures for 100 GHz silicon IMPATT diodes. 17th Europ. Microwave Conf. Rome (1987) pp. 820–825
2.12 L.H. Holway, Jr., S.L.G. Chu: Theory and measurement of back bias voltage in IMPATT diodes. IEEE Trans. MTT-31, 916–922 (1983)
2.13 A.S. Tager: The avalanche-transit diode and its use in microwaves, Sov. Phys.-Usb. **9** 892–912 (1967)
2.14 D.J. Coleman, Jr., S.M. Sze: A low-noise metal-semiconductor-metal (MSM) microwave oscillator. Bell Syst. Tech. J. **50**, 1695–1699 (1971)

2.15 S.M. Sze, D.J. Coleman, A. Loya: Current transport in metal-semiconductor-metal (MSM) structures. Solid State Electron. **14**, 1209–1218 (1971)

2.16 J. Nishizawa, Y. Watanabe: Sci. Rep. Res. Inst. Tohoku Univ., B (Comm.) **10**, No. 2, 91 (1958)

2.17 M.E. Elta, G.I. Haddad: Mixed tunneling and avalanche mechanisms in p-n junctions and their effects on microwave transit-time devices. IEEE Trans.-ED **25**, 694–702 (1978)

2.18 G.G.P. van Gorkum, A.M.E. Hoebrechts: An efficient silicon cold cathode for high current densities. Philips J. Res. **41**, 343–384 (1986)

2.19 J.F. Luy, R. Kühne: Tunneling assisted IMPATT operation. IEEE Trans.-ED-**36**, 589–595 (1989)

2.20 V.P. Kesan, D.P. Neikirk, B.G. Streetman, P.A. Blakey: A new transit-time device using quantum well injection. IEEE Trans.-EDL-**8**, 129–130 (1987)

2.21 J.-F. Luy, J. Büchler: 90 GHz SIMMWIC. MIOP Conf., Sindelfingen (1989) Paper 3A.2

2.22 T. Misawa: Negative resistance in PN junctions under avalanche breakdown conditions. P. I. IEEE Trans.-ED **13**, 137–143 (1966)

2.23 A. Schlachetzki: *Halbleiterbauelemente der Hochfrequenztechnik* Teubner Stuttgart (1984)

2.24 M. Claassen, W. Harth: Large-signal operation of pin IMPATT diodes for pulsed oscillators at millimeter-wave frequencies. Electron. Lett. **18**, 737–739 (1982)

2.25 J.-F. Luy: IMPATT operation below the avalanche frequency. Electron. Lett. **26**, 1960–1962 (1990)

2.26 P.A. Rolland, C. Dalle, M.R. Friscourt: Physical understanding and optimum design of high-power millimeter-wave pulsed IMPATT diodes. IEEE Trans.-EDL-**12**, 221–223 (1991)

2.27 C.-C. Chen, R.K. Mains, G.I. Haddad: High-power generation in IMPATT devices in the 100–200 GHz range. IEEE Trans.-ED-**38**, 1701–1705 (1991)

Section 2.3

2.28 W. Harth, M. Claassen: *Aktive Mikrowellendioden*, (Springer Beslin, Heidelberg 1981)

2.29 P.A. Blakey, J.R. East, G.I. Haddad, VLSI Electronics: *In Microstructure Science*, ed. by N.G. Einspruch, (Academic New York 1981)

2.30 P.A. Blakey, J.R. Giblin, A.J. Seeds: Large-signal time-domain modeling of avalanche diodes. IEEE Trans.-ED-**26**, 1718–1728 (1979)

2.31 D.L. Scharfetter, H.K. Gummel: Large signal analysis of a silicon read diode oscillator. IEEE Trans.-ED-**16**, 64–77 (1969)

2.32 R.L. Wierich: Computer simulation of avalanche diodes with special regard to noise. Ph.D. dissertation, Univ. of London, 1971

2.33 S.P. Yu, W. Tantraporn: A computer simulation scheme for various solid-state devices. IEEE Trans.-ED-**22**, 515–522 (1975)

2.34 G.R. Thoren: The effects of delayed secondary avalanche phenomena on the high efficiency operation of GaAs millimeter wave IMPATT diodes. Ph.D. dissertation, Cornell University (1981)

2.35 P.A. Blakey, R.K. Froelich, R.O. Grondin, R.K. Mains, G.I. Haddad: Millimeter-wave IMPATT diode modelling. 8th Cornell Conf. (1981)

2.36 D. Lippens, E. Constant: Effect of energy relaxation on injected current pulse in high-frequency IMPATT diodes. Electron. Lett. **17**, 878–879 (1981)

2.37 D. Lippens, J.L. Nieruchalski, C. Dalle, P.A. Rolland: Comparative studies of Si, GaAs, and InP millimeter-wave IMPATT diodes. Int'l J. Infrared and Millimeter Waves **7**, 771–783 (1986)

2.38 L.-C. Chang, D.-H. Hu: Large-signal analysis of Lo-Hi-Lo double-drift silicon IMPATT diodes at 50 GHz. IEEE Trans.-ED-**25**, 1137–1140 (1978)

2.39 C. Dalle, P.-A. Rolland: Read versus flat doping profile structures for the realization of reliable high-power, high-efficiency 94 GHz IMPATT sources. IEEE Trans.-MTT-**38**, 366–372 (1990)

Section 2.4

2.40 V.P. Kodali: Skin effect in microwave semiconductor diodes Electron. Lett. **4**, 67–68 (1968)
2.41 B.C. DeLoach, Jr.: The skin IMPATTs. IEEE Trans.-MTT-**18**, 72–74 (1970)
2.42 J.-F. Luy, R. Kühne: Current distribution in oscillating Mm-wave diodes. Solid State Electron. **29**, 471–476 (1986)
2.43 I. Ahmad, S. Ahmad: Skin effect consideration in metallised substrates of injection controlled transit time effect devices at mm-wave frequencies. Solid State Electron. **33**, 993–998 (1990)
2.44 I. Ahmad, S. Ahmad: Influence of a non-uniformly doped semiconductor region on skin effect resistance of ohmic contacts in mm-wave IMPATTs. Solid State Electron. **35**, 883–889 (1992)

Section 2.5

2.45 H.S. Caeslaw, J.C. Jaeger: *Conduction of Heat in Solids*, 2nd edn. (Oxford Univ. Press, London 1959)
2.46 J.-F. Luy, J. Schmide: Temperature distribution in cylinder symmetric Mm-Wave Devices. IEEE Trans.-MTT-42, 1–6 (1994)
2.47 A. Haiji-Sheikh: Peak temperature in high power chips. IEEE Trans.-ED-**37**, 902–907 (1990)
2.48 Landolt-Börnstein: *Zahlenwerte und Funktionen aus Naturwissenschaften und Technik*, Neue Serie, Band 17 Halbleiter, Teilband c Technologie von Si, Ge und SiC, (Springer Berlin, Heidelberg 1984)
2.49 M. Hazewinkel, R. Mattheij, E. van Groesen: Proc. 1st Europ. Symp. on Mathematics in Industry, ESMII, (Teubner, Stuttgart, 1985; and Kluwer, Dordrecht 1988)
2.50 C. B. Swan: Improved performance of silicon avalanche oscillators mounted on diamond heat sinks. Proc. IEEE **55**, 1617–1618 (1967)
2.51 G. Gibbons, T. Misawa: Temperature and current Distribution in an avalanching p-n junction. Solid State Electron. **11**, 1007–1014 (1968)
2.52 G. Gibbons: Transient thermal heatsink resistance. Solid State Electron. **133**, 799–826 (1969)
2.53 T.T. Fong, H.J. Kuno: Millimeter-wave pulsed IMPATT sources. IEEE Trans.-MTT-27, 492–499 (1979)
2.54 W. Behr, J.-F. Luy: High-power operation mode of pulsed IMPATT diodes. IEEE Trans.-EDL-**11**, 206–208 (1990)
2.55 D. Tjapkin, R. Ramovic, D. Stojanovic, D. Borcic: Semiempirical determination of avalanche Breakdown temperature parameters in p-n junctions. Solid State Electron. **27**, 407–411 (1984)
2.56 R.H. Haitz, H.L. Stover, N.J. Tolar: A method for heat flow resistance measurements in avalanche diodes. IEEE Trans-ED-16, 438–444 (1969)
2.57 J. Freyer, S. Ahmad: Measurement of heat-flow resistance in BARITT diodes. Electron. Lett., **12**, 527–528 (1976)

Section 2.6

2.58 P.A. Blakey, T.D. Linton: A-J plane analysis: A technique for active diode design. Cornell Conf. (1987)pp. 373–380
2.59 G.I. Haddad, J.R. East, C. Kidner: Tunnel transit-time (TUNNETT) diodes for terahertz sources. Microwave and Optical Technology Lett., 4, 23–29 (1991)

Section 2.7

2.60 E. Kasper, H. Kibbel, F. Schäffler: An industrial single-slice Si-M.B.E. apparatus. J. Electrochem. Soc., **136**, 1154–1158 (1989)

2.61 H. Jorke, H.J. Herzog: Carrier spilling in spreading resistance analysis of Si layers grown by molecular-beam epitaxy. J. Appl. Phys. **60**, 1735–1739 (1986)

2.62 H.-E. Sasse: Unpublished results (1989)

2.63 L.E. Terry, R.W. Wilson: Metallisation systems for silicon integrated circuits, Proc. IEEE **57** 1580–1586 (1969)

2.64 H.-E. Sasse, U. König: Preparation and characterization of nickel silicide, in *Polycrystalline Semiconductors* ed. by J.H. Werner, H.J. Möller, H.P. Strunk, Springer Proc. Phys., Vol. 35 (Springer, Berlin, Heidelberg 1989)

2.65 S.M. Sze: *Physics of Semiconductor Devices* (Wiley, New York, 1981)

2.66 D.D. Khandelwal: Beam lead impatts – a new dimension. Microwave J. **25**, No. 3, 81–85 (1982)

2.67 J. Freyer, R. Pierzina: Encapsulation techniques for millimeter-wave impatt diodes, Arch. Elekt. **40**, 321–325 (1986)

2.68 W. Behr: MM-Wave IMPATT diodes with integrated gold straps. Int'e Conf. on Infrared and Mm-Waves. Orlando (1985)pp. 13–14

Section 2.8

2.69 J.-F. Luy, F. Schäffler, M. Schlett: 17.6% conversion efficiency at 60 GHz with IMPATT diodes? 22nd EuMC, Helsinki (1992) pp. 485–490

2.70 P.A. Blakey, R.K. Froelich: Fundamental high-frequency performance limit for IMPATT mode operation. Electron. Lett., **21**, 28–29 (1985)

2.71 T. Ishibashi, M. Ohmori: 200-GHz 50-mW cw oscillation with silicon SDR IMPATT diodes. IEEE Trans.-MTT-24, 858–859 (1976)

2.72 J. Wenger: 140 GHz 70 mW cw output power with n-type silicon single-drift Impatt diodes. Electron. Lett. **19**, 908–909 (1983)

2.73 C. Chao, R.L. Bernick, E.M. Nakaji, R.S. Ying, K.P. Weller, D.H. Lee: Y-band (170–260 GHz) tunable cw IMPATT diode oscillators. IEEE Trans.-MTT-25, 985–991 (1977)

2.74 K.P. Weller, R.S. Ying, D.H. Lee: Millimeter IMPATT sources for the 130–170-GHz range. IEEE Trans.-MTT-24, 738–743 (1976)

2.75 M. Heitzmann, M. Boudot: New progress in the development of a 94 GHz pretuned module silicon IMPATT diode. IEEE Trans.-ED-20, 759–763 (1983)

2.76 T.A. Midford, R.L. Bernick: Millimeter-wave cw IMPATT diodes and oscillators. IEEE Trans.-MTT-27, 483–491 (1979)

2.77 E.J. Haugland: NASA seeking high-power 60-GHz Impatt diodes. Microwaves & RF, **23**, 100–107 (1984)

2.78 Y.E. Ma, E.M. Nakaji, W.F. Thrower: V-band double-drift Read silicon IMPATTs, Proc. IEEE MTT Sgmp. 167–168 (1984)

2.79 C.K. Pao, J.C. Chen, R.K. Rolph, A.T. Igawa, M.I. Herman: Millimeter-wave double drift hybrid read profile Si IMPATT diodes. Proc. IEEE MTT symp. 927–930 (1990)

2.80 P.N. Förg, J. Freyer: Baritt diodes advance to Ka band. Microwaves, No. 4, 61–65 (1981)

2.81 U. Güttich: 60 GHz BARITT diodes as self-oscillating mixers. Electron. Lett., **22**, 629–630 (1986)

2.82 J.-F. Luy, A. Casel, W. Behr, E. Kasper: A 90-GHz double drift IMPATT diode made with Si MBE IEEE Trans.- ED-34, 1084–1089 (1987)

2.83 J.J. Goedbloed, M.T. Vlaardingerbroek: Noise in IMPATT-diode oscillators. Acta Electronica **17**, 151–163 (1974)

2.84 K. Kurokawa: Noise in synchronized oscillators, IEEE Trans.-MTT-16, 234–240 (1968)

2.85 J. Wenger, S. Huber: Low-noise D-band IMPATT oscillators. Electron. Lett., **23**, 475–476 (1987)

2.86 R. Adler: A study of locking phenomena in oscillators. Proc. IRE **34**, 351–357 (1946)

2.87   W. Harth, D. Leistner, J. Freyer: FM noise measurement of W-band IMPATT diodes with a
       quasi-optical direct detection system. Electron. Lett., **18**, 355–356 (1982)
2.88   HP Product Note 11729C-2: Phase Noise Characterization of Microwave Oscillators – Fre-
       quency Discriminator Method" and HP Application Note 385: "Millimeter Measurements
       Using the HP 3048A Phase Noise Measurement System"
2.89   J.J. Goedbloed: Noise in IMPATT diodes. Philips Res. Rep. Suppl., No. 7 (1973)
2.90   H. Statz, R.A. Pucel, H.A. Haus: Velocity fluctuation noise in metal-semiconductor-metal
       diodes. Proc. IEEE **60**, 644–645 (1972)
2.91   D.M. Brookbanks, A.M. Howard, M.R.B. Jones: Si impatts exhibit low noise at MM-waves.
       Microwaves & RF, No. 2, 68–72 (1983)
2.92   I.G. Eddison: Noise in millimetre-wave oscillators. J. Instit. Electron. Radio Eng. **55**, No. 5,
       177–182 (1985)

Section 2.9

2.93   A. Christou, N.A. Papanicolaou: Summary abstract: Molecular beam epitaxy GaAs integrated
       IMPATT structure. J. Vac. Sci. Technol. B **32**, 791–792 (1985)
2.94   K.L. Wang, G.P. Li: A proposed high-frequency high-power silicon-silicide multilayered
       device. IEEE Trans.-EDL-4, 444–446 (1983)
2.95   S. Luryi, R.F. Kazarinov: Optimum BARITT structure. Solid State Electron. **25**, 943–945
       (1982)
2.96   H. Jorke, J.-F. Luy: Double drift BARITT operation mode in silicon, to be published
2.97   R.L. Kuvas, A.A. Immorlica, B.W. Ludington, F.J. Szalkowski: Heterojunction IMPATT
       diodes: Theoretical performance and material development studies. Proc. 6th Cornell Conf.
       (1977) pp. 247–256
2.98   J.-F. Luy, H. Jorke, H. Kibbel, A. Casel: Si/SiGe heterostructure MITATT diode, Electron.
       Lett. **24**, 1386–1387 (1988)

3 Schottky Contacts on Silicon

Jürgen H. Werner and Uwe Rau

Max-Planck-Institut für Festkörperforschung, Heisenbergstr. 1, 70569 Stuttgart, Germany

Experiments on rectifying contacts started in 1874 with the pioneering work of *Braun* who observed asymmetries in transport of electrical current across metal/semiconductor interfaces [3.1]. The following decades brought out a variety of technical applications, but it took more than sixty years until *Schottky* [3.2] and, independently, *Mott* [3.3] gave microscopic concepts to describe these so-called *Schottky contacts*. Today, in 1994, these interfaces are still not completely understood.

We do not intend to give a complete overview on the work on Schottky contacts. Several excellent texts [3.4–6] report on their electronic properties, and the development of the understanding until about 1989 was reviewed by *Mönch* [3.7a] who edited also a compilation of classic Schottky-contact publications [3.7b]. However, within the last few years, new insight has been gained on the fundamental role of interface *crystallography* in the formation of the Schottky-barrier height [3.8]: The epitaxy at the metal/semiconductor interface plays a major role. The most important observations were made on epitaxial silicides on Si [3.9]. The work on structural properties of *polycrystalline* silicides was recently reviewed by *Chen* and *Tu* [3.10], growth and electrical properties of *single crystalline* silicides by von *Känel et al.* [3.11], *Derrien et al.* [3.12], *Tung* [3.13], and *Mantl* [3.14].

The present chapter reviews information about Schottky contacts on silicon. We compile the structural and electronic properties of these interfaces that are important in basic research as well as for new device concepts. Chapter 3.1 presents some of the theoretical concepts that try to explain the formation of the Schottky Barrier Height (SBH). We summarize experimentally observed correlations for Schottky contacts on Si. The concepts of Metal-Induced Gap States (MIGS) and of charge transfer controlled by electronegativity differences explain the overall trend of SBHs. However, there are strong deviations from this combined *MIGS/electronegativity* model. Chapter 3.2 compiles barrier-height measurements for *epitaxial* contacts on Si. The *crystallography* at metal/Si interfaces is probably responsible for the deviations from the predictions of the MIGS/electronegativity model. These deviations are not really understood at present. Nevertheless, they are important, because many metals are either expected or even observed to grow epitaxially on Si. Chapter 3.3 gives a short review on the transport properties of Schottky contacts, and points out the importance of electrical *inhomogeneities* at the metal/semiconductor interface. In many cases, these electronic imperfections cannot be avoided. Finally, Chapter 3.4 lists methods for SBH measurements and compiles experimentally observed values for the SBH of a wide range of polycrystalline metals on Si.

Springer Series in Electronics and Photonics, Vol. 32
Silicon-Based Millimeter-Wave Devices, Eds.: Luy et al.
© Springer-Verlag Berlin Heidelberg 1994

3.1 Schottky-Barrier Models

3.1.1 The Schottky Mott Model

According to *Schottky* [3.2, 15] and *Mott* [3.3] the height of the barrier, as shown in Fig. 1 between a metal and a n-type semiconductor, is given by

$$\Phi_b^n = \Phi_M - \chi_s. \tag{3.1}$$

Here Φ_M is the work function of the metal and χ_s represents the electron affinity of the semiconductor [3.4–6]. Both quantities are measured with respect to the vacuum level, and (3.1) therefore assumes that the vacuum level is continuous across the interface.

Figure 3.1 illustrates the band diagram of a metal and a n-type semiconductor before (Fig. 3.1a) and after (Fig. 3.1b) the contact. We assume spatially homogeneous interfaces and therefore use a one-dimensional band diagram. The Schottky barrier heights for n-type and p-type material, i.e., the barrier Φ_b^n for electrons and the barrier Φ_b^p for holes, are then expected to add up to the band-gap energy E_g of the semiconductor according to

$$\Phi_b^n + \Phi_b^p = E_g, \tag{3.2}$$

as shown in Fig. 3.1c. It was realized early [3.15, 16] that most metal/semiconductor contacts do not follow (3.1), which predicts that the barrier height increases linearly as $S_\Phi = d\Phi_b/d\Phi_M = 1$, with the work function of the metal (for a review, see, for example, Ref. [3.17]. Experiments reveal deviations from (3.1), i.e. $S_\Phi < 1$, a finding which is usually ascribed to interface states. Most of the research on

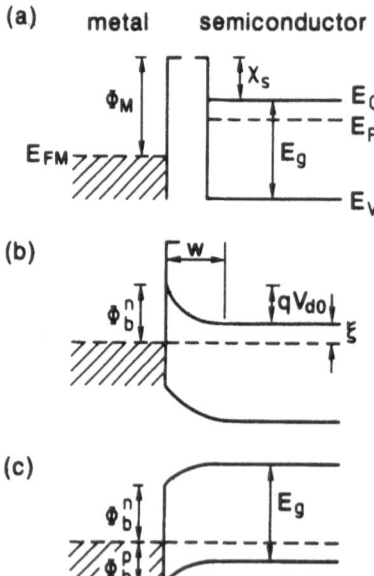

Fig. 3.1. Band diagram (a) of a metal and an n-type semiconductor before the contact; (b) after intimate contact; (c) intimate contact for a p-type semiconductor. The symbols E_{FM} and E_F denote the Fermi level within the metal and the semiconductor. The space charge region of width w creates the band bending V_{d0} at zero-bias within the semiconductor of band gap $E_g = E_C - E_V$

Schottky contacts within the last forty years has been devoted to giving either experimental or theoretical evidence for such states [3.8].

In (3.2) we distinguished between the barrier on n-type and on p-type material. In the following we restrict ourselves to the discussion of contacts on n-type Si, and use the symbol Φ_b for the height of the barrier.

3.1.2   The Space Charge Region

The electronic properties of Schottky diodes are not controlled by the Schottky barrier Φ_b directly, but by the space charge region of width w and the concomitant band bending V_d within the semiconductor. Within the space-charge region, the semiconductor is depleted from free carriers; the electrostatic potential $\Psi(z)$ for electrons is calculated from Poisson's equation which reads as

$$\frac{\partial^2}{\partial z^2}\Psi = \frac{-qN_D}{\varepsilon_s}. \tag{3.3}$$

Here q is the elementary charge, N_D the donor concentration, ε_s represents the permittivity of the semiconductor, and z is the coordinate perpendicular to the interface. In (3.3) we have neglected the contribution of the free carriers to the charge within the space charge region. Within this so-called *depletion approximation* [3.18], (3.3) yields then a parabolic solution for $\Psi(z)$ according to

$$-\Psi(z) = \frac{qN_D}{2\varepsilon_s}(w-z)^2. \tag{3.4}$$

Within the same depletion approximation, the electric field $E(z)$ increases linearly from the edge of the space charge region up to the maximal value

$$E_{max} = \frac{qN_D}{\varepsilon_s}w \tag{3.5}$$

right at the metal/semiconductor interface at $z = 0$. The width w is related to the band bending or diffusion potential V_d according to

$$w = \sqrt{\frac{2\varepsilon_s}{qN_D}V_d} \tag{3.6}$$

with

$$qV_d = \Phi_b - \xi - qV, \tag{3.7}$$

where $\xi = kT\ln(N_c/N_D)$ is the energy difference between the Fermi energy E_F in the semiconductor and the conduction band edge E_c. The quantity N_c denotes the effective density of states in the conduction band and V is the applied bias voltage. The total charge $Q_{sc} = qN_Dw$ per unit area in the depletion layer can be expressed as

$$Q_{sc} = \sqrt{2qN_D\varepsilon_s V_d}. \tag{3.8}$$

Within the Schottky-Mott model, this charge Q_{sc} within the depletion region is entirely counterbalanced by charges Q_M of opposite sign at the metallic side of the interface: i.e. it holds that $Q_M + Q_{sc} = 0$.

3.1.3 Bardeen's Model

In 1947 *Bardeen* proposed the first model for interface states in order to explain the deviations from the Schottky-Mott theory [3.19]. *Bardeen* took into account that the semiconductor surface without a metal may already have a certain density of *surface states* within the forbidden gap. The net charge of these surface states should be zero if the states were filled up to the so-called *neutrality level* Φ_0 [3.19]. In case the Fermi level position E_F at the semiconductor's surface does *not* coincide with Φ_0, a surface charge Q_{ss} should build up. Even without being in contact with the metal, the bands of the semiconductor could therefore be bent upwards, as shown in Fig. 3.2a. The total charge in the surface states

$$Q_{ss} = q D_{it}(q V_{d0} + \xi - E_g + \Phi_0) \tag{3.9}$$

depends on D_{it}, the density of surface states per unit area and unit energy. In (3.9) we have introduced the band bending V_{d0} for zero bias conditions, i.e. $V_{d0} = V_d(V = 0)$. To keep overall charge neutrality, the charge within the surface states must be counterbalanced by the electronic charge in the depletion region of the semiconductor: i.e. it holds that $Q_{ss} + Q_{sc} = 0$.

In the limit of a large density of surface states D_{it}, the expression in the bracket of (3.9) must go to zero. In this case the Fermi level position at the semiconductor surface, $E_F = q V_{d0} + \xi$, corresponds then approximately to the energy location of the neutrality level Φ_0: i.e., the Fermi level $E_F(= \Phi_0)$ at the surface is *pinned* by the large density of surface states D_{it}.

For a semiconductor like silicon, a density $D_{it} = 10^{12}$ cm^{-2} eV^{-1} would be sufficient to pin the Fermi level at the neutrality level Φ_0 [3.19]. If a metal is brought into contact with the semiconductor (Fig. 3.2b), the Fermi level remains

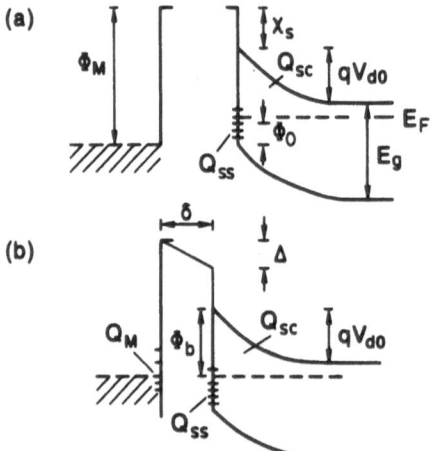

Fig. 3.2. Band diagram of a metal and an n-type semiconductor with surface states (a) before the contact and (b) after the contact with an interfacial layer of width δ. The interface states are assumed to have a charge neutrality level Φ_0

pinned as long as there is still a finite distance δ between the charges at the semiconductor and the metal surface. The density of surface states (which now have become *interface states*) needed to fix the Fermi level at Φ_0 even if a metal is present, amounts to about $D_{it} = 10^{13}$ cm^{-2} eV^{-1} [3.19]. Within this Bardeen model, the Schottky barrier height is given as

$$\Phi_b^n = E_g - \Phi_0. \tag{3.10}$$

Hence, in the Bardeen limit, the Schottky-barrier height is independent of the chemical nature of the metal, and depends only on the neutrality level Φ_0 of the surface states on the semiconductor. Within recent years, the charge neutrality levels derived by *Tejedor, Flores* and *co-workers* [3.20–23], *Cardona* and *Christensen* [3.24] have revived this concept of Fermi-level pinning.

3.1.4 Linear Models

The model of *Schottky-Mott* and that of *Bardeen* represent two extremes: The first one assumes the Schottky-barrier height to vary *linearly* with ($S_\Phi = 1$) the work function Φ_M of the metal, while the second one assumes the barrier height to be *independent* of the metal, i.e. $S_\Phi = 0$. For most semiconductors, both models do not predict the observed behavior. However, the experimental data may often be approximated by

$$\Phi_b^n = c_2 \Phi_M + c_3 \tag{3.11}$$

with $c_2 < 1$. The Schottky-Mott model assumes $S_\Phi \equiv c_2 = 1$ and $c_3 = -\chi_s$, whereas the Bardeen model holds for $c_2 = 0$ and $c_3 = E_g - \Phi_0$. A model which yields a basis for (3.11) was put forward by *Cowley* and *Sze* [3.25, 26]. Theories that assume the SBH to vary *linearly* with a bulk property of the metal (work function Φ_M, electronegativity χ_M, etc.) are termed linear models.

a) The Model of Cowley and Sze

The model of *Cowley* and *Sze* describes the dependence of the Schottky-barrier height on the metal work function Φ_M and the density of surface states D_{it}, thus connecting the Bardeen model with the Schottky-Mott model [3.25, 26]. The theory is based on the assumption of an interfacial layer of thickness δ between the metal and the semiconductor, as shown in Fig. 3.2b. The density of interface states D_{it} is assumed to be independent of energy E within the band gap. The charge Q_{ss} in these states is determined by the position of the Fermi-level E_F with respect to the neutrality level Φ_0. Combining (3.9 and 7) for the applied voltage $V = 0$ yields the charge

$$Q_{ss} = -q D_{it}(E_g - \Phi_0 - \Phi_b) \tag{3.12}$$

in the interface states. The voltage drop Δ across the interfacial layer is given by

$$\Delta/q = Q_M \delta/\varepsilon_i \tag{3.13}$$

where ε_i is the permittivity of the interfacial layer, and Q_M is the charge density at the metallic side of the interface. In contrast to the Schottky-Mott model, the Schottky barrier height

$$\Phi_b = \Phi_M - \chi_s - \Delta \tag{3.14}$$

depends here on the voltage drop Δ/q across the interfacial layer. Charge neutrality

$$Q_M + Q_{ss} + Q_{sc} = 0 \tag{3.15}$$

finally results in [3.25, 26]

$$\Phi_b = c_2(\Phi_M - \chi_s) + (1 - c_2)(E_g - \Phi_0)$$
$$\equiv c_2\Phi_M + c_3. \tag{3.16}$$

The constants c_2 and c_3 are defined as

$$c_2 = \frac{1}{1 + q^2\delta D_{it}/\varepsilon_i} \tag{3.17a}$$

and

$$c_3 = (1 - c_2)(E_g - \Phi_0) - c_2\chi_s \tag{3.17b}$$

For $D_{it} = 0$, the Schottky-Mott limit follows with $c_2 = 1$ and $c_3 = -\chi_s$, and (3.1) is recovered. In contrast, for $D_{it} \to \infty$, one obtains $c_2 = 0$ and $c_3 = E_g + \Phi_0$, i.e. the expression for the Bardeen model, namely (3.10).

Figure 3.3 represents a plot of the SBHs on n-type silicon for a variety of metals, as it was used by *Bené* and *Walser* [3.27]. A least square fit to the line results in $c_2 = 0.175$ and $c_3 = -0.12$ eV. From (3.17a and b) we obtain $D_{it} \approx 5.17 \times 10^{13}$ cm^{-2} eV^{-1} for the density of surface states, and $\Phi_0 \approx 0.4$ eV for the position of the neutrality level above the valence band edge E_V. Here we assume $\delta = 5\text{Å}$ for the thickness of the interfacial layer and a permittivity ε_i equal to the free space value, i.e. $\varepsilon_i = \varepsilon_0$.

A least-square fit to the original data of *Cowley* and *Sze* [3.25, 26] yields $c_2 = 0.27 \pm 0.05$ eV and $c_3 = -0.55 \pm 0.22$ eV. These data correspond to $D_{it} \approx (2.7 \pm 0.7) \times 10^{13}$ cm^{-2} eV^{-1} for the density of surface states and to $\Phi_0 \approx 0.30 \pm 0.36$ eV for the position of the neutrality level. Evaluation of data from other covalent semiconductors yields comparable results [3.28]. The experimental value for the charge neutrality level of *Cowley* and *Sze* [3.25] agrees well with the theoretical result for the charge neutrality level of *Tersoff* [3.23], and the dielectric midgap energy of *Cardona* and *Christensen* [3.24] which were later calculated on the basis of the band structure of Si. In contrast, some scatter is observed for the *slope parameter*

$$S_\Phi \equiv c_2 = \frac{d\Phi_b}{d\Phi_M}. \tag{3.18}$$

Cowley and *Sze* reported $S_\Phi = 0.27$, *Bené* and *Walser* $S_\Phi = 0.17$, and *Mönch* found later $S_\Phi = 0.15$ for contacts that were prepared in ultra-high vacuum (UHV) [3.17, 29]. For silicides on Si, $S_\Phi = 0.17$ was reported by *Wu*, et al. [3.30],

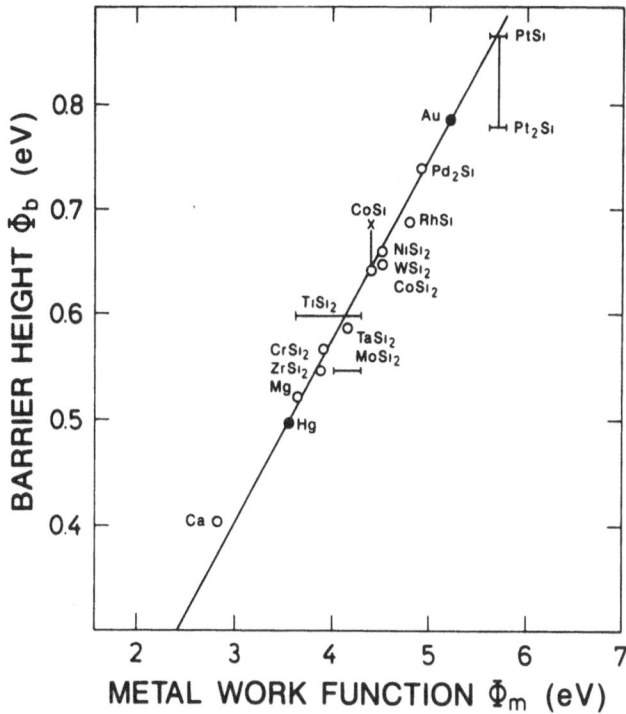

Fig. 3.3. Experimental data for the barrier heights versus the work function of the metal (taken from *Bené* and *Walser* [3.27]). We find a slope parameter $S_\Phi \equiv c_2 = 0.175$, and the intercept $c_3 = -0.12$ eV from these data

$S_\Phi = 0.17 - 0.22$ by *Tove* [3.31], and $S_\Phi = 0.19$ by *Bucher* et al. [3.32]. Recently, *Hwu* et al. analyzed Ag/Cs/Si interfaces and obtained a slope parameter of $S_\Phi = 0.06$ [3.33].

b) Electronegativity

The approach of *Cowley* and *Sze* [3.25] represented a first quantitatively satisfactory model for the prediction of the SBH Φ_b. However, the error bars in plots of SBHs Φ_b versus the work function Φ_M were very large not only due to the uncertainty in measurements of the SBH but also because of the difficulty of choosing the right value for the work function Φ_M. In the meantime the work functions of elemental metals were measured by *Michaelson* [3.34], and those of some silicides were measured by *Bucher* et al. [3.32]. A list for some silicides was also given in [3.35]. There is some scatter among these data, and it is not trivial to choose a value which is appropriate for the description of metal/semiconductor contacts.

The work function of a solid consists of a volume and a surface contribution. The Schottky-Mott description assumes implicitly that the work function of the metal and the electron affinity of the semiconductor (or, at least their difference) both remain unchanged when the two solids are brought into contact [3.36]. Instead of work functions, the electronegativity scales of L. Pauling or A.R. Miedema are therefore often used in the discussion of SBH dependences on metal properties. An empirical correlation of metal work functions Φ_M and their Pauling electronegativities X_M was given by *Gordy* and *Thomas* [3.37] as

$$\Phi_M \approx 2.27 X_M + 0.34 \text{ eV}. \tag{3.19a}$$

This correlation was revised on the basis of *Michaelson's* [3.34] work-function data [3.38] to

$$\Phi_M \approx 1.79 X_M + 1.11 \text{ eV}. \tag{3.19b}$$

The 'slope parameters' of experimentally measured barrier heights are therefore related to each other by

$$S_X = \frac{d\Phi_b}{dX_M} = A \frac{d\Phi_b}{d\Phi_M} = A S_\Phi \tag{3.19c}$$

with $A = 1.79$ or $A = 2.27$.

Kurtin, et al. analyzed the slope parameter S_X for different semiconductors [3.39]. They reported that for *ionic* semiconductors with a large electronegativity difference ΔX of anions and cations, the slope parameter S_X should be larger when compared to S_X for *covalent* semiconductors. Ionic semiconductors should follow the Schottky-Mott rule, (3.2), rather than the Bardeen limit. A transition between covalent (Bardeen limit) and ionic (Schottky-Mott behavior) was reported for $\Delta X = 0.7$. The ionicity of the semiconductor should therefore play a role in Fermi-level pinning. *Schlüter* re-investigated the data of *Kurtin* et al. [3.39] and took into account the large error bars for the SBHs and the electronegativities of the metals [3.40]. As a result, he stated only a trend instead of a sharp transition for the slope parameters. Recently, *Mönch* showed that the model of *Kurtin* et al. fails completely if one includes the SBH data of Xe [3.17, 41]. The overall behavior of the slope parameters for different semiconductors is better understood within the model of Metal-Induced Gap States (MIGS) and a plot of S_Φ versus dielectric constant, as will be discussed below [3.7a, 17, 41].

3.1.5 Barrier Height Correlations

Several empirical correlations for the SBH appeared in the literature on silicide/Si contacts. These contacts are important in device technology because they can be formed in a controlled and reliable manner by chemical reaction between a deposited metal and the Si substrate. In contrast to the assumption of the models of *Cowley* and *Sze*, these contacts are abrupt without an interfacial layer (e.g., by a native oxide) between the metal and the semiconductor. The correlations discussed here were based on the investigation of *polycrystalline* silicides on Si. These correlations explain just a subset of experimental results and do not stem from a microscopic theory of Schottky-barrier formation. Nevertheless, they were milestones for a better understanding of the key parameters controlling Schottky-barrier formation. Presently, it is obvious that further understanding of SBHs cannot be achieved without considering *epitaxial, single crystalline* metal/semiconductor interfaces.

a) Correlations with Silicide Bulk Properties

Heat of Formation. *Andrews* and *Phillips* [3.42, 43] attempted to correlate the SBHs of several polycrystalline Transition Metal (TM) silicide/Si diodes with the work functions Φ_M of the (unreacted!) metals (i.e., with the one of Pt, Pd etc. for PtSi, Pd_2Si, etc.) and with their Pauling-Gordy electronegativities X_M; neither correlation was satisfactory. Instead they found an empirical relationship between the *molar heat of formation* of the silicide and the SBH; their results are represented in Fig. 3.4 [3.42, 43]. With the exception of PtSi, their data can be represented as

$$\Phi_b = -0.18\Delta H_f + 0.83 \text{ eV}, \tag{3.20}$$

with ΔH_f being the heat of formation of the silicide normalized *per metal atom* per formula unit.

Andrews and *Phillips* explained their correlation with the *charge transfer* in the metallic covalent bonding between the TM atoms and the Si atoms within the bulk of the silicide. In fact, within Pauling's picture of strong (*d*-orbital controlled) bonding of TMs, one would expect that the charge transfer between the metal and Si varies proportionally with to the *square root* $(-\Delta H_f)^{1/2}$. However, for the TM silicides such *d*-orbitals seem to be of secondary importance. Instead, with increasing $-\Delta H_f$ more sp^3 orbitals are mixed into the TM *d*-orbitals. *Andrews* and *Phillips* expected the degree of this admixture (and therefore also the charge transfer) to be *linear* in $-\Delta H_f$ [3.42, 43].

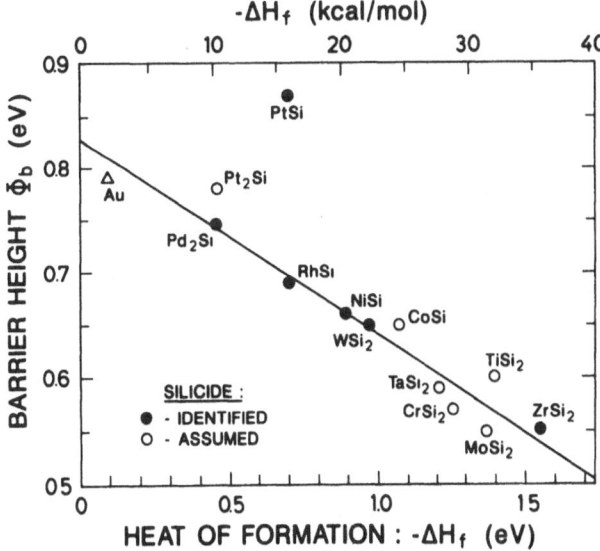

Fig. 3.4. The barrier height of several Transition Metal (TM) silicide/Si interfaces varies linearly with the heat of formation (normalized per TM atom) of the bulk silicide. The *y*-axis intercept is close to the value for the free Si surface. The *x*-axis intercept is close to the cohesive energy of Si. (Redrawn after [3.42])

Within their picture [3.42, 43], the SBH varies therefore *linearly* with the charge transfer between the TM atoms and the Si atoms *within the bulk* of the silicide. It remained open how this bulk charge transfer should be related to the charge transfer across the interface that makes up the SBH between the silicide and the Si. It would be interesting to re-investigate these early ideas within the presently available models that ascribe SBHs to the charge transfer *across* the metal/semiconductor interface and a combination of Metal-Induced Gap States (MIGS) and electronegativity concepts [3.41].

The y-axis intercept of Fig. 3.4 was explained by *Andrews* and *Phillips* in the following way: At zero heat of formation (i.e., for weak bonding and weak charge transfer), the barrier height should be close to that of a free Si surface. Indeed, the free-Si-surface value of 0.83 eV is close to their [3.42] Schottky barrier value (0.79 eV) for the 'weakly bonded' Au on Si. From the point of view of the present knowledge, it is interesting to note that the SBHs for the polycrystalline contacts of the weakly bonded Ag, Al and Pb on Si lie roughly in the same energy range. These data also coincide with the arguments of *Andrews* and *Phillips*. The charge neutrality level due to *Tersoff* [3.23] and the dielectric midgap energy due to *Cardona* and *Christensen* [3.24] range within the same energy regime when measured with respect to the conduction-band edge.

The x-axis intercept (and therefore also the slope) in Fig. 3.4 can be understood by recognizing that the 'barrier height' between two n-type Si samples should be zero [3.44]. The intercept at $-\Delta H_f = 4.8$ eV at the abscissa is therefore close to the cohesive energy of Si at 4.67 eV. In crystalline Si the bonds are completely sp^3 hybridized; i.e., the charge distribution is not spherically symmetrical. *Andrews* and *Phillips* assumed that the formation of the silicide results in a charge transfer between the Si and the TM atom, which goes hand in hand with a *de-hybridization* of Si bonds [3.42].

The deviation for PtSi from the anticipated behavior in Fig. 3.4 was explained by interface states which should be the result of epitaxial growth of PtSi on Si [3.42]. In this case, Pt and Si atoms were thought to be arranged in layers parallel to the interface which should result in *ionic* covalent rather than metallic covalent bonding. The influence of the local epitaxy, which seems to be the key to the understanding of SBHs [3.9, 45, 46], was therefore recognized in this early publication of *Andrews* and *Phillips*, too [3.42].

Core-Level Shift and d-Band Occupancy.   Chemical bonding properties in silicides were not much investigated at the time when the paper of *Andrews* and *Phillips* [3.42] was written in 1975. Recently, more light was shed by *Hirose* et al. [3.47], as well as by *Hara* and *Ohdomari* [3.48], on the correlation between heat of formation and SBHs [3.42]. These researchers used X-ray Photoelectron spectroscopy (XPS) and investigated the shift of core-levels [3.49] of the TMs upon the formation of silicides [3.47, 48]. They revealed the correlation between the SBH Φ_b and the chemical shift ΔE_{core} of core-levels of the bulk TMs on silicide formation in the form

$$\Phi_b = a\Delta E_{core} + b \tag{3.21}$$

with a = 0.15, b = 0.63 eV in [3.47] and a = 0.205, b = 0.570 eV in [3.48]. The different coefficients in the two publications stem obviously from different values for the SBHs of $MoSi_2$, $TiSi_2$, and WSi_2 on Si; the origin of this disparity is not clear. Figure 3.5 represents the correlation of *Hara* and *Ohdomari* [3.48].

The shift ΔE_{core} of the core-levels in the TM atoms is proportional to a shift of the *d-bands* of the TMs [3.48]. Therefore, this *d*-band shift correlates also with the SBH. The *d*-bands are characteristic for chemical bonding in TM silicides: As suspected by *Andrews* and *Phillips* [3.42], the sp^3 hybridization of the Si atoms is removed in the silicide. The main chemical bond in the TM silicides is made up of interactions between Si $3p$ and TM *d*-orbitals [3.48, 50, 51], with the occupancy of the *d*-bands increasing when going from refractory-metal silicides to near-noble-metal silicides [3.52]. The *shift* of *d*-bands upon the formation of the silicide reflects a change of the occupancy of the TM *d*-bands and a probable charge transfer between the valence electrons of the TM and the Si.. If one follows *Mailhiot* and *Duke*, this charge transfer should then be responsible for the SBH [3.53].

According to *Hara* and *Ohdomari*, not only the core-level shift ΔE_{core} but also the heat of formation ΔH_f correlates with the *d*-level occupancy [3.48]. Therefore, the correlation of *Andrews* and *Phillips* between the heat of formation ΔH_f of silicides and their barrier height on Si reflects also a *change* of the TM's *d*-band occupancy when the silicide is formed. In contrast, *Hara* and *Ohdomari* seemed to have measured this change via measurements of the core-level shift ΔE_{core}. The core-level shift ΔE_{core} and the heat of formation ΔH_f [3.42] are both

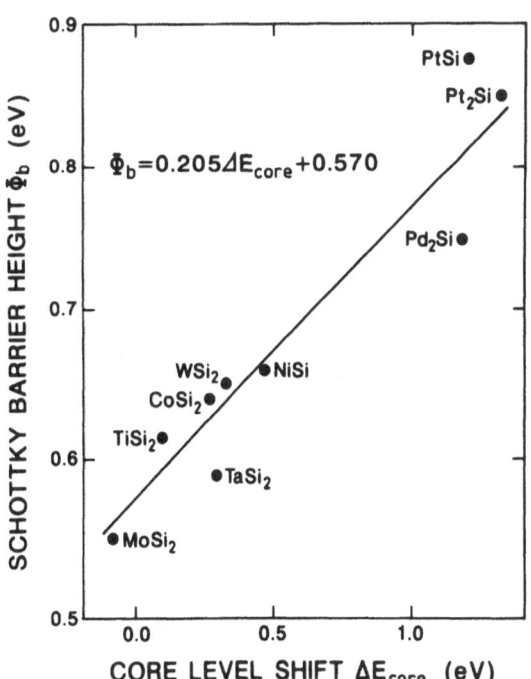

Fig. 3.5. The SBH of several transition metals silicides correlates with the core-level ΔE_{core} of the metals upon silicide formation. For example, the shift ΔE_{core} of the $4f$ 7/2 core level of Pt is measured when PtSi is formed. (Redrawn after [3.48])

correlated with the SBH because (i) both quantities reflect the change of d-level occupancy which is related to charge transfer between the metal atoms and the Si upon silicide formation, (ii) ΔE_{core} and ΔH_f are therefore correlated with each other, and (iii) the charge transfer within the *bulk* of the silicide seems to be a measure of the charge transfer at the *interface* between the silicide and the Si which makes up the Schottky barrier.

Eutectic Temperature   In 1980, *Ottaviani* et al. [3.54] had an *"uneasy feeling"* about the correlation of *Andrews* and *Phillips* [3.42] because it relates the interfacial SBH to a bulk property of the silicide. Besides, they criticized the deviations of PtSi and IrSi [3.55]. *Ottaviani* et al. therefore used the eutectic temperatures T_e for a correlation with the SBH [3.54]. Their correlation for silicide/Si contacts is represented in Fig. 3.6.

The reliability of the correlations in [3.54] is not clear. For example, in [Ref. 3.54, Fig. 1] for the metal/Si systems with a simple eutectic phase diagram (Au, Al, Ag), a SBH value for Be on Si is obviously used erroneously. It seems that this SBH value holds for GaAs and stems from [Ref. 3.4, Fig. 2.20]. In addition, the SBH values for contacts with and without an interfacial oxide which seem to stem originally from [Ref. 3.4, Fig. 2.16], and [Ref. 3.56, Fig. 6] were interchanged. Nevertheless, the SBH data in [Ref. 3.54, Fig. 1] for Au, Al, and Ag on Si with an etched surface (and therefore probably *with* an interfacial oxide) fall on a straight line when plotted versus eutectic temperature. The line predicts a SBH of zero at the melting point of Si.

The correlation for *silicides* (i.e., for reactive interfaces) from [3.54] is represented in Fig. 3.6. The choice of the correct eutectic temperature T_e is not easy.

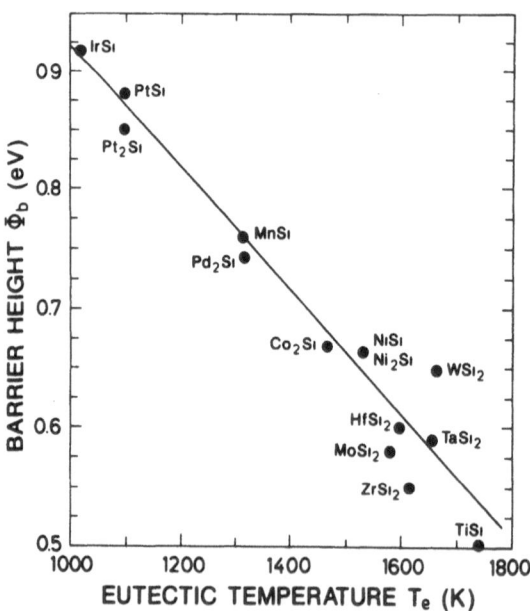

Fig. 3.6. Correlation of the SBH of silicide/Si contacts with the eutectic temperature T_e which is selected depending on the growth mechanism of the silicide. (Redrawn after Ref. [3.54])

Ottaviani et al. ascribed the observed linear correlation to a postulated interfacial layer between the silicide and the Si. A similar model was proposed by *Bené* and *Walser* [3.27]. This layer could differ from the two solids in contact in *"either structure or composition or both"* [3.54]. For those silicides whose growth is dominated by diffusion of the metal, the interfacial layer was expected to be enriched by metallic atoms; hence they [3.54] used the eutectic temperature of the closest eutectic on the metal side. For those silicides dominated by Si diffusion, the interfacial layer was expected to be enriched by Si atoms; hence they [3.54] chose the closest eutectic on the Si side. Figure 3.6 demonstrates that the silicides with a low eutectic temperature T_e yield the highest barriers on n-type Si. Additional work on rare-earth silicides fits also approximately in the scheme of Fig. 3.6, though the eutectic temperature was not known for such systems [3.57, 58]. Despite the successful correlation of *Ottaviani* et al., the underlying physical mechanism which leads to the SBH formation could not be revealed in [3.54].

b) Effective Work Function

Despite the difficulties in choosing the right work function for metals, *Freeouf* [3.59, 60] found that several silicide/Si Schottky contacts follow *exactly* the predictions of the Schottky-Mott model if one assumes the metal (M) silicides' work function $\Phi_{silicide}$ of composition $M_m Si_n$ is given as the *geometric mean*

$$\Phi_{silicide} = \sqrt[m+n]{\Phi_M^m \Phi_{Si}^n} \tag{3.22}$$

of the metal work function Φ_M and the silicon work function Φ_{Si}. *Freeouf* used the experimentally measured work functions of *Michaelson* [3.34] for the metals, the work functions of the silicides were not reported yet at the time of *Freeouf's* work, but were measured later *Bucher* et al. [3.32]. In addition, to calculate the work functions for the silicides from the stoichiometry-weighted average of (3.22), *Freeouf* had to assume that $m = 1$ and $n = 4$ holds just at the interface between the silicide and the Si. In other words, a silicon-rich layer of composition MSi_4 had to be postulated for all kinds of different silicide/Si diodes. Unfortunately, despite the attractive features of this effective work function approach, such compositions have not been found experimentally. In addition, the work functions which are derived from (3.22) do not agree with *experimental* silicide work functions which were determined later [3.32].

3.1.6 Advanced Models: Charge Transfer Across the Interface

a) Introduction

The correlations for silicides discussed above gave no microscopic explanation for the SBH. However, they revealed that the properties of the *bulk* of the metal and the semiconductor (and not just of the interface) play a role. Understanding of the charge transfer across the interface and the build-up of an interface dipole is the key for a quantitative model of SBHs. In our opinion, this transfer probably depends on, at least, four parameters:

(i) *Bulk* properties of the solids: The ability of the semiconductor and the metal to attract charges as, for example, described by the work function, or by Pauling's or Miedema's electronegativity scale.

(ii) *Intrinsic interface states*: The change of the bulk band structures at the interface by Metal-Induced Gap States (MIGS) that decay into the band gap of the semiconductor.

(iii) *Defects by crystallographic or chemical mismatch*: Charges that are introduced by point defects, like vacancies, antisites, impurity segregation, compound formation, and broken bonds as well as by lattice defects like dislocations and grain boundaries.

(iv) *Crystallography and epitaxy*: The relative orientation of atoms, their coordination, the relative areal atomic concentration, bond angles, bond lengths and the strain at the interface.

At present there is no model available which considers all of the listed four parameters. The Schottky-Mott model considered just bulk parameters [3.2, 3, 15]. The Bardeen model was the first to include interface states [3.19]. The models of *Heine* [3.61], as well as those of Tejedor et al. [3.20–22], *Tersoff* [3.23], and *Cardona* and *Christensen* [3.24] are more or less either extensions or elaborate combinations of these two classic models. The unified defect model of *Spicer* et al. favours defects to explain SBHs on compound semiconductors and neglects MIGS [3.62, 63]. *Ludeke* et al. developed a model that is based on the interaction of defect levels *and* MIGS [3.64–67]. Within *Ludeke's* delocalization model, stationary and localized defect levels at the semiconductor surface become resonance states when a metal is deposited. The interface dipole is then built up by a charge transfer between the defects and metallic states [3.64–67]. In contrast, *Mönch* has shown that the overall behavior of SBHs is understood *without* defects by the combined effects of electronegativity differences and MIGS. Interface defects were proposed to explain just *deviations* from a more general behavior [3.7, 8, 17, 41]. In this sense, *Mönch's* models seem to be the most advanced one for the description of polycrystalline or jellium-like Schottky contacts. However, the pioneering work of *Tung* has shown that *crystallography* is probably the most important parameter for further understanding [3.9]. Some of these models which neglect crystallography will be outlined next.

b) Metal-Induced Gap States (MIGS) Models

Heine's [3.61] classical publication represents not only an extension of *Bardeen's* surface-state model [3.19] and of the phenomenological work of *Cowley* and *Sze* [3.25], but he also introduced the concept of Virtual Gap States (VIGS, VGS) and Metal-Induced Gap States (MIGS) into the discussion of interface states. His model presents the starting point for a whole class of theories that are summarized in Table 3.1. *Heine* showed that the localized surface states that were proposed by *Bardeen* cannot exist when a metal is brought into intimate contact with the

Table 3.1. Acronym-zoo for interface dipole models that are based on bulk properties only

Acronym	Model	Researches
VIGS, VGS	Virtual Gap States	*Heine* [3.61]
MIGS	Metal Induced Gap States	
IDIS	Induced Density of States	*Tejedor* and *Flores* [3.20, 21, 22]
CNL	Charge Neutrality Level	*Tersoff* [3.23]
DME	Dielectric Midgap Energy	*Cardona* and *Christensen* [3.24]

semiconductor surface. Instead, "*virtual or resonance surface states can exist which behave for practical purposes in the same way. They are really the tails of the metal wave functions rather than separate states*" [3.61]. The MIGS are actually metallic states that decay exponentially into the semiconductor's forbidden gap. The existence of such states below the Fermi level at the interface is required by the matching of the wave functions. These states play the same role as *Bardeen's* surface states: The Fermi level is relatively insensitive to the work function of the metal because the build-up of the SBH is mainly caused by charge transfer between these MIGS and the metal. Those tailing states are actually virtual gap states (VIGS, VGS) of the semiconductor complex band structure. Pinning of the Fermi level by such MIGS was later confirmed theoretically by *Cohen* and *co-workers* [3.68, 69].

Tersoff applied Heine's hypothesis of virtual gap states (VIGS) and derived the concept of a Charge Neutrality Level (CNL) [3.23, 70]. This CNL defines the branch point, i.e. the energy within the gap of each semiconductor for which the character of the VIGS changes from valence-like to conduction-like. Although representing a property of the semiconductor surface, the CNL can be calculated from the band structure of the bulk semiconductor only. For Si the CNL is located at an energy of 0.36 eV above the valence-band edge E_V. In the spirit of *Heine's* work, *Tersoff* argued that the charge transfer between the metal and the semiconductor occurs almost completely between the MIGS and the metal [3.23, 70]. The resulting dipole is the origin of the Schottky barrier. *Local* charge neutrality at the interface is approximately maintained by this interfacial dipole, and the band bending within the semiconductor is mainly a consequence of doping. The large density of states of the MIGS which decay only several Å into the semiconductor (3.0 Å for Si) pins the Fermi level at the interface just at the CNL. The energetic position of the CNL fixes therefore the SBH. Indeed, the experimental SBHs for Au on several semiconductors are in good agreement with the CNL position [3.23, 70].

Following *Tersoff*, the band offsets of *semiconductor heterojunctions* are also predicted along the same line: The offset at the interface between two semiconductors is given by the energy differences of the two CNL. The charge neutrality level which depends on bulk properties of the semiconductor only, serves therefore as a natural reference, similar to the electron affinity in the *Schottky-Mott* and *Anderson* [3.71] theories. Continuity of the CNL – instead of the vacuum level – across the interface replaces here the electron-affinity/work-function rule. Similar ideas were published *earlier* by Tejedor and Flores [3.20–22], who showed that for a *one-dimensional* semiconductor the CNL coincides with the midgap.

Cardona and *Christensen* discussed the concept of the midgap energy further, and introduced the so-called *Dielectric* Midgap Energy (DME) into the discussion of SBHs and band offsets at heterostructures [3.24]. They argued that screening effects on the hydrostatic deformation potentials can be calculated by using the average energy position between the conduction bands and the valence bands. For example, the Penn gap (i.e., the optical gap), which describes the dielectric constant of semiconductors in a simple model, amounts to 4.77 eV for Si [3.72, 73]. The position of the DME within this optical gap was calculated at the first Baldareschi point of the band structure for several group IV elements and III–V and II–VI semiconductors. The DME was then projected into the fundamental band gap (1.12 eV). For Si one finds the DME to be located 0.23 eV above the valence band edge E_V [3.24]. For all materials considered, the DME values of *Cardona* and *Christensen* [3.24] were very close to the CNL of *Tersoff*. The ideas of *Cardona* and *Christensen* for the evaluation of screening effects on electron-phonon interaction can be used to calculate the screening at the heterointerface between two dissimilar semiconductors as well as at metal/semiconductor interfaces. The charge transfer between the two materials and the interface dipole is simply predicted by the adjustment of the DME positions [3.24].

The CNL model of *Tersoff*, and the DME model of *Cardona* and *Christensen* predict that the SBH for a particular semiconductor depends on bulk properties only and is *independent of the metal*. For example, all SBHs on n-type Si should amount to about 0.8 eV. This prediction is certainly incorrect if one considers the experimental SBH values between about 0.2 eV (for Hg [3.74]) to 0.95 eV (for IrSi [3.75]) on n-type Si. Nevertheless, these bulk models give a rough estimate not only for the absolute height of the SBHs, but also for their pressure and temperature dependences [3.76, 77].

Recently, *Flores* and *co-workers* [3.78, 79] extended their model of the midgap energy and the *intrinsic* CNL [3.20–22] and included *extrinsic* effects. They showed (for the one-dimensional case) that the CNL is only located at midgap "*if the metal band is assumed to be very broad, having a featureless density of states.*" Under such conditions i.e. of a jellium model for the metal, the "*Induced Density of Interface States*" (IDIS) in the semiconductor is independent of the chemical identity of the metal [3.20–22]. The Fermi level is then predicted to be pinned at the intrinsic CNL by MIGS/IDIS. For deviations from the jellium model, the IDIS/MIGS/CNL depend on the metal. The SBHs for a particular semiconductor depend then not only on the electronegativity of the metal atoms but also on their atomic position at the interface [3.79]. The extrinsic CNLs fluctuate around the intrinsic value. For the electro*positive* alkali metals K, Na, and Li the deviations from the intrinsic CNL are small for contacts on GaAs, whereas the difference is larger for the more electro*negative* Ag [3.79].

Instead of using *extrinsic* charge neutrality levels, *Mönch* combined the predictions of the *intrinsic* VIGS/MIGS/IDIS/CNL model with the concept of the work function/electronegativity, to account for the chemical properties of the metal [3.7, 8, 17, 41]. Similarly to *Schmid* [3.80], *Mönch* used the *Miedema* scale [3.81] for the metal's electronegativity. For metals like Ag and Au with an electronegativity similar to that of Si (4.7 eV on the *Miedema* scale), the barrier height at the metal/semiconductor interface indeed, agrees, with the value of about 0.8 eV that is expected form the VIGS/MIGS/IDIS/CNL/DME models of *Heine, Flores, Tersoff*, and *Cardona* and *Christensen*. *Mönch* argued therefore: *"For a metal exhibiting the same electronegativity as the semiconductor, the Fermi level should coincide with the branch point of the virtual gap states of the complex band structure of the semiconductor or, equally, with the Charge-Neutrality Level (CNL) in the continuum of the Metal-Induced Gap States* [3.7a].*"* The barrier height on n-type material will then be given by the difference between the conduction band edge and the intrinsic CNL. *"When the electronegativities of the metals are smaller or larger than that of the semiconductor, the heights will also be smaller and larger, respectively"* [3.7, 8], than this value.

Figure 3.7 supports this MIGS model of SBH formation. However, one has to make certain non-trivial assumptions. In this figure, *Mönch* [41] analyzed the slope parameter $S_\Phi = d\Phi_b/d\Phi_M$, see (3.17a), (3.18), for various semiconductors. The linear behavior is understood as follows:

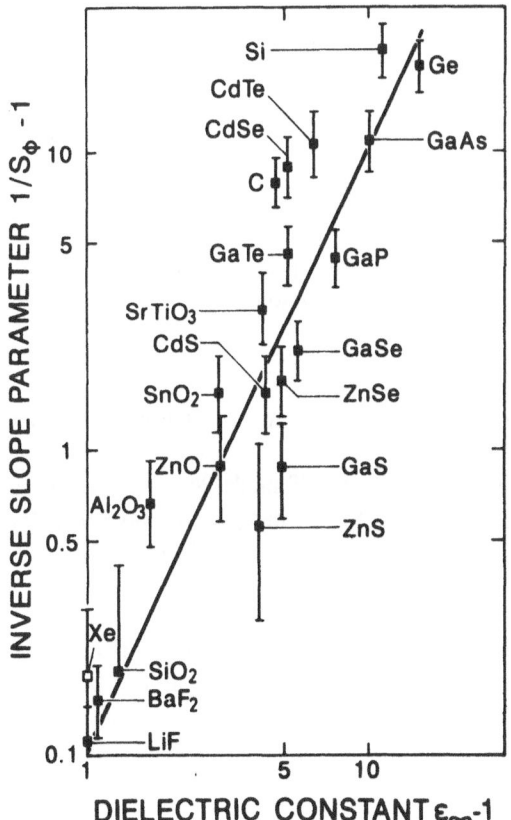

Fig. 3.7. Slope $S_\Phi = d\Phi_b/d\Phi_M$ versus the electronic contribution ε_∞ of the dielectric constant for twenty-one different semiconductors. The linear behavior gives strong support for the MIGS model. (Redrawn after [3.41])

(i) For a particular semiconductor, the slope

$$S_\Phi = \frac{1}{1 + q^2 D_{it} \delta / \varepsilon_0} \tag{3.23}$$

depends on the interface state density D_{it} of the MIGS around the CNL, the elementary charge q, the (vacuum) dielectric constant ε_0, and the width δ of the interface dipole, which is related to the decay length of the MIGS [3.17].

(ii) For a *one-dimensional* semiconductor, the product $D_{it}\delta$ in (3.23) depends, in turn, on the effective band gap $\langle E_g \rangle$ according to [3.17]

$$D_{it}\delta = \frac{c^2}{\langle E_g \rangle^2} \tag{3.24}$$

with c being a constant. Equation (3.24) predicts that the decay length δ is smaller for larger band gaps, a finding that is confirmed by the calculations of *Louie* et al. [3.69, 82]. Similar to the one-dimensional case one may assume for three dimensions the dependence

$$D_{it}\delta = \frac{c^{2n}}{\langle E_g \rangle^{2n}}. \tag{3.25}$$

Here $\langle E_g \rangle$ should then be identified with the average gap (i.e., the Penn gap) of the semiconductor.

(iii) The Penn gap $\langle E_g \rangle$ is related to the bulk plasmon energy $\hbar\omega_p$ of the valence electrons and the electronic contribution ε_∞ of the dielectric constant according to

$$\frac{1}{\langle E_g \rangle^2} = \frac{\varepsilon_\infty - 1}{(\hbar\omega_p)^2}. \tag{3.26}$$

For the semiconductors in Fig. 3.7, the plasmon energy $\hbar\omega_p$ is approximately constant and amounts to 16.5 eV [3.17].

(iv) If one assumes that the constant C is the same for all semiconductors considered, then one obtains from (3.23 to 26) the equation

$$\frac{1}{S_\Phi} - 1 = \frac{q^2 C^{2n}}{\varepsilon_0 (\hbar\omega_p)^{2n}} (\varepsilon_\infty - 1)^n. \tag{3.27}$$

From a fit of (3.27) to Fig. 3.7, Mönch found $n = 2$ and a value of 0.1 for the factor in front of $(\varepsilon_\infty - 1)^n$ [3.17].

Mönch's fit in Fig. 3.7 which is based on the MIGS model gave a better explanation for the chemical trend of the slope parameter S_Φ than previous investigations of *Kurtin*, et al. [3.39] that were already criticized earlier by *Schlüter* [3.40]. Correlations of the dielectric constant ε (or ε_∞) and the slope parameter S_Φ had been investigated by *Tersoff* [3.83] and *Phillips* [3.84], but they were less successful. The correlation of *Mönch* gives strong support for the MIGS concept. However, it remains unclear why just $n = 2$ should hold in (3.25). In addition, the extraction of a slope parameter S_Φ from experimental barrier data is not very well defined because there is a considerable scatter of data when the experimental SBHs for a particular semiconductor are plotted versus the work function or electronegativity of the metals.

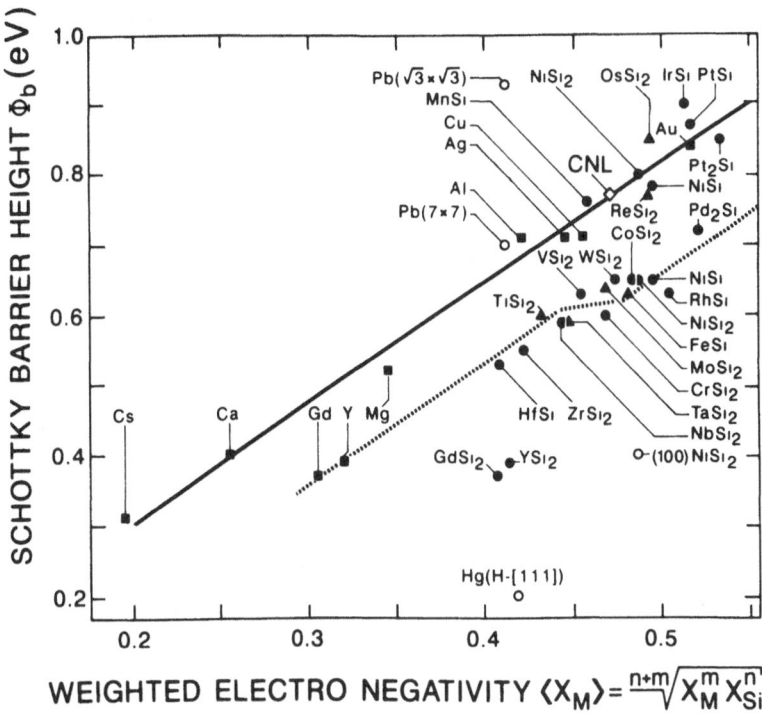

Fig. 3.8. The barrier height of different metals versus their electronegativity from (3.28). Heavy line gives strong support for the MIGS model. *Mönch* ascribed the dashed line to defects [3.41]. In addition to his data [3.41], we have added the data for Hg on H-terminated Si (111) [3.74], for epitaxial Pb on (7 × 7) or $\sqrt{3} \times \sqrt{3}$ reconstructed Si (111) [3.89], and for epitaxial NiSi$_2$ on Si (100)

Mönch ascribed the difficulty to derive a slope parameter S_Φ to defects at the metal-semiconductor interface. Figure 3.8 shows the SBHs for several metals on Si versus their *mean* electronegativity

$$\langle X_{\mathrm{M}} \rangle = \sqrt[m+n]{X_{\mathrm{M}}^m X_{\mathrm{Si}}^n} \tag{3.28}$$

for the metal of composition $M_m Si_n$ and the *electronegativity* X_{M} for the metal and X_{Si} for Si from the Miedema scale [compare to (3.22); *Freeouf* used the work functions for the definition of a mean value]. The *full* line in Fig. 3.7 is a least-square fit to fifteen data points and given by

$$\Phi_b = 0.17 \langle X_{\mathrm{M}} \rangle - 0.04 \text{ eV} \tag{3.29}$$

that goes right through the energy of *Tersoff's* Charge Neutrality Level CNL. Following *Mönch* [3.41], this finding suggests that those SBHs which define the straight line are determined by the VGS of silicon. Charge is transferred into the MIGS when the electronegativity of the metal and Si differ. The value of $S_X = 0.17$ for the 'slope parameter' in (3.29) was later confirmed: *Chang* et al. deduced a factor 0.22 from photoemission spectroscopy at interfaces of Si with the metals Au, Cu, Ni, Al, and Ag [3.85]. The same group measured later the change of the barrier height versus work function Φ_M (instead of electronegativity X_M) changes at Ag/Cs/Si interfaces and obtained a slope parameter of $S_\Phi = 0.06$ [3.33]. The possibility to employ fits according to (3.29) support therefore the MIGS model. Nevertheless, many data points in Fig. 3.8 deviated from the heavy line. Following *Mönch*, the data points on the *dashed* line in Fig. 3.8 form a second group. These data should be ascribed to *defects* with a concentration of around 10^{14} cm^{-2} at the metal-semiconductor interface [3.41].

It is interesting to note that earlier *Bucher* et al. had already realized that *silicide*/Si contacts fall in two groups [3.32]. The lower barriers for the TMs (those close to the dashed line, which *Mönch* ascribed to defects) were shown to correlate with the experimentally determined *work function* (instead of Miedema's electronegativity). The higher barriers for near noble metal silicides (i.e., those close to the heavy line in Fig. 3.7) did not correlate with the work function [3.32].

The elaborate and detailed analysis of *Mönch* gave deep insight into the role of MIGS and electronegativity differences in Schottky barrier formation [3.8]. However, it is unlikely that the dashed line in Fig. 3.8 can be ascribed to *defects*. Rather, the deviations from the MIGS model as well as the overall scatter of the data seem to be related to *epitaxy*. For example, the defect postulation of *Mönch* for the lower barriers on the dashed line were mainly based on the controversy of type-A and-B NiSi$_2$ SBHs [3.9, 86]. Defects at the type-A barrier were made responsible for the SBH value of 0.63 eV, whereas the SBH at type-B interfaces amounts to 0.78 eV [3.41, 86]. Unfortunately, such defects were not found in electrical measurements [3.87]. Instead, theory confirmed that the low barrier for type-A indeed is an intrinsic property of this defect-free NiSi$_2$ interface [3.46, 45]. The lower value for type-A than that of type-B is also not related to strain as previously suspected [3.76]. The low barriers for CoSi$_2$, YSi$_2$, and GdSi$_2$ in Fig. 3.8 may also be related to epitaxy [3.88]. In fact, most of the metals that are compiled in Fig. 3.8 grow either single crystalline or polycrystalline *epitaxial* on Si (see below). In addition, the deviations of OsSi$_2$, CrSi$_2$, ReSi$_2$, and FeSi from the MIGS line may be the result of semiconducting properties or of semiconducting traces within these materials.

In addition to the original data of *Mönch* [3.41], we also present in Fig. 3.8 later results for Hg on HF-treated (111) Si [3.74], epitaxial Pb on (111) Si [3.89], and epitaxial NiSi$_2$ on (100) Si [3.90] that deviate strongly from both lines. Additional dipoles due to *epitaxy* instead of defects may better explain the deviations in Fig. 3.8. Next, we show that almost all metals of Fig. 3.8 were observed to grow at least locally *epitaxially* on Si.

3.2 Epitaxial Diodes on Si

Charge transfer at the interface between the metal and the semiconductor is responsible for the formation of the Schottky barrier. From the discussion in the last section it is intuitively clear that the *relative* orientation of atoms at the interface must play a role. However, the role of *crystallography* was only recognized within the last few years. The pioneering work of *Tung* on single crystalline NiSi$_2$/Si interfaces has considerably increased the data base that is necessary for further fundamental understanding [3.9].

In the literature one is often confused because the terms *epitaxial* and *single crystalline* are used as synonyms. Epitaxial relationships between the grains of a polycrystalline metal and Si have often been observed, despite metal grain sizes on the submicrometer scale [3.10]. In this sense, probably *most* metal contacts on Si are locally epitaxial, but do not grow large single crystals on Si. We distinguish therefore clearly between *single crystalline* epitaxial and *polycrystalline* epitaxial contacts. Most of the metals that grow single crystalline on Si are silicides. Work on their preparation was reviewed by *Tung* [3.13, 91], *von Känel* [3.11], *Chen* and *Tu* [3.10], *Derrien* et al. [3.92], *Weitering* [3.93], and *Mantl* [3.14].

3.2.1 Single-Crystalline Schottky Contacts

Single-crystalline epitaxial metal/semiconductor interfaces represent a new basis for investigations of the SBH formation. In addition, this type of contacts allows for the fabrication of novel devices as, for example, permeable-base transistors or metal-base transistors. Only few metals have been grown single crystalline on Si, despite the large number that would be appropriate from the point of view of their lattice match [3.94] (see also the list of 'unconventional metals' [3.95]). Table 3.2 compiles the SBHs that were measured on crystalline metal interfaces to Si. A dependence of the barrier height on local epitaxy has been reported for $NiSi_2$, $CoSi_2$, and Pb on Si; the situation is not quite clear for $CoSi_2$. For two different orientations of

Table 3.2. Single crystalline metals and their measured Schottky barrier heights to n-type Si. Values in brackets are derived by us from the original data for p-type Si and the Si band gap of 1.12 eV

Metal	Structure	Si surface	Φ_b [eV]	Reference	Remark
$NiSi_2$	CaF_2	(111) A	0.65	3.9	7 fold
		(111) B	0.79		7 fold
		(100)	0.40	3.90	6 fold
		(100)	0.65		(111) A facets
		(110)	0.65	3.100	(111) A facets
$CoSi_2$	CaF_2	(111) A	0.78	3.102	7 fold, IV, $n = 1.4$
		(111) B	0.67	3.13	8 fold (?)
		(111) B	0.5		7 fold (?)
		(100)	0.67	3.102	Co rich
		(100)	0.78		8 fold (?)
		(110)	0.7	3.13	8 fold (?)
Pb	fcc	(111) (7 × 7)	0.70	3.89	CV barrier
		(111) ($\sqrt{3} \times \sqrt{3}$)	0.93		CV barrier
Al	fcc	(111) A	0.71	3.110	no orientational
		(111) B	0.71		dependence
α-$FeSi_2$	tetragonal	(111)	0.84	3.123	IV barrier, $n = 1.3$
YSi_2	hexagonal	(111)	0.36	3.122	
$ErSi_2$	hexagonal	(111)	0.74p (0.36)	3.117	measured on p-type
$GdSi_2$	orthorh.	(100)	0.67p (0.45)	3.119	measured on p-type

Al on Si (111), no difference for the SBHs was found. The metals α-FeSi$_2$, YSi$_2$, ErSi$_2$, and GdSi$_2$ have just been grown with *one* type of interface orientation on Si. Some others like CaSi$_2$ and Ag were grown in single crystalline form, but data for their SBHs on Si seem not to be available for these special samples.

a) NiSi$_2$

The metal NiSi$_2$, with CaF$_2$ structure and a lattice mismatch of -0.46%, is ideally suited for single crystalline growth on Si. The innovative work of *Tung* on NiSi$_2$/Si interfaces showed the SBH to depend on crystallography [3.9, 96]. Using ultra high vacuum techniques, *Tung* produced single crystalline NiSi$_2$ of two different orientations on (111) oriented Si surfaces: The A-type NiSi$_2$ has the same orientation

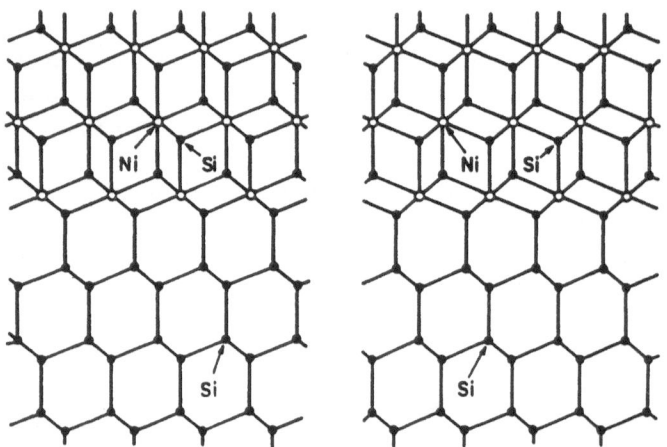

Fig. 3.9. Ball and sticks model of the single crystalline *A-type* (left part) and *B-type* (right part) NiSi$_2$/Si (111) interface. The Ni atom at the interface are sevenfold coordinated ([3.13].)

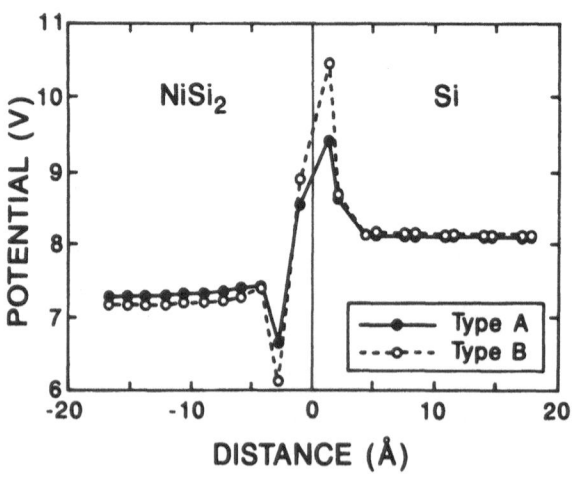

Fig. 3.10. Calculated potential distribution at each Si site across the NiSi$_2$/Si (111) interfaces of type A and B. The dissimilar barrier height is seen as the shift of the two curves on the metal side. The Schottky-barrier height is the difference between the potential deep in the semiconductor and the value deep in the metal (After Ref. [45])

as the Si substrate, whereas the B-type silicide shares the ⟨111⟩ direction but is
rotated 180° about this axis, as shown in Fig. 3.9 [3.96]. For both, type-A and type-
B, the Ni is 7-fold coordinated at the interface, whereas in bulk $NiSi_2$ every Ni has
eight Si neighbours. The SBHs of these two twin-related $NiSi_2$/Si interfaces differ
by about 140 meV [3.9] as shown in Table 3.2. This difference was confirmed
by theory [3.45, 46, 97]. Figure 3.10 displays that the potentials at the interfaces
are indeed different, a fact which results from different charge transfer across
the interface between the $NiSi_2$ and the Si [3.45]. The difference between the
two interface dipoles is found to be largely due to states *below* the valence band
maximum of the semiconductor and not due to different MIGS *within* the band
gap of Si. This finding makes the model of MIGS questionable for these interfaces
[3.13]. At least, the MIGS should depend on orientation.

Recently, *Tung* et al. found also two different SBHs for (100)-oriented Si
surfaces which are metallized by crystalline $NiSi_2$ [3.98, 99]: Flat interfaces on
Si (100) have a SBH of 0.40 eV on n-type Si, whereas those with inclined
facets exhibit a barrier of 0.65 eV. These facets at the $NiSi_2$/Si(100) interface
are simply sections of type-A $NiSi_2$/Si(111) interfaces and have therefore the same
barrier as type-A on Si (111) (Table 3.2). The Ni atoms are 6-fold coordinated
at the flat $NiSi_2$/Si(100) interface [3.98]. The type-A $NiSi_2$/Si(111) facets and the

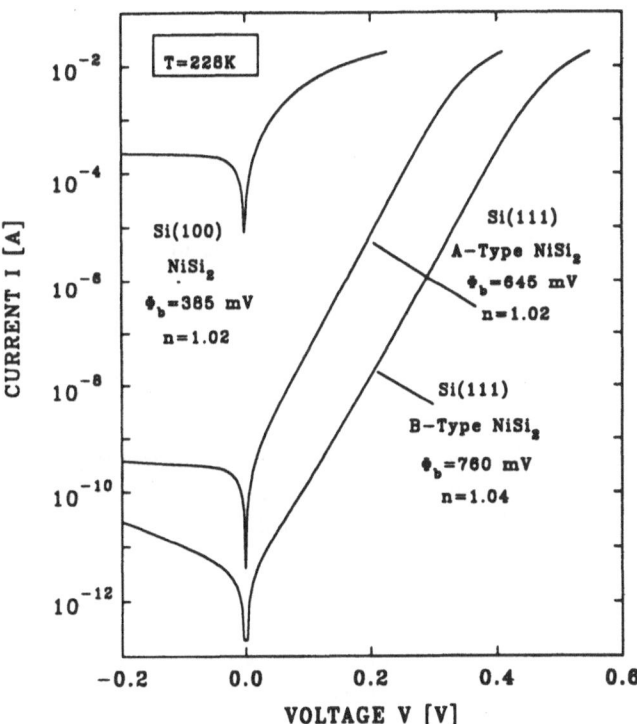

Fig. 3.11. Current/voltage curves at 228 K for different $NiSi_2$/Si interfaces. The sample on Si (100)
has a flat interface without facet bars. The diodes have an area of 1×10^{-2} cm^{-2} [3.101]

corresponding barrier occur also if NiSi$_2$ is grown on (110) surfaces [3.13, 100]. Figure 3.11 demonstrates that the different NiSi$_2$/Si interfaces differ drastically in their current/voltage curves [3.101].

b) CoSi$_2$

The metal CoSi$_2$ with a -1.2% lattice mismatch to Si was also grown single crystalline on Si substrates. However, the dependence of the SBH on interface structure and orientation is less transparent and confirmed than for the case of NiSi$_2$. On Si (111), the type-B CoSi$_2$ is the dominant epitaxial orientation. Growth techniques which involve the deposition of Si produce therefore pure type-B [13]. Only ion implantation has also been able to produce type-A interfaces [3.14, 102]. A SBH of 0.78 eV for these type-A-CoSi$_2$/Si(111) interfaces with a 7-fold coordinated interface structure was deducted from current-voltage (I-V) curves with an ideality $n = 1.3-1.5$ [3.14, 102]. For the type-B-CoSi$_2$/Si(111) the barrier height seems to depend on the *coordination* of the Co at the interface: SBHs between 0.65 and 0.70 eV were reported for samples grown around 600°C [3.103-105]. Tentatively, *Tung* ascribed an 8-fold coordination to these samples [13]. Type-B-CoSi$_2$/Si(111) that is grown at lower temperatures shows lower SBHs around 0.5 eV, which could be the result of a 7-fold or 5-fold coordinated interface [3.13]. The situation is not clear at present.

The data for CoSi$_2$ on Si (100) are also not coherent. *White* et al. reported a barrier value of 0.64 eV [3.106], and *Tung* measured 0.7 eV [3.13]. *Schüppen* et al. found two different values which seem to depend on coordination [3.102]: For the top diode of implanted Si (100)/CoSi$_2$/Si (100) interfaces these researchers reported a SBH of 0.67 eV (ideality $n = 1.05$), while the bottom diode had a higher SBH of 0.78 eV ($n = 1.15$). The different barriers were ascribed to a *"chemical dipole"*: A higher density of Co at the upper interface should lower the Schottky barrier [3.102]. For CoSi$_2$ on (110) oriented Si, a SBH of 0.7 eV was reported [3.13]. The data are compiled in Table 3.2.

c) Al

For Al, the mismatch of the lattice constant to Si is 25%; nevertheless single crystalline Al films can be prepared by epitaxy on Si (111) even by vapour deposition of Al at room temperature [3.107, 108]. Matching translational symmetry [3.94] on both sides of the interface is the reason for such surprising behavior: In the Al(111)/Si(111) system (as well as in the Ag(111)/Si(111) system), *four* Al (111) lattice planes match to *three* Si (111) planes with only 0.55% (0.33%) effective mismatch. Single crystals of Al were prepared by sputtering [109] and electron beam evaporation [3.110] on Si (111). Epitaxial Al/CaF$_2$/Al/Si (111) structures were also reported recently [3.111].

The SBHs of type-A and type-B Al films on n-type Si (111) were investigated by *Miura* et al., who deduced a value of 0.71 eV from IV-curves and 0.72–0.73

from capacitance-voltage (CV)-curves [3.110]. No difference in SBH for these two different interfaces was therefore observed, a finding which is in strong contrast to the results for NiSi$_2$ on Si (111).

d) Pb

The system Pb on Si is of particular interest because no chemical reactions (similar to Ag) or interdiffusions occur at this interface. After NiSi$_2$, the metal Pb on Si (111) is the second system for which a dependence of the barrier height on interface crystallography has definitely been established [3.89, 93]. *Heslinga* et al. deposited Pb on a metastable Si (111) (7 × 7) surface; measurements of IV-curves revealed a barrier height of 0.62 eV, while CV measurements gave 0.70 eV [3.89]. The second interface investigated, Pb on the stable Si (111) ($\sqrt{3} \times \sqrt{3}$)$R30°$, had a higher SBH of 0.90 eV from IV measurements and of 0.93 from CV measurements [3.89]. The different interface structures lead therefore to a barrier difference of at least 230 meV. Similar differences were reported by *Le Lay* et. al. [3.112]. Any theory that is solely based on bulk properties cannot explain this difference.

e) ErSi$_2$

The hexagonal structure AlB$_2$ of the Rare-Earth silicides (RE) yields a good match to the Si (111) surface with the lattice mismatch varying from 0.05% for TbSi$_2$ to −2.55% for LuSi$_2$ [3.113, 114]. The structure is Si deficient by about 20% yielding hexagonal RESi$_{1.6-1.7}$. Apart from Hg on H-terminated Si [3.74], these RE silicides – in their polycrystalline form – were shown to yield the highest barriers to p-type Si [3.58, 115]. Recently, single crystalline ErSi$_2$ was grown on Si (111) surfaces; a very low value of the minimum yield $\chi_{min} = 2.2\%$ was found in channelled Rutherford backscattering spectrometry (RBS) [3.116]. Single crystals of ErSi$_2$ on Si (111) with values of $\chi_{min} = 5\%$ were characterized in IV, CV and Photoresponse (PR) measurements [3.117]. The IV barrier height on p-type Si amounted to 0.73 eV, and the PR barrier was 0.74 eV [3.117]. Both barrier data are in good agreement with earlier values obtained on *polycrystalline* samples [58, 115] but lower than the 0.8 eV of *Wu*, et al. [3.30]. The diodes on n-type Si were ohmic at room temperature; at lower temperatures they had a degraded ideality [3.117]. Therefore, only a CV barrier of 0.29 eV and a PR barrier of 0.28 eV were given for $T = 77K$ [3.117]. The atomic structure of the crystalline ErSi$_2$/Si interface was investigated by *d'Anterroches* et al. [3.118].

f) GdSi$_2$

Not only a hexagonal AlB$_2$ structure, but also an *orthorhombic* epitaxial structure of GdSi$_2$ on Si was recently reported [3.119, 120]: Evaporation of Gd on Si and subsequent annealing allow both phases to be formed epitaxially by solid phase reaction. On (100)-oriented Si, *orthorhombic* GdSi$_2$ with a lattice mismatch of 4% grows for annealing at 500°C [3.120] and 600°C [3.119], while on (111) oriented Si the usual *hexagonal* GdSi$_{1.7}$ with a mismatch of around 1% was observed for annealing between 350 and 600°C [3.121]. The hexagonal phase was also reported later for Si (100) [3.119]. Single crystalline GdSi$_2$ layers were obtained for the orthorhombic phase on Si (100) when the evaporated Gd was 50 nm thick [3.120]. For these single-crystal GdSi$_2$/Si samples, a SBH of 0.67 eV was measured on p-type Si (100), while the values for textured orthorhombic, polycrystalline orthorhombic, and polycrystalline hexagonal samples ranged between 0.58 and 0.64 eV [3.119]. All these values are lower than the 0.71 eV of *Tu* et al. for the *hexagonal* polycrystalline GdSi$_2$ on p-type Si [3.58]. It is unlikely that the different barrier values are related to different doping concentrations, as suspected by Kovaczs et al. [3.119]. Rather, the different interface structure with different epitaxial relationships and structures (orthorhombic instead of hexagonal) could play a role.

g) YSi$_2$

Epitaxial films of hexagonal YSi$_2$ on Si (111) using a straightforward vacuum furnace annealing technique were prepared by *Gurvitch* et al.; the minimum channelling yield in RBS measurements amounted to χ_{min} = 8% [3.122]. The SBH of 0.36 eV for these single crystals [3.122] on n-type Si is in good agreement with the earlier value for polycrystalline films [3.58, 115]. Recently, *Fujitano* and *Asano* investigated the band structure for such epitaxial contacts and predicted a SBH of 0.56 eV which is 0.2 eV higher than the experimental value [3.88]. If one assumes that the distance between the interface Si and Y layers is 0.1Å smaller than in the bulk, then the theory predicts a lower barrier of 0.46 for n-type Si [3.88].

h) FeSi$_2$

The silicides β-FeSi$_2$, ReSi$_2$, and CrSi$_2$ are semiconductors. The work on their single-crystalline and polycrystalline growth by deposition techniques was reviewed by *Derrien* et al. [3.12]. *Mantl* reviewed the ion beam synthesis of CrSi$_2$, FeSi$_2$, and several other silicides [3.14]. In the framework of Schottky-barrier work, just the *metallic*, tetragonal α-FeSi$_2$ instead of the semiconducting, orthorhombic β-FeSi$_2$ is of interest. Single crystalline Schottky contacts of α-FeSi$_2$/Si were prepared by *Radermacher* et al. by ion implantation of Fe into (111) oriented n-type Si and subsequent rapid thermal annealing [3.123]. Current-voltage measurements yielded a barrier height of 0.84 eV at room temperature; Richardson plots of the saturation current gave 0.83 eV for these α-FeSi$_2$/Si diodes. However, the higher values for the SBH of 0.94–1.05 eV obtained from capacitance measurements as well as the large idealities n = 1.3–1.5 of the IV curves [3.123] indicate that these Schottky contacts probably were rather inhomogeneous [3.124]. This conclusion is also consistent with the high χ_{min} = 75% [3.123] in channeled RBS measurements, which indicates relatively poor matching with the consequence of spatial SBH fluctuations at the α-FeSi$_2$/Si interface.

3.2.2   Other Orientational Dependences

The single crystalline Schottky contacts of either NiSi$_2$ or Pb on Si clearly proved the dependence of the SBH on crystallographic relationships at the metal/silicon interface. Similar orientational dependences were observed on crystalline Sb and Yb contacts on GaAs [3.125, 126]. Apart from these data of crystalline contacts on Si and GaAs, there are at least three further indications that crystallographic orientations may play a role in the formation of the SBH. However, for these systems listed in Table 3.3, little information on the atomic structure at the interface is available.

a) Ir and IrSi

The first indication for a dependence of the barrier height on crystallographic orientation goes back to 1978: *Ohdomari* et al. found that the SBHs of *polycrystalline* Ir and IrSi on n-type Si (100) are about 30 meV higher than on Si (111), as shown in Table 3.3; an explanation for this finding was not given [55]. Unfortunately, the barriers of around 0.9 eV were derived from IV curves only [3.55]. On the one hand, it is unlikely that these 30 meV differences could be ascribed to the ideality n and an artefact in extrapolating the IV curves towards zero in order to extract the saturation current density j_0. For example, the samples on Si (111) (with the lower SBH) had a slightly better ideality n than the samples on Si (100) (with the higher SBH). This finding indicates that the result of a lower SBH on Si (111) cannot be ascribed to an inhomogeneous interface [3.124]. On the other hand, without a systematic comparison to other measurement techniques it cannot be excluded that the different IV-barriers stem, for example, from different Richardson constants A^* for the (100) and (111) direction at the investigated interfaces. At present, it also seems unknown if local epitaxial relationships exist at the interface between IrSi

Table 3.3. Orientational dependences of the SBH for several metals on Si. The structure of these interfaces was not always investigated. Values in brackets are derived by us for n-type Si from the measured values on p-type Si and the Si band gap of 1.12 eV

Metal	Si surface	Φ_b [eV]	Reference	Remark
Ir	(111)	0.883	3.55	randomly oriented Ir
	(100)	0.914		grains, 10 nm size
IrSi	(111)	0.888	3.55	randomly oriented
	(100)	0.922		IrSi
PtSi	(111)	0.22p (0.9)	3.127	2 to 20 nm thick
	(100)	0.3p (0.82)	3.127	PtSi, single crystalline?
Sb	(111)	0.6p (0.52)	3.129	Sb thickness in
	(100)	0.19p (0.93)		monolayer regime

and Si: None are listed in a recent review of *Chen* and *Tu* [3.10]. Recent investigations of *Wittmer* revealed crystallinity up to the interface but no local epitaxy between IrSi and Si [3.75].

b) PtSi

Recently, *Pellegrini*, et al. reported that the SBH of PtSi on p-type Si was about 0.1 eV lower on a Si (100) surface than on a (111) surface [3.127]. These results were derived from Richardson plots of the saturation current density j_0 as well as from photoresponse measurements. Structural investigations of the metal/semiconductor interface as well as a physical explanation for the different SBHs were not given in [3.127], but it is obvious that the barrier difference could be related to *single* crystalline growth of PtSi on Si: The thickness of the evaporated Pt on Si was varied between 1 and 10 nm [3.127]. Upon annealing, these Pt thicknesses should then yield about 2 to 20 nm PtSi on Si. At such small thicknesses the PtSi could grow single crystalline over the whole contact area. Then, the SBH could be determined by the different epitaxy on the differently oriented surfaces. *Local* epitaxial growth of *thick* layers of PtSi on (111) and Si (100) was reviewed by *Tung* et al. [3.128] and *Chen* and *Tu* [3.10].

c) Sb

Hricovini et al. recently gave results on synchrotron-radiation core-level spectroscopy for Sb and Pb on Si. For p-type Si, they deduced a SBH of 0.6 eV for Sb on Si (111) and of 0.19 eV for (100) oriented Si [3.129]. This difference of more than 0.4 eV between the SBH for the two Sb orientations is even larger than the one between the B-type $NiSi2/Si(111)$ and the flat $NiSi_2/Si(100)$ interface.

3.2.3 CaSi$_2$ and Ag

a) CaSi$_2$

$CaSi_2$ has a lattice mismatch of only 0.4% to Si, a value that compares favourably with the 1.2% and 0.4% mismatches of $CoSi_2$ and $NiSi_2$. Atomically abrupt and smooth interfaces of $CaSi_2$ on Si (111) were prepared by *Morar* and *Wittmer* [3.130, 131]. Investigations by transmission electron microscopy revealed that the epitaxial $CaSi_2$ assumes the trigonal/rhombohedral phase of bulk $CaSi_2$. Measurements of the SBH have not been carried out [3.130, 131].

b) Ag

Ag is one of the few metals which form an abrupt, inert interface with Si. Similar to Al, the mismatch of the Ag lattice constant to Si is 25%. Nevertheless, the effective mismatch of Ag on Si (111) is only 0.33%. Epitaxial relationships of Ag on Si (111) were investigated by *LeGoues*, et al. [3.107] as well as by *Park* et al.

[3.132]. Systematic SBH investigations on crystalline interfaces seem not to be available. The work on polycrystalline Ag/Si interfaces was recently reviewed by *Cros* and *Muret*; these authors also gave SBH results for Cu and Au [3.133].

3.2.4 Polycrystalline Epitaxial Contacts on Si

It is likely that many metals on Si display some epitaxial relationships between the metal grains and the Si. In particular, such a behavior is expected for the reactive TMs. There are probably only few metal/Si contacts with an amorphous transition region made up by a mixture of metal and Si atoms. For the TMs, an early review of epitaxial relationships to Si was given by *Tung* et al. for the metals $CoSi_2$, $NiSi_2$, Pd_2Si, and PtSi [3.128]. The structure and properties of many more silicides were later reviewed by *Nicolet* and *Lau* [3.35] and *Murarka* [3.134]. Since then, many more epitaxial relationships were observed between TM silicides and Si. Work on the following metals on Si was recently reviewed by *Chen* and *Tu* [3.10]:

(i) the *near-noble* silicides $CoSi_2$, $NiSi_2$, the systems Pd/Si, Pt/Si, $FeSi_2$;
(ii) the *refractory* silicides $TiSi_2$, $ZrSi_2$, $HfSi_2$, VSi_2, $NbSi_2$, $TaSi_2$, $CrSi_2$, $MoSi_2$, WSi_2;
(iii) the *platinum group* silicide systems Ir/Si, Rh/Si, Ru/Si, Os/Si;
(iv) the systems Mn/Si, Re/Si.

Epitaxial growth of the *rare-earth* silicides on Si (111) was investigated by *Knapp* and *Picraux* [3.113].

3.2.5 Unconventional Metals with Small Lattice Mismatch to Si

Recently, *Murarka* tabulated metals with a good lattice match (i.e., with a similar lattice constant) to Si, GaAs and InP [3.95]. Many metals seem appropriate (but have not been used yet) for the growth of *single crystalline* interfaces on Si; we list them here with their lattice mismatch in brackets [3.95]: $NbAl_3$ (0.02%), $TiAl_3$ (0.02%), Zr_4Al_3 (0.04%), Co_4Gd (0.07%), Yb (0.07%), $TaAl_3$ (0.2%), Rh_2B (0.2%), Ce_3Al (0.4%), $NiSi_2$ (0.46%), Cr_5B_3 (0.5%), Pd_3B (0.6%), $TiPd_3$ (0.9%), $CoSi_2$ (1.2%), Cu_4Fe (1.5%), $GdNi_4$ (1.5%), Rh_2P (1.5%), Ir_2P (2%), Mo_2B (2%), W_2B (2.4%), CoP (2.9%). From these twenty metals, just two ($NiSi_2$ and $CoSi_2$) have been investigated for crystalline Schottky contacts on Si. For future investigations of orientational dependences, the element Yb with an almost perfect lattice match (and an fcc structure) seems especially interesting for deposition on differently oriented surfaces of Si.

3.2.6 Summary

Practically *all* metals of those indicated in Fig. 3.8 (and many more) were observed to grow (at least locally) epitaxially on Si. For special single crystalline metals ($NiSi_2$, Pb etc.) it was proven that their SBH to Si depends on the *local* chemical and atomic configuration at the interface. It is therefore suggested that similar dependences exist also for those metals that were observed to grow epitaxially (but not in large area single crystalline form) on Si. For these *polycrystalline* epitaxial metals, the SBH deduced from electrical measurements represents then a macroscopic average over many microscopic, spatially distributed SBHs. In this case, the measured average barrier value depends on the lateral distribution of the microscopic barriers, which, in turn, depends on the texture, local epitaxy, and the size distribution of grains in the metal. The *crystallography* is therefore the key issue for further understanding of Schottky barrier formation and probably also responsible for the deviations from models that take only into account MIGS and electronegativity differences between the metal and the semiconductor (Fig. 3.8).

At polycrystalline contacts, the dependence of the SBH on local crystallography necessarily results in fluctuations of the SBH and consequently also of the Fermi level position with respect to the band edges. These inhomogeneities raise the serious question, whether the concept of *Fermi level pinning* can be used at all: If the Fermi level was pinned, spatially inhomogeneous Schottky contacts could not occur. In this sense, terms like *Fermi level pinning* that intend to describe deviations from an anticipated linear dependence of the SBH on bulk properties like work function (or electron affinity, or electronegativity) seem no longer tenable. This matter was also discussed by *Tung* [3.13].

3.3 Electrical Transport Properties

Here we give only a short description of the current transport mechanisms and refer the reader to standard textbooks [3.46]. An n-type semiconductor is assumed in the summary of the transport processes. Figure 3.12 indicates the transport processes for majority carriers only:

 (a) Emission of electrons over the top of the barrier,
 (b_1) thermally assisted tunneling,
 (b_2) tunneling at the bottom of the conduction band in the quasi-neutral region.

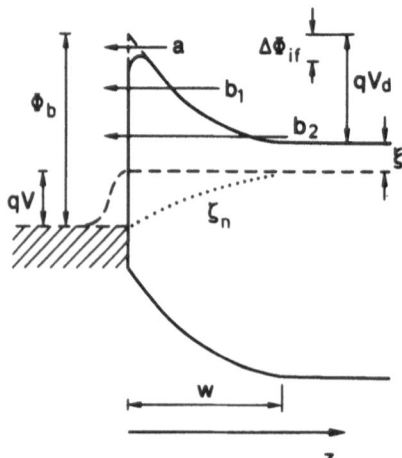

Fig. 3.12. Band diagram of a Schottky contact under forward bias. Transport paths for electrons are indicated by arrows. (a) thermionic emission, (b_1) thermionic field emission, and (b_2) pure field emission. The curves for the quasi Fermi level ζ_n for electrons are indicated for thermionic emission (dashed line) and diffusion (dotted line), respectively

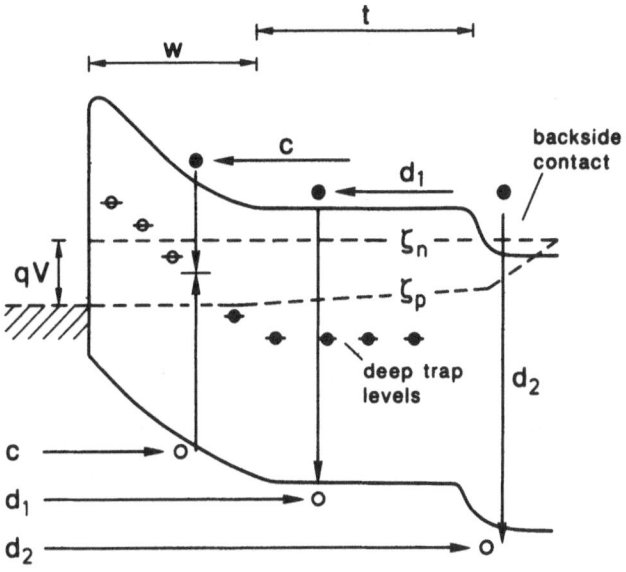

Fig. 3.13. Band diagram indicating the transport mechanisms involving minority carriers (holes, indicated as hollow circles) *and* majority carriers (electrons, filled circles): (c) recombination via deep trap levels, (d_1) recombination in the quasi neutral region, and (d_2) recombination at the back contact (assumed here as an nn$^+$-transition). The quasi Fermi-levels ζ_n and ζ_p for electrons and holes are split across the quasi neutral region of the sample

Processes involving also minority carriers are sketched in Fig. 3.13:

(c) recombination of holes and electrons within the space charge region,

(d_1) recombination in the quasi-neutral region,

(d_2) recombination at the back-contact.

The transport processes (a–d) are connected in parallel. Thus, the total diode current is the sum of the contributions of the individual current paths. It is important to note that the usual models for these transport processes describe electronic transport only in one dimension and assume that the Schottky barrier is spatially uniform. During the last years, the consideration of spatial *fluctuations* of the barrier has given new impact on the understanding of electronic transport across Schottky barriers [3.98, 99, 135, 136]. In the following, we briefly describe the 'classical' *one-dimensional*.models for the transport mechanisms (a–d) of homogeneous contacts before we discuss recent work on the electrical transport of inhomogeneous Schottky barriers.

3.3.1 Emission Over the Barrier

a) Thermionic Emission

Emission over the barrier is the most important current path across metal/semiconductor interfaces, especially for Schottky contacts on silicon. The

theory is based on *Bethe's* model of vacuum tubes [3.137] which is modified to
describe the current density j across metal/semiconductor interfaces by

$$j = j_0 [\exp\left(\frac{qV}{kT}\right) - 1].$$ (3.30a)

Here, V denotes the voltage, k the Boltzmann constant, and T is the temperature.
The saturation current density j_0 is given by

$$j_0 = A^{**} T^2 \exp\left(\frac{-\Phi_b}{kT}\right),$$ (3.30b)

where A^{**} is the effective Richardson constant. For Schottky diodes on silicon, the values for the effective Richardson constants as computed by *Andrews*
and *Lepselter* are usually used [3.138]: $A^{**} = 112$ A cm^{-2}K^{-1} for n-type and
$A^{**} = 32$ A cm^{-2} K^{-1} for p-type Si. Thermionic emission implicitly assumes
that the drop of the quasi Fermi level ζ_n for the majority carriers occurs immediately at the metal/semiconductor interface instead of decreasing continuously across
the space-charge region of the contact, as it is described by the diffusion theory
of *Wagner* [3.139] and of *Schottky* and *Spenke* [3.140]. Theory and experiment
showed that diffusion processes within the space-charge region play no major role
in the current transport across Schottky contacts on high-mobility semiconductors
like silicon under reverse or moderate forward bias [3.141–143].

In most cases, (3.30a, b) do not describe the experimentally observed behavior
correctly. Instead, these equations have to be modified in order to fit the experimental data. This finding results from the fact that the barrier Φ_b^j which controls
the current *deviates* from the value Φ_b that describes the *electrostatic* properties
of the barrier. The classic example for such a deviation is image force lowering.

b) Image Force Lowering

A charge carrier traversing the barrier is attracted by its image force in the metal.
This Schottky effect lowers the SBH which controls the current by an amount
[3.144] of

$$\Delta\Phi_{if}/q = \left[\frac{q^3 N_D}{8\pi^2 \varepsilon_s^3}\left(V_d - \frac{kT}{q}\right)\right]^{1/4}.$$ (3.31)

The barrier lowering for a semiconductor like silicon is small (typically of the order
of 10–20 mV for moderately doped n-type silicon). Nevertheless, this lowering is
noticeable in current-voltage characteristics under reverse bias. To include the effect
of image force lowering into thermionic emission theory, the barrier $\Phi_b^j = \Phi_b - \Delta\Phi_{if}$ instead of Φ_b has to be inserted into (3.30b) for the description of IV curves.

c) The Ideality Factor n

The decrease of the SBH Φ_b^j by image force lowering according to (3.31) depends
on the applied voltage. As will be seen in the following subsections, the SBH which

controls the current can also depend on the voltage for other reasons. Usually, such deviations from the ideal diode behavior are *hidden* in the diode ideality factor n. Instead of using (3.30a, b) with a voltage dependent barrier height $\Phi_b^j(V)$, the ideality n is added to the voltage dependent term of (3.30a) and the saturation current density j_0 is expressed by a voltage *independent* SBH Φ_{b0}. Then, the current/voltage relation for voltages $V > 3kT/q$ reads as

$$j = A^{**}T^2 \exp\left(\frac{-\Phi_{b0}}{kT}\right) \exp\left(\frac{qV}{nkT}\right) \tag{3.32}$$

with the zero-bias Schottky barrier Φ_{b0}. The ideality factor n is related to the voltage dependence of the current Schottky barrier Φ_b^j by the relation [3.145]

$$\frac{1}{q}\frac{d\Phi_b^j}{dV} = 1 - \frac{1}{n}. \tag{3.33}$$

The ideality becomes larger than unity if the SBH Φ_b^j increases with increasing forward bias.

According to the ideal thermionic emission theory, under reverse bias the current density approaches asymptotically the saturation current density j_0. If the barrier height entering in Eq. (30b) depends on the applied voltage, the reverse current density cannot saturate. The reverse conductance G_{rev} for voltages $V < -3kT/q$ is obtained from the derivative (3.30b) as

$$G_{rev} \equiv \frac{dj}{dV} = \frac{|j|}{kT}\frac{d}{dV}\Phi_b^j(V). \tag{3.34a}$$

If the voltage dependence of Φ_b^j (V) is linear and is the same for reverse and forward bias, the ideality n, which describes the *forward* characteristic, is related to the *reverse* conductance by

$$\frac{1}{|j|}\frac{dj}{dV} = \left(1 - \frac{1}{n}\right)\frac{q}{kT}. \tag{3.34b}$$

It is therefore also possible to determine the ideality factor n from the reverse bias behavior.

Ideality factors larger than one and 'soft', non-saturating reverse characteristics are two often-observed features in the current-voltage relationship of Schottky diodes. Equations (3.32–34) are a formal description of these phenomena in terms of a voltage dependent SBH $\Phi_b^j(V)$ or with the help of the ideality factor n, respectively. However, both descriptions are phenomenological, replacing one term by another. The remainder of this section will be concerned with possible *physical* reasons for such deviations from ideal thermionic emission theory.

3.3.2   Tunneling Through the Barrier

Thermionic emission theory treats the charge carriers as classical particles. Transmission of a carrier across the Schottky-barrier is only possible if its energy exceeds the barrier height. As a result (3.30) is obtained. However, a microscopical particle can be *reflected* by a potential barrier even if its energy *exceeds* the height of the barrier (quantum mechanical reflection) and it can *tunnel* through the barrier with a certain probability if it has an energy *below* the top of the barrier. The quantum mechanical reflection of carriers with energies above the SBH can be taken into account by a modification of the effective Richardson constant A^{**} [3.146]. However, consideration of the tunneling of carriers with energies *below* the maximum of the SBH results in a change of the overall voltage dependence of the current.

Padovani and *Stratton* [3.147] distinguished between:

(i) *Field emission*, i.e., tunneling of electrons at an energy close to the Fermi level.
(ii) *Thermionic field emission*, i.e., tunneling of thermally activated carriers through the top of the barrier.

Thermionic field emission can be applied for n-type Si with a doping concentration of $N_D = 10^{19}$ cm^{-3} in the temperature range $T > 180$ K [3.148]. Only for even higher doping levels pure field emission must be considered. Disregarding the case of highly doped, degenerate semiconductors, we restrict ourselves to the case of thermionic field emission. Expressions for the current-voltage relationship in the case of thermionic field emission were derived by *Padovani* and *Stratton* [3.147] and by *Crowell* and *Rideout* [3.149]. Following [3.147], the current density for *forward* bias voltages $V > 3kT/q$ is given by

$$j = a_f \exp\left[\left(\frac{1}{n_{\text{TFE}}} - 1\right)\frac{\xi}{kT}\right] \exp\left(\frac{-\Phi_b}{n_{\text{TFE}}kT}\right) \exp\left(\frac{qV}{n_{\text{TFE}}kT}\right), \qquad (3.35)$$

where the prefactor a_f depends only slightly on the temperature T and on the applied voltage V. The ideality factor n_{TFE} for thermionic field emission is given by

$$n_{\text{TFE}}(T) = \frac{E_{00}}{kT}\coth\left(\frac{E_{00}}{kT}\right). \qquad (3.36a)$$

The parameter

$$E_{00} = \frac{q\hbar}{2}\left(\frac{N_D}{m^*\varepsilon_s}\right)^{1/2} \qquad (3.36b)$$

has the dimension of an energy and is the most important quantity in tunneling theory [3.147]. In (3.36b) m^*, is the effective tunneling mass of the electrons in the semiconductor.

According to (3.36a), the ideality n_{TFE} decreases with decreasing temperature T. If one neglects the first term in (3.35) one can compare the result with the phenomenological (3.32). Then, the IV curves can be described by the ideality n_{TFE} and a *temperature* dependent zero-bias SBH

$$\Phi_{b0}(T) = \frac{\Phi_b}{n_{\text{TFE}}(T)}. \qquad (3.37)$$

The zero-bias SBH Φ_{b0} decreases with temperature T in the same way as does the quantity $1/n_{\text{TFE}}$. The decrease of the SBH Φ_{b0} as well as the increase of the ideality factor n_{TFE} by thermionic field emission are small: For n-type silicon, typical values for E_{00} range between 3×10^{-4} eV (for a doping $N_D = 10^{15}$ cm^{-3}) and 3×10^{-2} eV (for $N_D = 10^{19}$ cm^{-3}). A noticeable effect of thermionic field emission on the ideality n_{TFE} at room temperature occurs only at high doping: We find $n_{TFE} = 1.16$ for $N_D = 10^{19}$ cm^{-3} and $T = 300$ K. In this case, the zero-bias SBH Φ_0 is lowered with respect to Φ_b by a factor of $1/n_{TFE} = 0.86$. In contrast, for moderately doped silicon there is no observable effect on the ideality n: One obtains $n_{TFE} = 1.0007$ for a doping concentration of $N_D = 10^{15}$ cm^{-3} even at 77 K.

For *reverse* bias voltages with $V < -3kT/q$ the current density j obtained by *Padovani* and *Stratton* [3.147] is given by

$$j = -a_r \exp\left(-\frac{\Phi_b}{n_{TFE}kT}\right) \exp\left[-\frac{qV}{kT}(n_{TFE}-1)\right] \tag{3.38}$$

where a_r depends only weakly on temperature T and voltage V. Equation (3.38) describes a non-saturating reverse characteristic for values of $n_{TFE} > 1$. However, this effect is again negligible for Schottky contacts on moderately doped silicon as for the example considered above with $N_D = 10^{15}$ cm^{-3} and $n_{TFE} = 1.0007$.

3.3.3 Generation/Recombination in the Space Charge Region

Schottky contacts are usually regarded as pure majority-carrier devices. Nevertheless, minority carriers may cause an additional current component across the interface. Here, the most important process is the generation or recombination of minority carriers in the depletion region of the contact. As can be seen in Fig. 3.13, the quasi Fermi levels ζ_n for electrons and ζ_p for holes are split across the space charge region of the contact. This situation is equivalent to the situation in the depletion region of a pn-junction under bias. Thus, the theory of generation-recombination currents as developed for pn-junctions [3.150] applies for Schottky diodes in the same way.

Under forward bias, the product of the concentrations n of electrons and p of holes exceeds its equilibrium value and a *recombination* current flows across the interface. According to *Sah* et al. [3.150], recombination of holes with electrons is most efficient via recombination centers with energies located close to the intrinsic Fermi level. With capture cross-sections σ_n for electrons and σ_p for holes assumed to be equal ($\sigma_n = \sigma_p = \sigma$) the recombination current density j_r is given by [3.151]

$$j_r = \frac{qw}{2}\sigma v_{th} N_t n_i \exp\left(\frac{qV}{2kT}\right) \tag{3.39a}$$

where the quantity N_t denotes the density of traps, v_{th} the carriers' thermal velocity, and n_i is the intrinsic carrier concentration. With the effective lifetime $\tau_e = (\sigma v_{th} N_t/2)^{-1}$ for minority carriers one obtains from (3.39a)

$$j_r = \frac{qw}{\tau_e}\sqrt{N_v N_c} \exp\left(\frac{E_g}{2kT}\right) \exp\left(\frac{qV}{2kT}\right), \tag{3.39b}$$

where N_v and N_c are the effective densities of states in the valence band and in the conduction band, respectively. To derive (3.39b), we have used $n_i^2 = N_v N_c \exp[E_g/kT]$ for the intrinsic carrier concentration.

Under reverse bias, the current via deep traps is due to the *generation* of holes and electrons. With similar arguments as above, one obtains for the reverse bias (generation) current density

$$j_g = \frac{qw}{\tau_e} \sqrt{N_v N_c} \exp\left(\frac{E_g}{2kT}\right). \tag{3.40}$$

Due to its dependence on the width w of the space charge region which, in turn, depends on the applied voltage V, the generation current density does not saturate with increasing reverse bias voltage.

According to (3.39b) and (40) the generation-recombination current is thermally activated with an activation energy of $E_g/2$ and has an ideality of two in forward bias direction. In contrast, the majority carrier current arising from thermionic emission over a Schottky barrier has a SBH $\Phi_b > E_g/2$ and an ideality close to one. Thus, the generation/recombination current component is expected to be important only under reverse or low forward bias and at low temperatures.

3.3.4 Minority Carrier Injection

Minority-carrier injection is usually neglected in the description of current transport in Schottky diodes. However, the tacit disregard of minority carriers in the quasi-neutral region of the semiconductor is justified only for reverse or moderate forward bias, and for semiconductors with a relatively low minority carrier lifetime. In contrast, for long-lifetime semiconductors like silicon, minority carrier injection may play an important role preferably for the electronic transport under ac conditions and under high forward bias [3.87, 152]. But even in dc-experiments on low-barrier Schottky diodes, minority carriers may show up because they modulate the resistance of the substrate [3.153]. This effect is particularly severe in case of thin epitaxial layers, but also for thick substrates there may still be some influence of minority carriers on dc-experiments [3.154–158].

As long as the voltage drop across the quasi-neutral region is negligible, the minority carrier flow into the quasi-neutral region of the semiconductor is purely diffusive and corresponds completely to the situation encountered in a pn diode. If the distance between the edge of the depletion region and the *back-side contact* is comparable to the minority carriers' diffusion length, the recombination properties of this contact influence minority-carrier injection at the front Schottky contact. The problem of the back-contact was originally (and since then practically exclusively) discussed by *Scharfetter* [3.159]. He calculated the injection of holes from a Schottky contact into an n-type bulk semiconductor and demonstrated a sensitive dependence on the properties of the back-contacts: An imperfect back-contact extracts minority carriers via recombination and increases therefore the injection of minority carriers at the front Schottky diode considerably.

According to *Scharfetter* the influence of the back-contact is more severe when the sample thickness is smaller than or comparable to the minority carrier diffusion length because under these circumstances the recombination at the back-contact outstrips bulk recombination. As an example, consider a Schottky contact on an epitaxial layer of a thickness $t = 5$ μm and a diffusion length $L_D = 200$ μm. Here, the injection ratio γ, i.e., the ratio between the minority-carrier and the majority-carrier current densities, can vary by several orders of magnitude depending on the recombination velocity at the backside contact. The absolute values of γ, however, are restricted to values between 5×10^{-3} and 3×10^{-6} for a typical Schottky contact with a barrier height $\Phi_b = 0.8$ eV on n-type silicon with a doping of $N_D = 10^{15}$ cm^{-3}. For silicon one should always expect an influence of the back-contact since the typical diffusion length around 100 μm is *always* comparable to sample thickness.

The second source for an increase of minority-carrier injection comes from the dependence of the injection ratio γ on the *area* of the Schottky diode. Even when one somehow prevents excess minority-carrier injection due to recombination at the back-contact, there may still be substantial injection. The minority carrier current can even *equal* the majority-carrier current when the diameter of the Schottky contact is small compared to the diffusion length: *Clarke* et al. [3.160] showed that under such geometrical conditions the injection ratio can reach values of up to one because the Schottky diode serves as a point source and the bulk wafer as a three-dimensional diffusion sink. This area dependence and three-dimensionality of the problem should therefore be studied in systematic experiments.

3.3.5   Inhomogeneities in Schottky Contacts

Up to this point we have used one-dimensional theories for Schottky contacts and *tacitly* disregarded the fact that one-dimensional band diagrams like the one in Fig. 3.1 represent only the *spatial average* of electronic and thermodynamic properties of the contact over the entire diode area. The shortcoming of such a crude (but in most cases successful) model becomes immediately apparent if we consider that the band bending of a Schottky contact is not built up by a continuously distributed charge density, but by discrete, randomly distributed impurity atoms. For a contact on n-type silicon with a donor concentration of $N_D = 10^{18}$ cm^{-3} and a band bending $V_d = 0.5$V, the space-charge region is built up by about 4 layers of dopant atoms.

The influence of randomly distributed dopant atoms on the electronic transport is usually neglected. This assumption seems justified as long as one regards the *electrostatic* properties of the diode. For instance, a consideration of the *mean* potential is sufficient to describe the SBH when measured by capacitance voltage measurements. In contrast, within the last few years it has been realized that the electronic *transport* is strongly altered under the influence of *spatial potential fluctuations*. For example, it was shown recently that only minute deviations from electronic homogeneity result in noticeable changes of the ideality n in current-voltage curves: The current density j depends exponentially on the local value of the Schottky-barrier height; consequently, inhomogeneities which cover only a fraction of 10^{-3} or 10^{-4} of the total interface area suffice to make the contact to appear inhomogeneous for the electronic transport [3.136].

There are several reasons for such potential inhomogeneities at the interface of metal/semiconductor contacts:

(i) The dopant atoms are randomly distributed within the semiconductor.

(ii) Atomic steps and lattice defects at the interface modulate the barrier height even at single crystalline epitaxial contacts.

(iii) The dependence of the SBH on the relative orientation of semiconductor and metal atoms results in barrier height fluctuations over the area of *polycrystalline* Schottky contacts.

(iv) At polycrystalline contacts, grain boundaries in the metal may modify the SBH.

(v) Interface roughness results in spatially varying *effective* SBHs by local barrier lowering due to field emission even for nominally homogeneous contacts.

(vi) In the particular case of reactive contacts (for example, silicide/Si), different phases of the metal yield different barrier heights.

(vii) The metal atoms can diffuse into the semiconductor and (when charged) after the distribution of dopant atoms in the vicinity of the interface.

(viii) Contact *edges* are often sites of charges which locally modify the Schottky barrier height. The contact appears therefore also inhomogeneous.

Only very few of the effects of our list have been discussed in the literature; however, fluctuations due to the statistical distribution of dopant atoms were already considered by *Chang* and *Sze* [3.161]. They discussed tunneling currents across Schottky contacts in terms of a tunnelling probability $T(E)$ and used a *Gaussian distribution* to model concentration fluctuations of the dopant atoms leading to fluctuations in the electrical field at the interface. In the presence of such fluctuations, the tunnelling current deviates from a simple one-dimensional model by 10% up to 100% for average doping concentrations of $N_D = 10^{18}$ cm^{-3} and $N_D = 10^{19}$ cm^{-3}, respectively. These values hold for a Schottky contact on n-type silicon with a SBH $\Phi_b = 0.8$V and for a temperature of $T = 300$ K. The deviations from the one-dimensional model become even more pronounced at lower temperatures [3.161]. *Van Schilfgaarde* [3.162] calculated the fluctuations of the electrostatic potential at the interface due to doping statistics. His result can be expressed in terms of a standard deviation

$$\sigma_\Phi = \frac{3q}{5\varepsilon_s}\left(\frac{4\pi}{3}N_D\right)^{1/3} \tag{3.41}$$

of the SBH. For moderately doped silicon with $N_D = 10^{15}$ cm^{-3}, (3.41) yields a value of $\sigma_\Phi = 15$ meV for the fluctuations.

Up to now, current transport across inhomogeneous Schottky contacts has been described, basically, by two approaches: The first approach models inhomogeneities either by two different discrete values [3.163] or by a continuous distribution $P(\Phi_b)$ of barrier heights [3.6, 124, 164–166]. The current transport is then modelled by a *parallel connection* of the different barrier heights, each of which is described by a one-dimensional theory (usually the thermionic emission model). The second type of theories, the *geometric models*, assume specific interface geometries at the contacts. These models focus on either a numerical or analytical solution of Poisson's equation for the space charge region of the inhomogeneous diode [3.135, 167–169]. Convenient geometries are chosen for lower-barrier regimes (like circular patches or semi-infinite stripes) within an otherwise homogeneous contact. The size and energetic depth of these lower-barrier patches are then varied in order to study their influence on the SBH that is effective in current transport [3.135].

a) Parallel Connection Models

This type of model assumes that the interface is composed of a *distribution* $P(\Phi_b)$ of barrier heights Φ_b over the contact area. Within each of the contact patches, the current transport is described by the theory that is usually assumed for a *homogeneous* contact. The total current is then obtained by an integration over the barrier distribution. Several researchers assumed a *Gaussian barrier distribution*

$$P_G(\Phi_b) = \frac{1}{\sigma_\Phi \sqrt{2\pi}} \exp\left(\frac{(\bar{\Phi}_b - \Phi_b)^2}{2\sigma_\Phi^2}\right) \tag{3.42}$$

which is described by its first two momenta: the mean value $\bar{\Phi}_b$ of the barrier height and the standard deviation σ_Φ [3.16, 124, 164–166]. Within thermionic emission theory, the total current across the entire contact area is then calculated by an integration of the saturation current

$$j_0 = A^* T^2 \exp\left(-\frac{\Phi_b}{kT}\right) \tag{3.43}$$

for a single barrier over the barrier distribution according to

$$j_0 = A^* T^2 \int \exp\left(-\frac{\Phi_b}{kT}\right) P_G(\Phi_b) d\Phi_b. \tag{3.44}$$

The integration reveals that the usual equation for the current across a *homogeneous* contacts, (3.32), can be maintained for an *inhomogeneous* contact if one just replaces the barrier Φ_b by an *effective* barrier [3.124, 166]

$$\Phi_b^j = \bar{\Phi}_b - \frac{\sigma_\Phi^2}{2kT}. \tag{3.45}$$

Equation (3.145) predicts a decrease of the effective current Schottky-barrier height Φ_b^j with inverse temperature T^{-1}. *Güttler* and *Werner* demonstrated that the temperature dependence of Φ_b^j obtained from different kinds of silicide/silicon Schottky contacts follows this rule [3.124, 166]. The mean barrier height $\bar{\Phi}_b$ can independently be measured by capacitance-voltage curves. The difference of Φ_b^j and $\bar{\Phi}_b$ allows one therefore to determine the standard deviation σ_Φ of the barrier distribution. Values of $\sigma_\Phi = 50 - 70$ mV were found for silicide/Si contacts [3.124]. Recently it was shown that the standard deviation σ_Φ is also accessible by the curved behavior of Richardson plots of the saturation current j_0 [3.170]. Similarly to current/voltage curves, the noise properties of Schottky diodes depend also sensitively on the standard deviation σ_Φ which gives a measure for the contact inhomogeneity [3.166, 171].

The ideality factor n of current/voltage curves is also understood within the model of potential inhomogeneities [124]: The factor n describes the homogenization of the barrier distribution under the application of a bias voltage V. Within this model, a temperature dependence of the ideality n according to

$$n^{-1} - 1 = -\rho_2 + \rho_3 \frac{q}{2kT} \tag{3.46}$$

is predicted. Here, the coefficients ρ_2 and ρ_3 describe the linear voltage dependence of the mean barrier height $\overline{\Phi_b}$ and of the variance σ_Φ^2. The temperature dependence of the ideality n of a variety of Schottky contacts in terms of this barrier-fluctuation model was recently investigated on the basis of (3.46) [3.124]. For most of the contacts considered, mechanisms other than inhomogeneities which could also cause the temperature dependence of the ideality factor n could be excluded [3.124].

b) Geometric Models

The integration over a distribution $P(\Phi_b)$ of barrier heights, as described above, represents a parallel connection of a series of *one-dimensional* models. These parallel connection models cannot be regarded as a real *three-dimensional* theory. Instead, in principle, all transport processes such as thermionic emission, diffusion, tunnelling, etc. should be re-inspected by a solution of the relevant equations in *three* dimensions.

The potential fluctuations at the metal/semiconductor interface result also in a spatial variation of the potential within the space-charge region of the semiconductor. In the solution of Poisson's equation for this region, one cannot expect that interfacial inhomogeneities of arbitrary size result in a profile within the space charge region, that is given by a parallel connection of simple parabolic potential curves. Either numerical [3.167] or analytical solutions [3.135, 136, 168] of Poisson's equation are therefore necessary to account for the three dimensionality of inhomogeneities at Schottky contacts.

Numerical solutions of *Poisson's* equation for the space charge region of an inhomogeneous contact were computed by *Freeouf* et al. [3.167]. It turned out that parallel connection models risk significant errors in the case of barrier modulations with a size smaller than the width of the space charge region [3.167]. Such numerical studies have to assume special geometries for the interfacial inhomogeneities in order to obtain boundary values for the solution of the partial differential equation [3.167]. These solutions are therefore hardly applicable to the quantitative *evaluation* of experimental data on contacts with an unknown interface geometry.

Systematic numerical *and* analytical studies of the electrostatic potential within the space-charge region of inhomogeneous Schottky contacts were recently published by *Tung* et al. [3.99, 135, 168, 169]. The idea of *Tung* et al. is based on the consideration of small regions of a locally lowered SBH embedded in a much larger area of a higher, uniform SBH. For the low SBH regions, special geometric configurations like circular patches were assumed [3.135, 168]. For such cases, the electrostatic potential in the vicinity of an inhomogeneity can be approximated well by the superposition of the usual one-dimensional parabolic potential and the potential caused by an *electrostatic dipole* located in the centre of the low-barrier patch. The strength of this dipole depends on the size of the patch and on the difference Δ between the SBH of the low-barrier patch and the SBH of the embedding area. The specific geometry has little influence on the dipole strength [3.168].

As an alternative to the dipole approach of *Tung* et al., inhomogeneities can also be modelled by a sinusoidal modulation of the barrier height with an amplitude Δ around a mean value Φ_b^a [3.136]. Figure 3.14 illustrates the resulting potential distribution obtained from an analytical solution of Poisson's equation with a periodic boundary condition. As can be seen from Fig. 3.14 the electrostatic potential displays a *sandle-point barrier*, i.e., a *minimum* value of the electrostatic potential along the axis *lateral* to the interface plane and a *maximum* value along the axis *perpendicular* to the interface. The energetic height of this saddle-point represents the saddle-point SBH Φ_b^{sp} which is relevant for the electronic transport across the inhomogeneous interface. The voltage dependence of this saddle point barrier results in a *"pinch-off"* [3.135] of lower barrier regimes when the voltage is increased.

The definition of this saddle-point barrier allows one to describe the current transport in terms of analytical expressions [3.135, 136, 168]. The effective barrier depends on the size of the inhomogeneities, the energy difference Δ, the semiconductor's doping and dielectric constant, and the applied voltage. The major impact of interfacial inhomogeneities on current transport is described in terms of *a distribution* of saddle point barriers attributed to a distribution of the size and the energetic amplitude of the inhomogeneities [3.136].

Tung et al. demonstrated the validity of their model by the evaluation of transport measurements at epitaxial NiSi$_2$/Si Schottky contacts with a well defined geometry for the inhomogeneities [3.98]. They used NiSi$_2$ on Si (100) with and

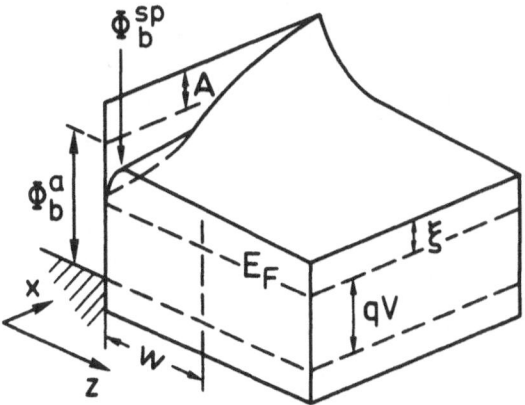

Fig. 3.14. Two-dimensional band diagram of a Schottky contact with an inhomogeneous barrier. The SBH is assumed to be modulated by a sine function within the interface area (x-direction). The average barrier height is denoted by Φ_b^a and the energetic amplitude of the modulation by the symbol Δ

without facet bars [3.98]. The description of barrier inhomogeneities in terms of these geometrical models explain also the occurrence of idealities $n > 1$ as well as the dependence of the measured Schottky barrier height on the doping of the semiconductor [3.99, 135, 168, 169].

3.3.6 Noise Properties

Noise measurements are a sensitive tool for the characterization of metal-semiconductor contacts. However, the interpretation of noise data requires a precise knowledge of the electronic transport mechanism across the interface. Experiments on noise at Schottky contacts should therefore always be supported by other characterization methods (IV, CV) in order to detect the main noise sources and to achieve an understanding of the electronic transport across the interface.

There are two conventional models for Schottky-barrier noise. *Hsu* explained the low-frequency $1/f$-noise by a modulation of the carrier *density* due to multistep tunnelling processes within the space charge region of the semiconductor [3.172]. *Luo* et al. ascribed noise to fluctuations of the carrier *mobility* within the space charge region [3.173]. The mobility fluctuations were explained by statistical scattering processes within the space-charge region and were analyzed using the framework of *Handel's* theory of infra-quanta emission [3.174, 175]. The fit of this theory to measured room-temperature noise of a commercially available diode yielded, however, a Hooge parameter α which deviated by at least an order of magnitude from the theoretically predicted value for silicon [3.175].

Despite adjusting the Hooge parameter, the model of *Luo* et al. [3.173] is – as is the model of *Hsu* [3.172] – unable to account for the observed temperature dependence of the noise signal. The experiments reported in [3.124, 166] showed a strong *increase* of the current noise upon cooling, whereas the theory of *Luo* [3.173] predicts a *decrease*. The model of *Hsu* [3.172] anticipates a noise which is independent of temperature T.

The increase of the noise upon cooling is correlated with the strength of spatial inhomogeneities at the metal/semiconductor interface [3.166]. *Güttler* and *Werner* presented temperature-dependent noise data using a particular delineation. They scaled the temperature axis with the value for the standard deviations σ_Φ of the spatial distribution of the potential distribution at the metal/semiconductor interface. This delineation revealed a universal behavior for all diodes investigated. The increase of noise upon cooling is obviously correlated with the increasing influence of the potential fluctuations at lower temperatures.

The effect of inhomogeneities can be understood semi-quantitatively by the consideration of the charging and decharging dynamics of trap levels close to the inhomogeneous interface [3.171]. These traps are able to modulate locally the barrier height and thereby cause current fluctuations. Upon decreasing the temperature, current flow becomes more and more concentrated in fewer low-barrier regions. Thus, at lower temperatures the total current flow becomes controlled by the stochastic charging and decharging of fewer and fewer trap states located close to these low-barrier regions. The result is a drastic increase of current noise. The quantitative model [3.136] predicts an exponential dependence of the current noise power density S_I on the variance σ_Φ^2 of the SBH according to

$$S_I \sim \exp\left(\frac{\sigma_\Phi^2}{(kT)^2}\right). \tag{3.47}$$

This model explains the dependence of the current noise on the inhomogeneity of the diodes as well as the drastic increase of noise upon cooling the diodes.

Equation (3.47) reveals also the empirical law, found earlier [3.166], that noise due to interfacial inhomogeneities is low as long as the 'strength' of potential fluctuations is small compared to the thermal energy, i.e. for $\sigma_\Phi < 2kT$. On the other hand, the noise increases drastically when the standard deviation σ_Φ of the potential fluctuations exceeds the critical threshold value of $2kT$. Low-noise Schottky diodes can therefore be fabricated as long as their inhomogeneity does not exceed this critical value.

3.3.7 Microwave Properties

Their use as microwave detectors and mixer diodes is one of the most important technological applications for Schottky diodes. One major advantage of Schottky diodes is the fact that they are comparatively easy to fabricate and to incorporate into integrated circuits. However, the main reason for their wide usage is that they do not exhibit large minority carrier effects resulting in large reverse recovery times and large diffusion capacitances. In general, it holds that Schottky diodes are preferably appropriate for applications where the nonlinear conductance of the diode is the one and only desired effect.

Basically, there are two principles for the application of Schottky contacts in Radio-Frequency (RF) receivers:

(i) The *homodyn* receiver directly converts the microwave signal into a dc signal by the nonlinearity of the diode characteristic.

(ii) The *heterodyn* receiver in its usual application converts a microwave signal to a signal centered at a lower intermediate frequency. The frequency conversion results from the mixing of the microwave frequency with a signal at another frequency obtained from a local oscillator. The resulting intermediate frequency is the difference between the signal and the local oscillator frequency.

Fundamental limitations to the sensitivity for both types of microwave receivers depend primarily on the diode junction properties and secondarily on the diode's electronic environment. In the following we briefly discuss the most important limiting quantities for the performance of a Schottky diode as a microwave receiver. These quantities can be predicted very accurately from measurable circuit parameters and from the diode properties [3.176].

Figure 3.15 depicts the design of two planar Schottky diodes together with the equivalent circuit. In the first case the semiconductor consists of a thin ($< 1\,\mu$m) epitaxial layer on a highly conducting substrate and the metalization is given by thin metal layer which establishes the Schottky barrier and a thicker overlayer

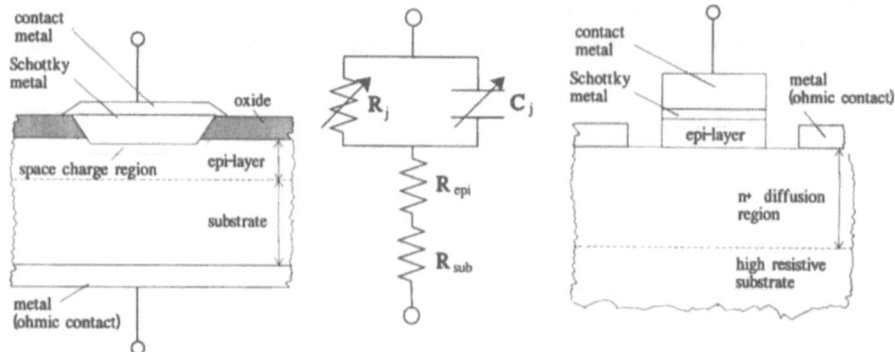

Fig. 3.15. Cross sections of two basic structures of Schottky barrier diodes for high frequency applications together with the equivalent circuit

providing the connection to the outer circuitry. For the second structure a *high resistive* semiconductor substrate is used with a heavily doped region on top. A lightly doped epitaxial layer together with the Schottky metal establishes here the Schottky barrier. Such structures are suitable for monolithic millimeter wave integrated circuits [3.177].

The elements of the equivalent circuit are given by the small signal resistance R_j and the capacitance C_j of the diode, and by the series resistance R_S. The series resistance R_S is split in to the contribution R_{epi} of the epitaxial layer and the contribution R_{sub} of the substrate. Note that in Fig. 3.15 we have omitted possible additional components of the equivalent circuit such as the parasitic capacitance contributed by the metal contact overlaying the passivating oxide layer.

The small-signal junction resistance R_j for a diode of area A at a given bias voltage V is obtained by differentiating the diode equation (3.30a) with respect to the voltage:

$$\frac{1}{R_j}(V) = A\frac{d}{dV}j(V) = Aj_0\frac{q}{nkT}\exp\left(\frac{qV}{nkT}\right).$$ (3.48)

The junction capacitance C_j of the diode is determined by the space charge capacitance C_{sc} of the Schottky contact according to

$$C_{SC}(V) = A\frac{dQ_{SC}}{dV}(V) = A\left(\frac{q^2\epsilon_S N_D}{2(\Phi_b - \zeta - qV - kT)}\right)^{1/2}.$$ (3.49)

This equation is obtained from the derivative of (3.8). Note that in (3.49) we have included the correction term $-kT$, which expresses the effect of the penetration of majority carriers from the volume of the semiconductor into the depletion region. The series resistance R_S of practical diodes depends much on the specific geometry. However, for the design considered on the left hand side of Fig. 3.15 (assuming axial symmetry) R_S may be calculated as $R_S = R_{epi} + R_{sub}$ with [3.178]

$$R_{epi} = \frac{\rho_{epi}}{2\pi} \left[\frac{t}{r^2 + t^2} + \frac{1}{r} \arctan\left(\frac{t}{r}\right) \right] \approx \frac{\rho_{epi} t}{\pi r^2} \qquad (3.50a)$$

and

$$R_{sub} = \frac{\rho_{sub}}{4r} \qquad (3.50b)$$

where ρ_{epi} and ρ_{sub} are the resistivities of the epitaxial layer and of the substrate, respectively. The thickness of the epitaxial layer is denoted by t and the radius of the contact by r. Note that the second equality in (3.50a) holds in the limit $r \gg t$.

For the high frequency performance of Schottky diodes the cut-off frequency

$$f_c = \frac{1}{2\pi R_S C_j} \qquad (3.51)$$

is used as a simple figure of merit. In order to maximize f_c, i.e. to minimize the series resistance R_S, the following conclusions may be drawn from (3.50a, b) [3.179, 180]. The doping of the epitaxial layer should be chosen high enough to minimize the series resistance but low enough to guarantee that the diode is working properly in the thermionic emission range. If the doping is too high this will cause tunneling to dominate the electronic transport and, therefore, increase the ideality factor n (Section 3.3.2). Typical doping concentrations used for Schottky contacts on n-type silicon are $N_D = 10^{16} - 10^{17}$ cm^{-3}. In contrast, the conductivity of the underlying substrate should be as high as possible in order to minimize the series resistance contribution R_{sub} of the substrate. The areal dependence of the cut-off frequency f_c for a circular contact of area $A = \pi r^2$ can be calculated from (3.50, 51) as

$$f_c = \frac{1}{2\pi C_{sc}} \frac{1}{\rho_{epi} t + \sqrt{A\pi} \rho_{sub}} \qquad (3.52)$$

where $C_{sc} = C_{sc}/A$ is the space charge capacitance per unit area. It follows from (3.52) that the thickness t of the epitaxial layer should be reduced to a value that is required to accommodate the depletion layer of the Schottky contact. Further maximation of f_c is achieved by reducing the diode area as much as possible.

The importance of high values for the cut-off frequency f_c is also reflected in expressions for the basic detector parameters which will be discussed in the following. We refer to the simple detector circuit displayed in Fig. 3.16 [3.181] where the diode is dc biased by the voltage V_0 via a load resistor R_L. Let the RF input signal with angular frequency ω_{RF} be $V_{RF} = V_1 \cos(\omega_{RF} t)$ and $V_1 < kT/q$ such that small signal expansion of the diode equation (3.30a) is allowed. Under this condition the detector is said to operate in the square law regime. The diode current as a function of time t is then given by [3.182, 180]

$$I = j_0 \left[\exp\left\{ \frac{q}{nkT}[V + V_1' \cos(\omega_{RF} t)] \right\} - 1 \right]$$

$$\approx j_0 A \left\{ \exp\left(\frac{qV}{nkT}\right) \left[1 + \frac{V_1' \cos(\omega_{RF} t)}{nkT/q} + \frac{1}{2}\left(\frac{V_1' \cos(\omega_{RF} t)}{nkT/q} \right)^2 + \cdots \right] - 1 \right\}$$

$$(3.53)$$

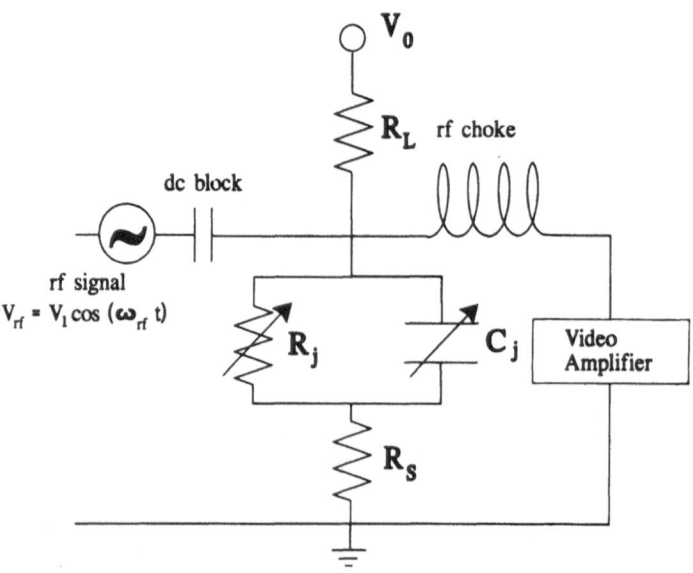

Fig. 3.16. Equivalent circuit of a Schottky detector diode connected to a detector circuit [3.181]

where V_1' is the portion of the RF voltage V_1 which drops immediately at the diode. From the temporal average of (3.53) the dc current across the diode can be calculated as

$$\langle I \rangle \approx j_0 A \left\{ \exp\left(\frac{qV}{nkT} \right) \left[1 + \frac{1}{4} \left(\frac{V_1'}{nkT/q} \right)^2 \right] - 1 \right\}. \tag{3.54}$$

It is straightforward from (3.54) and Fig. 3.16 to calculate the current sensitivity β and the voltage sensitivity γ of the detector [3.182, 183]. The current sensitivity is the ratio of the short-circuit current I_{sc} to the RF input power P_{RF} for the voltage controlled diode (i.e., $R_L \ll R_j$). It holds that

$$\beta = \frac{I_{sc}}{P_{RF}} = \frac{q}{2nkt} \frac{1}{1 + R_S/R_j + \omega_{RF}^2 C_j^2 R_S R_j}. \tag{3.55}$$

The voltage sensitivity is defined as the ratio of the open circuit voltage V_{oc} to the RF input power P_{RF} for the current controlled diode (i.e., $R_L \gg R_j$) and it holds that $\gamma = V_{oc}/P_{RF} = \beta(R_S + R_j)$. Thus, in the square law range the output voltage and the output current are both proportional to the input power. Maximum sensitivity can be achieved when the junction resistance equals the reactance of the diode, i.e. $R_j C_j \omega_{RF} = 1$. Then, the voltage sensitivity β becomes a function of the cut-off frequency alone and it holds for a given RF frequency $f_{RF} = \omega_{RF}/2\pi$ that

$$\beta = \frac{q}{2nkt} \frac{1}{1 + 2f_{RF}/f_c}. \tag{3.56}$$

Other parameters are also commonly used to measure the performance of detectors. One example is the Tangential Sensitivity (TSS) which is the input power

P_{TS} required to change the dc voltage output by an amount equal to the voltage-noise fluctuations. The TSS corresponds to a signal-to-noise ratio of approximately 2.5, i.e. for the open circuit situation: $P_{TS} = 2.5|\bar{V}_n^2|/\gamma$ where $\bar{V}_n^2$ is the mean square of the voltage noise at the terminals of the video amplifier [3.182, 183]. Further parameters depend on the receiver circuit as well as on the diode properties and we refer the reader to the standard literature [3.180, 182–184].

3.4 Schottky-Barrier Measurements

3.4.1 Current-Voltage Curves

The determination of the SBH from IV measurements is based upon the diode equation

$$I = A j_0 \left[\exp\left(\frac{qV}{nkT} \right) - 1 \right] \tag{3.57}$$

where A denotes the diode area and j_0 is the saturation current density. Under the assumption of thermionic emission, j_0 is given by (3.30b). The usual straight line fit to a semilogarithmic plot of the IV curve for voltages $V > 3nkT/q$ yields the saturation current $I_0 = A j_0$ from the y-axis intercept and the ideality n from the slope q/nkT. If the diode area A and the effective Richardson constant A^{**} are known, the Schottky barrier height Φ_{b0}^j can be obtained from (3.33). The index '0' indicates that Φ_{b0}^j is found by an extrapolation of the IV curves to zero bias. The need for an *extra*polation can be circumvented by a semilogarithmic plot of $1/[1 - \exp(qV/kT)]$ vs. V which should give a straight line not only for voltages $V > 3nkT/q$ but also for negative bias voltages [3.185]. The saturation current I_0 is then found by *interpolation*.

If IV data under higher forward bias are evaluated, a correction for the series is necessary. A proper separation between the effect of the series resistance R_S and of the ideality n can be achieved by an evaluation scheme [3.186] that makes use of the small signal conductance $G = dI/dV$: Under the assumption of a voltage independent ideality, a plot of G/I vs. G should yield a straight line with $1/R_S$ as the x-axis intercept and q/nkT as the y-axis intercept.

If the effective Richardson constant A^{**} is not known, Φ_{b0}^j can be obtained from an activation energy plot (Richardson plot) of $\ln(j_0/T^2)$ vs. $1/T$. From (3.33) it follows that

$$\ln(j_0/T^2) = \ln(A^{**}) - \frac{\Phi_{b0}^j}{kT}. \tag{3.58}$$

The slope of the Richardson plot yields Φ_{b0}^j, and from the y-axis intercept one obtains the effective Richardson constant A^{**}.

In (3.58) it is implicitly assumed that the barrier height Φ_{b0}^j is independent of temperature. However, the Richardson plot is insensitive to the dependence of Φ_{b0}^j

on the temperature T, if Φ_{b0}^j varies linearly with T according to

$$\Phi_{b0}^j(T) = \Phi_{b0}^j(0) + \alpha_\Phi T \quad (3.59)$$

where α_Φ is the temperature coefficient. In this case, the slope of the Richardson plot yields the barrier height $\Phi_{b0}^j(0)$ at zero temperature and from the y-axis intercept the temperature coefficient is found when A^{**} is known.

3.4.2 Capacitance Measurements

The determination of the SBH by the Capacitance-Voltage (CV) method is based upon the voltage dependence of the charge in the depletion region of the diode. The space charge capacitance C_{sc} of a Schottky contact of area A was derived in the preceding section. Equation (3.49) may be rewritten as

$$C_{sc}^{-2}(V) = \frac{2(\phi_b - \zeta - qV - kT)}{q^2 \epsilon_s N_D A^2}. \quad (3.60)$$

According to (3.60) a plot of the measured capacitance $C^{-2} = C_{sc}^{-2}$ versus the voltage V should yield a straight line, with an x-axis intercept at $V = V_i$. The SBH is then deduced from

$$\phi_b = qV_i + \zeta + kT. \quad (3.61)$$

The slope of the straight line yields the dopant concentration of the semiconductor

$$\frac{d}{dV}C^{-2} = -\frac{2}{qN_D\epsilon_s A^2}. \quad (3.62)$$

The CV method measures the electrostatic properties of the Schottky barrier and is insensitive to transport effects such as tunneling and image force lowering. For an inhomogeneous interface the CV method averages over the whole sample area and measures, therefore, the mean barrier height $\overline{\phi_b}$ [3.124]. A difference between the SBH deduced from IV and CV measurements can be attributed to the effect of inhomogeneities as far as this difference exceeds the value that is expected from the effect of image force lowering and tunneling.

3.4.3 Internal Photoemission (Photoresponse)

Internal photoemission is the most accurate and direct method for the determination of the SBH. Monochromatic light, incident on a metal/semiconductor contact, excites charge carriers at the Fermi level of the metal. If the quantum energy $h\nu$ of the photons exceeds the threshold $h\nu_0 = \phi_b - \Delta\phi_{if}$, some of these excited carriers cause a photocurrent in the external circuit. According to *Fowler* [3.187], this photocurrent I_p obeys the proportionality

$$I_p \sim (h\nu - h\nu_0)^2 \quad (3.63)$$

as long as $h(v - v_0) > 5kT$. Thus, a plot of $I_p^{1/2}$ versus hv should yield a straight line with an x-axis intercept at $hv_0 = \phi_b - \Delta\phi_{if}$. The advantage of the PhotoResponse (PR) method is given by the fact that it can be applied at an arbitrary but fixed bias and does not need an extrapolation to zero bias or to the flat band voltage. Thus, the photoelectric method can be used to measure the voltage dependence of the SBH directly.

3.4.4 External Photoemission

Photoelectron spectroscopy measures the energy distribution of the optically excited electrons which are emitted from a solid via the photoelectric effect. Depending on the wavelength of the monochromatic incident light, the method is denoted by X-ray Photoelectron Spectrometry (XPS) or Ultraviolet Photoelectron Spectrometry (UPS). Photoelectron spectrometry with synchroton radiation is termed as soft-XPS (for a discussion, see (3.188–191).

Some of the photoexcited electrons within a distance of about 20Å from the semiconductor surface can escape without loss of energy. The electrons with the maximal energy E_v in the photoemission spectrum of the semiconductor are those emitted from the valence band minimum at the surface. The determination of the Fermi level position at the semiconductor surface is achieved by measuring a metal in contact with the semiconductor. The maximum energy of the photoexcited electrons corresponds to the Fermi level in the metal E_F^M. The barrier height Φ_b is established by the relation $\Phi_b = E_g + E_v - E_F^M$.

The energy of photoelectrons from the core levels of atoms near the semiconductor surface can also be measured by photoemission spectroscopy. Core level electrons supply sharp and, for each element, well-defined peaks in the photoemission spectra. These peaks are preferably used to measure changes of the Fermi level position during the metallization of semiconductor surfaces.

Photoemission spectroscopy of a semiconductor surface can only be performed as long as photoemission from the metal overlayer does not mask the spectrum. This is only true for thin metal overlayers in the range of a few Ångstroms. Since photoelectron spectroscopy does not need electrical contacts, this method is appropriate for in-situ measurements of the early stage of metallization.

3.4.5 Results for Polycrystalline Contacts

Table 3.4 presents a compilation of barrier heights for a large number of *polycrystalline* metals on Si. Single crystalline contacts are listed in Table 3.2 and 3.3. A compilation of Schottky-barrier heights for the noble metals Au, Ag, Cu was recently published by *Cros* and *Muret* [3.133]. The barrier heights of some other elements and their silicides were reviewed in [3.192].

Table 3.4. Measured barrier heights for polycrystalline metals on n-type silicon. Values in brackets are derived by us from measured values on p-type Si

Metal	Φ_b^n [eV]	Method[a]	Reference	Remark
Ag	0.78	CV	3.193	(111), UHV cleaved, $T = 80$ K, influence of cleavage steps
Al	0.69	CV	3.193	(111), UHV cleaved, $T = 80$ K, influence of cleavage steps
Au	0.81	CV	3.193	(111), UHV cleaved, $T = 80$ K, influence of cleavage steps
Ca	0.40	IV	3.194	
Cd	0.78	IV, CV	3.195	electrochemically deposited, $n = 1.02$
Co	0.68	IV	3.196	(111), (100)
CoSi$_2$	0.64	IV	3.196	(111), (100), review in [192]
Cr	0.62	IV, CV, AE	3.197	review in [192]
Cs	0.67	CV	3.195	(111), UHV cleaved, $T = 80$ K
Cu	0.61	IV, CV	3.198	(111), (100); UHV samples give higher barrier [192, 133]
DySi$_2$	0.37	IV	3.58	$n = 1.05$, 0.73 eV on p-type
Er	0.68p (0.44)	IV, PR	3.115	CV value slightly higher
ErSi$_{1.7}$	0.76p (0.36)	IV, PR	3.115	
ErSi$_2$	0.39	IV	3.58	$n = 1.1$, 0.70 eV on p-type
Fe	0.71	CV	3.193	(111), UHV cleaved, $T = 80$ K
Ga	0.72	CV	3.193	(111), UHV cleaved, $T = 80$ K, influence of cleavage steps
GdSi$_2$	0.37	IV	3.58	$n = 1.05$, 0.71 eV on p-type
Hf	0.69p (0.43)	PR	3.199	
HoSi$_2$	0.37	IV	3.58	
In	0.66	IV	3.195	electrochemically deposited, $n = 1.03$
Ir	0.88	CV, PR	3.200	(100), (111)
IrSi	0.95	IV	3.75	(100), $n = 1.02$
K	0.73	CV	3.193	(111), UHV cleaved, $T = 80$ K
Mg	0.62	CV	3.193	(111), UHV cleaved, $T = 80$K, influence of cleavage steps, review in [3.192]
Mn	0.85p (0.27)	IV, CV	3.201	(111)
MnSi	0.76	PR	3.202	(111)
MnSi$_{1.7}$	0.72	PR	3.202	(111)
Mo	0.55	IV	3.203	(100), review in [3.192]
MoSi$_2$	0.63	IV, CV	3.204	(100), review in [3.192]
Na	0.70	CV	3.191	(111), UHV cleaved, $T = 197$ K
Ni	0.6			large scatter of data [3.192]
NiSi$_2$	0.65	PR	3.205	review in [3.192]
Os	0.85	IV, PR	3.206	
Pb	0.61	IV, PR	3.207	(111)
Pd	0.74	IV	3.208	(111), $n \approx 1.04$, review in [3.192]
Pd$_2$Si	0.71	IV, PR	3.205	(100), (111), review in [3.192]
Pt	0.84	PR	3.209	(111), review in [3.192]
PtSi	0.90	IV	3.210	(100), review in [3.192]
Re	0.77	IV, PR	3.206	(100)
Rh	0.72	IV	3.211	(111), $n < 1.1$
RhSi	0.70	IV	3.212	(111), $n = 1.12$
Ru	0.77	CV	3.213	

Table 3.4 (*Continued*)

Metal	Φ_b^n [eV]	Method[a]	Reference	Remark
Sb	0.30	CV, IV	3.198	(111), (100); 100–200 K; 0.73 eV for p-type
Sn	0.58	CV, IV	3.198	(111), (100), 100–200 K
TaSi$_2$	0.58	IV	3.214	(100), review in [3.192]
Tb	0.64p (0.48)	IV, PR	3.115	(100)
Ti	0.515	IV	3.215	(100), review in [3.192]
TiSi$_2$	0.60	IV	3.215	(100), review in [3.192]
TiN	0.55	IV	3.216	(100), review in [3.192]
V	0.61	XPS	3.217	(111)
VSi$_2$	0.64	XPS	3.217	(111)
		IV	3.57	(111), (100)
W	0.66	CV, IV PR	3.218	(111)
Y	0.68p (0.44)	IV, PR	3.115	(100)
YSi$_{1\,7}$	0.74p (0.37)	IV, PR	3.115	(100)
YSi$_2$	0.39	IV	3.58	
Yb	0.64p (0.48)	IV, PR	3.115	(100)
Zn	0.75	CV, IV	3.219	(111)
Zr	0.55p (0.57)	PR	3.202	(111)
ZrSi$_2$	0.55	IV	3.138	

[a] IV: = current-voltage curve, AE: Richardson plot, CV: capacitance-voltage curve, PR: photoresponse, XPS: X-ray induced photoemission spectroscopy

3.5 Conclusions

The presently available models for the Schottky barrier height do not explain all of the available data. The theoretical models that consider the pure bulk band structure of the semiconductor only predict a Schottky-barrier height of around 0.8 eV for all metals on Si [3.23, 24]. The consideration of metal induced gap states and electronegativity differences of the metal [3.9] allows for deviations from this intrinsic value. However, the experimentally observed barrier heights of single crystalline, epitaxial Schottky contacts cannot be explained by these theories. Only supercell calculations that take into account the interface crystallography [3.46] are able to model some of the experimental results. Unfortunately, these methods are restricted to single-crystalline, atomically abrupt, and defect free interfaces. Such idealized, perfect interfaces do not exist. On the contrary, even single crystalline interfaces appear electrically inhomogeneous [3.98, 124]. These electrical inhomogeneities are hardly understood within the concept of 'Fermi level pinning'. More experimental as well as theoretical work is thus required on such epitaxial contacts on Si. We have shown that many metals on Si seem suitable for these investigations.

Acknowledgements. The authors gratefully acknowledge the fruitful collaboration with H.H. Güttler and R.T. Tung, and thank H.J. Queisser for his steady support. We are obliged to C. Lenuzza, R. Plieninger and S. Trollinger for their assistance during the preparation of the manuscript and to J.K. Arch and F. Noll for a critical reading. Part of this work was supported by the German *Bundesministerium für Forschung und Technologie* under contract 0328962A.

References

Section 3.0

3.1   F. Braun: Über die Stromleitung durch Schwefelmetalle. Pogg. Ann. **153**, 556 (1874)
3.2   W. Schottky: Halbleitertheorie der Sperrschicht, Naturwissenschaften, **26**, 843 (1938)
3.3   N.F. Mott: Note on the contact between a metal and an insulator or semiconductor, Proc. Cambr. Philos. Soc. **34**, 568 (1938)
3.4   E.H. Rhoderick: *Metal-Semiconductor Contacts* (Clarendon, Oxford 1978)
      E.H. Rhoderick, R.H. Williams: *Metal-Semiconductor Contacts 2nd edn.* (Clarendon, Oxford 1988)
3.5   H.K. Henisch: *Semiconductor Contacts* (Clarendon, Oxford 1984)
3.6   H. Lüth: *Surfaces and Interfaces of Solids*, 2nd edn. (Springer, Berlin, Heidelberg 1993)
3.7   W. Mönch: On the physics of metal-semiconductor interfaces, Rep. Prog. Phys. **53**, 221 (1990)
      W. Mönch (ed.): *Electronic Structure of Metal-Semiconductor Contacts* (Jaca Book, Milano 1990)
3.8   W. Mönch: *Semiconductor Surfaces and Interfaces*, Springer Ser. Surf. Sci., Vol. 26 (Springer, Berlin, Heidelberg 1993)
3.9   R.T. Tung: Schottky barrier formation at single-crystal Metal-Semiconductor Interfaces, Phys. Rev. Lett. **52**, 461 (1984)
3.10 L.J. Chen, K.N. Tu: Epitaxial growth of transition-metal silicides on silicon, Mater. Sci. Rept. **6**, 53 (1991)
3.11 H. von Känel, J. Henz, M. Ospelt, J. Hugi, E. Müller, N. Onda, A. Gruhle: Epitaxy of metal silicides, Thin Solid Films **184**, 295 (1990)
3.12 J. Derrien, J. Chevrier, V. Le Thanh, J.E. Mahan: *Semiconducting silicide-silicon heterostructures: growth, properties and applications*, Appl. Surf. Sci. **56–58**, 382 (1992)
3.13 R.T. Tung: In *Atomic-Level Properties of Interface Materials*, ed. by D. Wolf, S. Yip (Chapman & Hall, London 1992)
3.14 S. Mantl: Ion beam synthesis of epitaxial silicides: Fabrication, characterization and applications, Mater. Sci. Rept. **8**, 1 (1992)

Section 3.1

3.15 W. Schottky: Abweichungen vom Ohmschen Gesetz in Halbleitern, Phys. Zeitschrift **41**, 570 (1940)
3.16 H. Schweikert: Über Selen – Gleichrichter, Verh. Phys. Ges. **3**, 99 (1939)
3.17 W. Mönch: *Festkörperprobleme (Advances in Solid State Physics)* **24**, 67 (Vieweg, Baunschweig 1986)
3.18 For an exact solution see, for example, [Ref. 3.46, Appendix B]
3.19 J. Bardeen: Surface states and rectification at a metal semiconductor contact, Phys. Rev. **71**, 717 (1947)
3.20 C. Tejedor, F. Flores, E. Louis: The metal-semiconductor interface: Si (111) and zincblende (110) junctions, J. Phys. C: **10**, 2163 (1977)
3.21 C. Tejedor, F. Flores: A simple approach to heterojunctions, J. Phys. C: **11**, L19 (1978)

3.22   F. Flores, C. Tejedor: Energy barriers and interfaces states at heterojunctions, J. Phys. C: **12**, 731 (1979)

3.23   J. Tersoff: Schottky barrier heights and the continuum of gap states, Phys. Rev. Lett. **52**, 214 (1984); Phys. Rev. B**30**, 4874 (1984)

3.24   M. Cardona, N.E. Christensen: Acoustic deformation potentials and heterostructure band offsets in semiconductors, Phys. Rev. B **35**, 6182 (1987)

3.25   A.M. Cowley, S.M. Sze: Surface States and barrier height of metal semiconductor systems, J. Appl. Phys. **36**, 3212 (1965)

3.26   S.M. Sze: *Physics of Semiconductor Devices*, 2nd edn. (Wiley, New York 1981) p. 272.

3.27   R.W. Bené, R.M. Walser: Effect of a glassy membrane on the Schottky barrier between silicon and metallic silicides, J. Vac. Sci. Technol. **14**, 925 (1977)

3.28   B.L. Sharma, S.C. Gupta: *Metal-Semiconductor Schottky Barrier Junctions*, Solid State Technology p. 97 (May 1980) p. 90 (June 1980),

3.29   W. Mönch: On metal – semiconductor surface barriers, Surf. Sci. **21**, 443 (1970)

3.30   C.S. Wu, D.M. Scott, S.S. Lau: The effect of an interfacial oxide layer on the Schottky barrier height of Er-Si contact, J. Appl. Phys. **58**, 1330 (1985)

3.31   P.A. Tove: Simple dipole model for barrier heights of silicide-silicon and metal-silicon barriers, Surf. Sci. **132**, 336 (1983)

3.32   E. Bucher, S. Schulz, M. Ch. Lux-Steiner, P. Munz, U. Gubler, F. Greuter: Work function and barrier heights of transition metal silicides, Appl. Phys. A **40**, 71 (1986)

3.33   Y. Hwu, M. Marsi, P. Alméras, G. Margaritondo: Microscopic Schottky-barrier control: Semiconductor on metal case, Phys. Rev. B**46**, 1835 (1992)

3.34   H.B. Michaelson: The work funktion of the elements and its periodicity, J. Appl. Phys. **48**, 4729 (1977); IBM J. Res. Dev. **22**, 72 (1978)

3.35   M.-A. Nicolet, S.S. Lau: VLSI *Electronics – Microstructure Science*, **6**, 329 (Academic, New York)

3.36   For a discussion, [Ref. 3.5, Pages 10 and 47]

3.37   W. Gordy, W.J.O. Thomas: Electronegativities of the elements, J. Chem. Phys. **24**, 439 (1956)

3.38   K.W. Frese' jr.: Simple method for estimating energy levels of solids, J. Vac. Sci. Technol. **16**, 1042 (1979)

3.39   S. Kurtin, T.C. McGill, C.A. Mead: Fundamental transition in the electronic nature of solids, Phys. Rev. Lett. **22**, 1433 (1970)

3.40   M. Schlüter: Chemical trends in metal-semiconductor barrier heights, Phys. Rev. B **17**, 5044 (1978)

3.41   W. Mönch: Role of virtual gap states and defects in metal semiconductor contacts, Phys. Rev. Lett. **58**, 1260 (1987)

3.42   J.M. Andrews, J.C. Phillips: Chemical bonding and structure of metal-semiconductor interfaces, Phys. Rev. Lett. **35**, 56 (1975);
        J.M. Andrews, J.C. Phillips: CRC Crit. Rev. Sol. St. Sci. **5**, 405 (1975)

3.43   J.C. Phillips: In *Thin Film Phenomena – Interfaces and Interactions*, ed. by J.E.E. Baglin, J.M. Poate (Electrochemical Society, Princeton 1978) p. 3

3.44   This argument assumes epitaxial matching without charges that depend on orientation. For a review on such grain boundary charges in Si, see J.H. Werner: Inst. Phys. Conf. Ser. **104**, 63 (1989)

3.45   H. Fujitani, S. Asano: Schottky barriers at NiSi$_2$/Si(111) interfaces, Phys. Rev. B**42**, 1696 (1990); Appl. Surf. Sci. **41/42**, 164 (1989)

3.46   G.P. Das, P. Blöchl, O.K. Andersen, N.E. Christensen, O. Gunnarsson: Electronic structure and schottky-barrier heights of (111) NiSi$_2$/Si A- and B-type interfaces, Phys. Rev. Lett. **63**, 1168 (1989); Phys. Rev. Lett. **65**, 2084 (1990)

3.47   K. Hirose, I. Ohdomari, M. Uda: Schottky barrier heights of transition – metal – silicide – silicon contacts studied by x-ray photoelectron spectroscopy measurements, Phys. Rev. B **37**, 6929 (1988)

3.48   S. Hara I. Ohdomari: Chemical trend in silicide electronic structure and Schottky barrier heights of silicide-silicon interfaces, Phys. Rev. B38, 7554 (1988)

3.49   For a review on chemical shifts and XPS see for example, W.M. Riggs, M.J. Parker: *Methods of Surface Analysis*, ed. by A.W. Czandra (Elsevier, Amsterdam 1975) Chap. 4, p. 115

3.50   G.W. Rubloff: Microscopic properties and behaviour of silicide interfaces, Surf. Sci. **132**, 268 (1983)

3.51   G.W. Rubloff: In *Dynamical Phenomena at Surfaces, Interfaces and Superlattices*, ed. by F. Nizzoli, K.H. Rieder, R.F. Willis, Springer Ser. Surf. Sci., Vol. 3 (Springer, Berlin, Heidelberg 1985) p. 220

3.52   J.H. Weaver, A. Franciosi, V.L. Moruzzi: Bonding in metal disilicides $CaSi_2$: Experimental and theory, Phys. Rev. B29, 3293 (1984)

3.53   C. Mailhiot, C.B. Duke: Many-electron model of equilibrium metal-semiconductor contacts and semiconductor heterojunctions, Phys. Rev. B33, 1118 (1986)

3.54   G. Ottaviani, K.N. Tu, J.W. Mayer: Interfacial reaction and Schottky barrier in metal-silicon systems, Phys. Rev. Lett. **44**, 284 (1980)

3.55   I. Ohdomari, K.N. Tu, F.M. d'Heurle, T.S. Kuan, S. Petersson: Schottky barrier height of iridium silicide, Appl. Phys. Lett. **33**, 1028 (1978)

3.56   M.J. Turner E.H. Rhoderick: Metal-silicon schottky barriers, Solid-St. Electron. **11**, 291 (1968)

3.57   R.D. Thompson, K.N. Tu: Comparison of the three classes (rare earth, refractory and near-noble) of silicide contacts, Thin Solid Films **93**, 265 (1982)

3.58   K.N. Tu, R.D. Thompson, B.Y. Tsaur: Low Schottky barrier of rare-earth silicide on n-Si, Appl. Phys. Lett. **38**, 626 (1981)

3.59   J.L. Freeouf: Silicide Schottky barriers: An elemental description, Solid State Commun. **33**, 1059 (1980)

3.60   J.L. Freeouf: Silicide interface stoichiometry, J. Vac. Sci. Technol. **18**, 910 (1981)

3.61   V. Heine: Theory of surface states, Phys. Rev. A **138**, 1689 (1965)

3.62   W.E. Spicer, R. Cao, K. Miyano, T. Kedelewicz, I. Lindau, E. Weber, Z. Lilienthal-Weber, N. Newman: From synchrotron radiation to IV measurements of GaAs Schottky barrier formation, Appl. Surf. Sci. **41/42**, 1 (1989)

3.63   W.E. Spicer, R. Cao, K. Miyano, C. McCants, T.T. Chiang, C.J. Spindt, N. Newman, T. Kendelewicz, I. Lindau, E. Weber, Z. Liliental-Weber: In *Metallization and Metal-Semiconductor Interfaces*, ed. by I.P. Batra (Plenum, New York 1989) p. 139

3.64   R. Ludeke, G. Jezequel, A. Taleb-Ibrahimi: Delocalization effects at metal-semiconductor interfaces, Phys. Rev. Lett. **61**, 601 (1988)

3.65   R. Ludeke, G. Jezequel, A. Taleb-Ibrahimi: Screening and delocalization effects in Schottky barrier formation, J. Vac. Sci. Technol. B6, 1277 (1988)

3.66   R. Ludeke, A. Taleb-Ibrahimi, G. Jezequel: Delocalization of defects: A new model for the Schottky barrier, Appl. Surf. Sci. **41/42**, 151 (1989)

3.67   R. Ludeke: In *Metallization and Metal-Semiconductor Interfaces*, ed. by I.P. Batra (Plenum, New York 1989) p. 39

3.68   S.G. Louie, M.L. Cohen: Electronic structure of a metal-semiconductor interface, Phys. Rev. B13, 2461 (1976)

3.69   S.G. Louie, J.R. Chelikowsky, M.L. Cohen: Ionicity and the theory of Schottky barriers, Phys. Rev. B15, 2154 (1977)

3.70   J. Tersoff: In *Heterojunction Band Discontinuities – Physics and Device Applications*, ed. by F. Capasso, G. Margaritondo (North-Holland, Amsterdam 1987) p. 3

3.71   R.L. Anderson: Experiments on Ge-GaAs heterojunctions, Solid-State Electron. **5**, 341 (1962)

3.72   See, for example, G. Burns: *Solid state physics*, (Academic Press, Boston, 1985) pp. 179–184; for a table of average Penn gaps see J.C. Phillips: *Bands and Bonds in Semiconductors*, (Academic, New York 1973) p. 42, see also K.W. Boer: *Survey of semiconductor physics*, (Van Nostrand Reinhold, New York 1990) p. 361

3.73   O. Madelung: *Introduction to Solid State Theory*, Springer Ser. Solid-State Sci., Vol. 2 (Springer, Berlin, Heidelberg 1978) p. 349

3.74   M. Wittmer, J.L. Freeouf: Ideal Schottky diodes on passivated silicon, Phys. Rev. Lett. **69**, 2701 (1992)

3.75   M. Wittmer: Current transport in high-barrier IrSi/Si Schottky diodes, Phys. Rev. **B42**, 5249 (1990)

3.76   J.H. Werner: Silicide/silicon Schottky barriers under hydrostatic pressure, Appl. Phys. Lett. **54**, 1528 (1989)

3.77   J.H. Werner, H.H. Güttler, Temperature dependence of Schottky barrier heights on silicon, J. Appl. Phys. **73**, 1315 (1993)

3.78   F. Flores, A. Munoz, J.C. Durán: Semiconductor interface formation: The role of the induced density of interface states, Appl. Surf. Sci. **41/42**, 144 (1989)

3.79   F. Flores, J. Ortega: Semiconductor interface formation: theoretical aspects, Appl. Surf. Sci. **56–58**, 301 (1992)

3.80   P.E. Schmid: Silicide-silicon Schottky barriers, Helvetica Physica Acta **58**, 371 (1985)

3.81   A.R. Miedema, P.F. de Chatel, F.R. de Boer: Cohesion in alloys – Fundamentals of a semi-empirical model, Physica **B100** 1 (1980)

3.82   For a discussion see also [Ref. 3.4a, ps. 32 and 87]

3.83   J. Tersoff: Schottky barriers and semiconductor band structures, Phys. Rev. **B32**, 6968 (1985)

3.84   J.C. Phillips: Microscopic theory of covalent-ionic transition at metal-semiconductor interfaces, Solid State Com. **12**, 861 (1973)

3.85   Y. Chang, Y. Hwu, J. Hansen, F. Zanini, G. Margaritondo: Nature of the Schottky term in the Schottky barrier, Phys. Rev. Lett. **63**, 1845 (1989)

3.86   M. Liehr, P.E. Schmid, F.K. LeGoues, P.S. Ho: Correlation of Schottky barrier height and microstructure in the epitaxial Ni silicide on Si(111), Phys. Rev. Lett. **54**, 2139 (1985)

3.87   J.H. Werner, A.F.J. Levi, R.T. Tung, M. Anzlowar, M. Pinto: Origin of the excess capacitance at intimate schottky contacts, Phys. Rev. Lett. **60**, 53 (1988)

3.88   H. Fujitani, S. Asano: Schottky barriers at epitaxial silicide/Si interfaces, Appl. Surf. Sci. **56–58**, 408 (1992)

3.89   D.R. Heslinga, H.H. Weitering, D.P. van der Werf, T.M. Klapwijk, T. Hibma: Atomic-structure-dependent Schottky barrier at epitaxial Pb/Si(111) interfaces, Phys. Rev. Lett. **64**, 1589 (1990)

3.90   R.T. Tung, J.P. Sullivan, F. Schrey: On the inhomogeneity of Schottky barriers, Mat. Sci. Eng. **B14**, 266 (1992)

Section 3.2

3.91   R.T. Tung: In *Silicon-Molecular Beam Epitaxy* ed. by E. Kasper and J.C. Bean (CRC Press, Boca Raton 1988) Vol. 2, p. 13

3.92   J. Derrien, J. Chevrier, A. Younsi, V. Le Than, J.P. Dussaulcy, N. Cherief: Structure and electronic properties of epitaxially grown silicides, Physica Scripta T**35**, 251 (1991)

3.93   H.H. Weitering: Epitaxial metal-semiconductor interfaces, Mat. Sci. Eng. **B14**, 281 (1992)

3.94   A. Zur, T.C. McGill, M.-A. Nicolet: Transition-metal silicides lattice-matched to silicon, J. Appl. Phys. **57**, 600 (1985)

3.95   S.P. Murarka: *Metallization – Theory and Practice for VLSI and ULSI* (Butterworth-Heinemann, Boston 1993) p. 28

Section 3.3

3.96   R.T. Tung, J.M. Gibson, J.M. Poate: Formation of ultrathin single crystal silicide films on Si: Surface and interfacial stabilization of Si-NiSi$_2$ epitaxial structures, Phys. Rev. Lett. **50**, 429 (1983)

3.97   G.P. Das: Electronic structure of epitaxial interfaces, Pramana – J. Phys. (India) **38**, 545 (1992)

3.98   R.T. Tung, A.F.J. Levi, J.P. Sullivan, F. Schrey: Schottky-barrier inhomogeneity at epitaxial NiSi₂ interfaces on Si(100). Phys. Rev. Lett. **66**, 72 (1991)

3.99   J.P. Sullivan, R.T. Tung, F. Schrey, W.R. Graham: Correlation of the interfacial structure and electrical properties of epitaxial silicides on Si, J. Vac. Sci. Technol. A **10**, 1959 (1992)

3.100   R.T. Tung, J.M. Gibson: Single crystal silicide silicon interfaces: Structures and barrier heights, J. Vac. Sci. Technol **A3**, 987 (1985)

3.101   H.H. Güttler, Elektrische Eigenschaften inhomogener Metall-Halbleiter-Grenzflächen Dissertation, University of Stuttgart (1991) p. 152

3.102   A. Schüppen, S. Mantl, L. Vescan, S. Woiwod, R. Jebasinski, H. Lüth: Permeable-base transistors with ion-implanted CoSi₂ gate, Mat. Sci. Eng. B**12**, 157 (1992)

3.103   R.T. Tung: Schottky barrier heights of single crystal silicides on Si(111), J. Vac. Sci. Technol. B**2**, 465 (1984)

3.104   E. Rosencher, S. Delage, F. Arnaud d'Avitaya: Transient capacitance study of epitaxial CoSi₂/Si ⟨111⟩ Schottky barriers, J. Vac. Sci. Technol. B**3**, 762 (1985)

3.105   Y.C. Kao, Y.Y. Wu, K.L. Wang: In *Proc. 1st Intern. Symp. Si MBE*, ed. by J.C. Bean (Electrochemical Society, Pennington 1985) p. 261

3.106   A.E. White, K.T. Short, K. Maex, R. Hull, Y.-F. Hsieh, S.A. Audet, K.W. Goossen, D.C. Jacobson, J.M. Poate: Exploiting Si/CoSi₂/Si heterostructures grown by mesotaxy, Nucl. Instr. Methods B**59**, 693 (1991)

3.107   F.K. LeGoues, W. Krakow, P.S. Ho: Atomic structure of the epitaxial Al-Si interface, Phil. Mag. A**53**, 833 (1986)

3.108   N. Thangaraj, K.H. Westmacott, U. Dahmen: Epitaxial growth of (001) Al on (111) Si by vapour deposition, Appl. Phys. Lett. **61**, 913 (1992)

3.109   H. Niwa, M. Kato: Single-crystal Al films grown by sputtering on (111) Si substrates, Appl. Phys. Lett. **60**, 2520 (1992)

3.110   Y. Miura, K. Hirose, K. Aizawa, N. Ikarashi, H. Okabayashi: Schottky barrier inhomogeneity caused by grain boundaries in epitaxial Al film formed on Si(111), Appl. Phys. Lett. **61**, 1057 (1992)

3.111   C.-C. Cho, H.-Y. Liu, H.-L. Tsai: Epitaxial growth of an Al/CaF₂/Al/Si (111) structure, Appl. Phys. Lett. **61**, 270 (1992)

3.112   G. Le Lay, M. Abraham, A. Kahn, K. Hricovini, J.E. Bonnet: Abrupt metal-semiconductor interfaces, Physica Scripta T**35**, 261 (1991)

3.113   J.A. Knapp, S.T. Picraux: Epitaxial growth of rare-earth silicides on (111) Si, Appl. Phys. Lett. **48**, 466 (1986)

3.114   A. Iandelli, A. Palenzona, G.L. Olcese: Valence fluctuations of ytterbium in silicon-rich compounds, J. Less-Common. Met. **64**, 213 (1979)

3.115   H. Norde, J. de Sousa Pires, F. d'Heurle, F. Pesavento, S. Petersson, P.A. Tove: The Schottky barrier height of the contacts between some rare-earth metals (and silicides) and p-type silicon, Appl. Phys. Lett. **38**, 865 (1981)

3.116   F.H. Kaatz, M.P. Siegal, W.R. Graham, J. van der Spriegel, J.J. Santiago: Epitaxial growth of ErSi2 on (111) Si, Thin Solid Films **184**, 325 (1990)

3.117   R. Arnaud d'Avitaya, P.-A. Badoz, Y. Campidelli, J.A. Chroboczek, J.-Y. Duboz, A. Perio, J. Pierre: Growth, characterization and electrical properties of epitaxial erbium silicide, Thin Solid Films **184**, 325 (1990)

3.118   C. d'Anterroches, P. Perret, F. Arnaud d'Avitaya, J.A. Chroboczek: High resolution electron microscopy study of the ErSi₂-Si (111) interface, Thin Solid Films **184**, 349 (1990)

3.119   B. Kovacs, Zs. J. Horvath, I. Mojzes, G. Molnar, G. Peto, M. Andrasi: Comparative electrical study of epitaxial and polycrystalline GdSi₂/(100) p-Si Schottky barriers, Mat. Res. Soc. Symp. Proc. **260**, 697 (1992)

3.120   I. Gerocz, G. Molnar, E. Jaroli, E. Zsoldos, G. Peto, J. Gyulai, E. Bugiel: Epitaxy of orthorhombic gadolinium disilicide on ⟨100⟩ silicon, Appl. Phys. Lett. **51**, 2144 (1987)

3.121   G. Molnár, I. Gerócs, G. Petó, E. Zsoldos, J. Gyulai, E. Bugiel: Epitaxy of GdSi$_{1.7}$ on ⟨111⟩ Si by solid phase reaction, Appl. Phys. Lett. **58**, 249 (1991)

3.122   M. Gurvitch, A.F.J. Levi, R.T. Tung, S. Nakahara: Preparation and characterization of epitaxial yttrium silicide on (111) silicon, MRS Proc. **91**, 457 (1987)

3.123   K. Radermacher, S. Mantl, Ch. Dieker, H. Lüth: Ion beam synthesis of buried α-FeSi$_2$ and β-FeSi$_2$ layers, Appl. Phys. Lett. **59**, 2145 (1991)

3.124   J.H. Werner, H.H. Güttler: Barrier inhomogenities at Schottky contacts, J. Appl. Phys. **69**, 1522 (1991)

3.125   K. Hirose, K. Akimoto, I. Hirosawa, J. Mizuki, T. Mizutani, J. Matsui: Microstructure and Schottky-barrier height of the Yb/GaAs interface, Phys. Rev. B. **39**, 8037 (1989)

3.126   K. Hirose, K. Akimoto, I. Hirosawa, J. Mizuki, T. Mizutani, J. Matsui: Relationship between interfacial superstructures and Schottky-barrier heights of Sb/GaAs contacts, Phys. Rev. B. **43**, 4538 (1991)

3.127   P.W. Pellegrini, C.E. Ludington, M.M. Weeks: The dependence of Schottky barrier potential on substrate orientation in PtSi infrared diodes, J. Appl. Phys. **67**, 1417 (1990)

3.128   R.T. Tung, J.M. Poate, J.C. Bean, J.M. Gibson, D.C. Jacobson: Epitaxial silicides, Thin Solid Films **93**, 77 (1982)

3.129   K. Hricovini, G. Le Lay, A. Kahn, A. Taleb-Ibrahimi, J.E. Bonnet, L. Lassabatère, M. Dumas: Structure effects on Schottky barrier heights of Pb/Si and Bi/Si interfaces, Surface Science **251**, 424 (1991)

3.130   J.F. Morar, M. Wittmer: Metallic CaSi$_2$ epitaxial films on Si (111), Phys. Rev. B **37**, 2618 (1988)

3.131   J.F. Morar, M. Wittmer: Growth of epitaxial CaSi$_2$ films on Si (111), J. Vac. Sci. Technol. A **6**, 1340 (1988)

3.132   K.-H. Park, H.-S. Jin, L. Luo, W.M. Gibson, G.-C. Wang, T.-M. Lu: Epitaxial growth of thick Ag/Si (111) films, MRS. Proc. **102**, 271 (1988)

3.133   A. Cros, P. Muret: Properties of noble-metal/silicon junctions, Mater. Sci. Rept. **8**, 271 (1992)

3.134   S.P. Murarka: *Silicides for VLSI Applications* (Academic Orlando 1983) p. 62

3.135   R.T. Tung: Electron transport of inhomogeneous Schottky barriers, Appl. Phys. Rev. Lett. **58**, 2821 (1991)

3.136   U. Rau, H.H. Güttler, J. H. Werner: The ideality of spatially inhomogeneous Schottky contacts, MRS. Proc. **260**, 245 (1992)

3.137   H.A. Bethe: Theory of the boundary layer of crystal rectifier, MIT Radiation Lab. Rep. 43–12 (1942)

3.138   J.M. Andrews, M.P. Lepselter: Reverse current-voltage characteristics of metal-silicide schottky diodes, Solid-St. Electron. **13**, 1011 (1970)

3.139   C. Wagner: Theory of current rectifiers, Phys. Zeitschr. **32**, 641 (1931)

3.140   W. Schottky, E. Spenke: Zur quantitativen Durchführung der Raumladungs- und Randschichttheorie der Kristallgleichrichter, Wiss. Veröff. a. d. Siemens Werken **18**, 225 (1939)

3.141   T. Arizumi, M. Hirose: Transport properties of metal-silicon schottky barriers, Jap. J. Appl. Phys. **8**, 749 (1969)

3.142   E.H. Rhoderick: Comments on the conduction mechanism in Schottky diodes, J. Phys. D: **5**, 1920 (1972)

3.143   For a discussion see also [Ref. 3.4a, ps. 102 and 107]

3.144   See [Ref. 3.4a, p. 36]

3.145   See [Ref. 3.4a, p. 99]

3.146   C.R. Crowell, S.M. Sze: Current transport in metal-semiconductor barriers, Solid-St. Electron. **9**, 1035 (1966)

3.147   F.A. Padovani, R. Stratton: Field and thermionic-field emission in schottky barriers, Solid-St. Electron. **11**, 695 (1966)

3.148   V.L. Rideout, C.R. Crowell: Effects of image force and tunneling on current transport in metal-semiconductor (Schottky barrier) contacts, Solid-St. Electron. **13**, 993 (1970)

3.149   C.R. Crowell, V.L. Rideout: Normalized thermionic-field (T-F) emission in metal-semiconductor (Schottky) barriers, Solid-St. Electron. **12**, 89 (1969)

3.150   C.T. Sah, R.N. Noyce, W. Shockley: Carrier-generation and recombination in P-N junctions and P-N characteristics, Proc. IRE **45**, 1228 (1957)

3.151   See [Ref. 3.26, p. 92]

3.152   M. Alavi, D.K. Reinhard, C.C.W. Yu: Minority-carrier injection in Pt-Si Schottky-barrier diodes at high current densities, IEEE Trans. Electron Dev. ED – **34**, 1134 (1987)

3.153   H. Jäger, W. Kosak: Modulation effect by intense hole injection in epitaxial silicon Schottky-barrier diodes, Solid-State Electron. **16**, 357 (1973)

3.154   H.C. Card, E.H. Rhoderick: The effect of an interfacial layer on minority carrier injection in forward-biased silicon Schottky Diodes, Solid-State Electron. **16**, 365 (1973)

3.155   J.C. Manifacier, H.K. Henisch: Minority-carrier injection into semiconductors, Phys. Rev. B**17**, 2640 (1978)

3.156   C.T. Chuang: On the minority charge storage for an epitaxial Schottky-barrier diode, IEEE Trans. ED – **30**, 700 (1983)

3.157   C.T. Chuang: On the current-voltage characteristics of epitaxial Schottky-barrier diodes, Solid-State Electron. **27**, 299 (1984)

3.158   B. Elfsten, P.A. Tove: Calculation of charge distributions and minority-carrier injection ratio for high-barrier Schottky diodes, Solid-State Electron. **28**, 721 (1985)

3.159   D.L. Scharfetter: Minority carrier injection and charge storage in epitaxial Schottky-barrier diodes, Solid-State Electron. **8**, 299 (1965)

3.160   R.A. Clarke, M.A. Green, J. Shewchun: Contact area dependence of minority-carrier injection in Schottky-barrier diodes, J. Appl. Phys. **45**, 1442 (1974)

3.161   C.Y. Chang, S.M. Sze: Carrier transport across metal-semiconductor barriers, Solid-St. Electron. **13**, 727 (1970)

3.162   M. van Schilfgaarde: Scattering from ionized dopants in Schottky barriers, J. Vac. Sci. Technol. B**8**, 990 (1990)

3.163   I. Ohdomari, K.N. Tu: Parallel silicide contacts, J. Appl. Phys. **51**, 3735 (1980)

3.164   Y.P. Song, R.L. van Meirhaeghe, W.H. Laflère, F. Cardon: On the difference in apparent barrier height as obtained from capacitance-voltage-temperature measurements on Al/p-InP Schottky barriers, Solid-St. Electron. **29**, 633 (1986)

3.165   A. Singh, K.C. Reinhardt, W.A. Anderson: Temperature dependence of the electrical characteristics of Yb/p-InP tunnel metal-insulator-semiconductor junctions, J. Appl. Phys. **68**, 3475 (1990)

3.166   H.H. Güttler, J.H. Werner: Influence of barrier inhomogenities on the noise at Schottky contacts, Appl. Phys. Lett. **56**, 1113 (1990)

3.167   J.L. Freeouf, T.N. Jackson, S.E. Laux, J.M. Woodall: Effective barrier heights of mixed phase contacts: Size effects, Appl. Phys. Lett. **40**, 634 (1982); J. Vac. Sci. Technol. **21**, 570 (1982)

3.168   R.T. Tung: Electron transport at metal-semiconductor interfaces: General theory, Phys. Rev. B**45**, 13509 (1992)

3.169   J.P. Sullivan, R.T. Tung, M.R. Pinto, W.R. Graham: Electron transport of inhomogeneous Schottky barriers: A numerical study, J. Appl. Phys. **70**, 7403 (1991)

3.170   J.H. Werner, H.H. Güttler, U. Rau: Barrier inhomogenities at Schottky contacts: Curved Richardson plots, idealities, and flat band barriers, MRS Proc. **260**, 311 (1992)

3.171   U. Rau, H.H. Güttler, J.H. Werner: Barrier inhomogenities dominating low-frequency excess noise of Schottky contacts, MRS Proc. **260**, 305 (1992)

3.172   S.T. Hsu: Flicker noise in metal semiconductor Schottky barrier diodes due to multistep tunneling process, IEEE Trans. ED – **18**, 882 (1971)

3.173   M.Y. Luo, G. Bosman, A. van der Ziel, L.L. Hench: Theory and experiments of 1/f noise in Schottky barrier diodes operating in the thermionic-emission mode, IEEE Trans. ED – **35**, 1351 (1988)

3.174   P.H. Handel: Quantum approach to 1/f noise, **Phys**. Rev. A **22**, 745 (1980)

3.175 G.S. Kousik, C.M. Van Vliet, G. Bosman, P.H. Handel: Quantum 1/f noise associated with ionized impurity scattering and electron-photon scattering in condensed mater, Adv. Phys. **34**, 663 (1985)

3.176 D.N. Held, A.R. Kerr: Conversion loss and noise of microwave and milimeter-wave mixes: Part 1 – Theory, IEEE Trans. MTT-**26**, 49 (1978)

3.177 K.M. Strohm, J.F. Luy, J. Büchler, A. Schaub: Planar 100 GHzsilicon detector circuits, Microelectronic Eng. **15**, 285 (1991)

3.178 M. Ida, Y. Sato, M. Uchida, K. Shimoda: GaAs Schottky barrier diodes for ultrahigh frequency communication systems, Rev. Elec. Comm. Lab. **21**, 800 (1973)

3.179 B.L. Sharma: in *Metal-Semiconductor Schottky Barrier Junctions and Their Applications*, ed. by B.L. Sharma (Plenum, New York 1984) p. 113

3.180 K. Chang (ed.,): *Handbook of microwave and optical components*, Vol. 2: *Microwave Solid State Components* (Wiley, Chichester 1990) p. 62

3.181 A.R. Kerr, Y. Anand: Schottky – diode MM detektor, Microwave J. **24**, 67 (1981)

3.182 G. Kesel, J. Hammerschmitt, E. Lange: *Signalverarbeitende Dioden* (Springer, Berlin, Heidelberg 1982) p. 148

3.183 Y. Anand: In *Metal-Semiconductor Schottky Barrier Junctions and Their Applications*, ed. by B.L. Sharma (Plenum, New York 1984) p. 219

3.184 See also J.F. Luy (Ch. 2), J. Buechler (Ch. 5), M. Claasen (Ch. 6), M. Willander (Ch. 8), this volume.

Section 3.4

3.185 See [Ref. 3.4a, p. 39]

3.186 J.H. Werner: Schottky barrier and pn-junction I-V plots – Small signal evaluation, Appl. Phys. A **47**, 291 (1988)

3.187 R.H. Fowler: The analysis of photoelectric sensitivity curves for clean metals at various temperatures, Phys. Rev. **38**, 45 (1931)

3.188 R.H. Williams, G.P. Srivastava, I.T. McGovern: Photoelectron spectroscopy of solids and their surfaces, Rep. Prog. Phys. **43**, 1357 (1980)

3.189 M. Prutton: *Surface Physics* (Clarendon, Oxford 1983)

3.190 B.K. Agarwal: *X-Ray Spectroscopy*, 2nd edn., Springer Ser. Opt. Sci., Vol. 15 (Springer, Berlin, Heidelberg 1991)

3.191 S. Hüfner: *Photoemission Spectroscopy*, Springer Ser. Solid-State Sci., Vol. 82 (Springer, Berlin, Heidelberg 1994)

3.192 K.K. Ng: In *Properties of Silicon, Barrier Heights and Contact Resistance: Metal/Si*, (INSPEC, London 1988) p. 799

Section 3.5

3.193 J.D. van Otterloo: Schottky barriers on clean-cleaved silicon, Surf. Sci. **104**, L205 (1981)

3.194 C.R. Crowell, H.B. Shore, E.E. LaBate: Surface-state and interface effects in Schottky barriers at n-Type silicon surfaces, J. Appl. Phys. **36**, 3843 (1965)

3.195 R.N. Mitra, S.B. Roy, K. Ghosh, A.N. Daw: Electrochemically deposited schottky contacts of In, Cd and In-Cd alloy, Solid-St. Electron. **23**, 793 (1980)

3.196 G.J. van Gurp: Cobalt silicide layers on Si. II. Schottky barrier height and contact resistivity, J. Appl. Phys. **46**, 4308 (1975)

3.197 A.M. Cowley, R.A. Zettler: Shot noise in silicon Schottky barrier diodes, IEEE Trans. ED – **15**, 761 (1968)

3.198 M. Hirose, N. Altaf, T. Arizumi: Contact properties of metal-silicon Schottky barriers, Jpn. J. Appl. Phys. **9**, 260, (1970)

3.199   M. Leppihalme, T. Tuomi: Photovoltage spectra of metal-silicon (p-Type) diodes near the indirect absorption edge of silicon, Phys. Stat. Sol. (a) **33**, 125 (1976)

3.200   J. de Sousa Pires, P. Ali, B. Crowder, F. d'Heurle, S. Petersson, L. Stolt, P.A. Tove: Measurements of the rectifying barrier heights of the various iridium silicides with n-Si, Appl. Phys. Lett. **35**, 202 (1979)

3.201   M.P. Ali, P.A. Tove, M. Ibrahim: Barrier height of evaporated-manganese contacts to silicon, J. Appl. Phys. **50**, 7250 (1979)

3.202   K.E. Sundström, S. Petersson, P.A. Tove: Studies of formation of silicides and their barrier heights to silicon, Phys. Stat. Sol. (a) **20**, 653 (1973)

3.203   K.T.-Y. Kung, I. Suni, M.-A. Nicolet: Electrical characteristics of amorphous molybdenium-nickel contacts to silicon, J. Appl. Phys. **55**, 3882 (1984)

3.204   A.K. Kapoor, M.E. Thomas, M.B. Vora: *A low-barrier schottky process using MoSi$_2$*, IEEE Trans. ED – **33**, 772 (1986)

3.205   P.E. Schmid, P.S. Ho, H. Föll, T.Y. Tan: Effects of variations of silicide characteristics on the Schottky-barrier height of silicide-silicon interfaces, Phys. Rev. B**28**, 4593 (1983)

3.206   A. Smirnov, P.A. Tove, J. de Sousa Pires, H. Norde: Barrier height of Re and Os contacts to n-silicon, Appl. Phys. Lett. **36**, 313 (1980)

3.207   A. Thanailakis: Contacts between simple metals and atomically clean silicon, J. Phys. **8**, 655 (1975)

3.208   M. Bartur, M.-A. Nicolet: Chromium as a diffusion barrier between Pd$_2$Si, or PtSi and Al, J. Electrochem. Soc. **131**, 1118 (1984)

3.209   N. Toyama, T. Takahashi, H. Murakami, H. Koriyama: Variation of the effective Richardson constant of Pt-Si Schottky diode due to annealing treatment, Appl. Phys. Lett. **46**, 557 (1985)

3.210   M. Wittmer: Conduction mechanism in Pt/Si Schottky diodes, Phys. Rev. B**43**, 4385 (1991)

3.211   U.A. Shakirov, M.S. Yunusov: Influence of X-irradiation on silicon Schottky diodes, Phys. Stat. Sol. (a) **37**, 681 (1976)

3.212   D.J. Coe, E.H. Rhoderick, P.H. Gerzon, A.W. Tinsley: Silicide formation in Rh-Si Schottky barrier diodes, Inst. Phys. Conf. Ser. **22**, 74 (1974)

3.213   H. Jäger, W. Kosak: Die Metall-Halbleiter-Kontaktbarrieren der Metalle aus der Nebengruppe I und VIII auf Silizium und Germanium, Solid-St. Electron. **12**, 511 (1969)

3.214   F. Neppl, F. Fischer, U. Schwabe: A TaSi barrier for low resistivity and high reliability of contacts to shallow diffusion regions in silicon, Thin Solid Films **120**, 257 (1984)

3.215   M.O. Aboelfotoh, K.N. Tu: Schottky-barrier heights of Ti and TiSi$_2$ on n-type and p-type Si (100), Phys. Rev. B**34**, 2311 (1986)

3.216   M. Finetti, I. Suni, M. Bartur, T. Banwell, M.-A. Nicolet: Schottky barrier height of sputtered TiN contacts on silicon, Solid-St. Electron. **27**, 617 (1984)

3.217   J.G. Clabes, G.W. Rubloff, T.Y. Tan: Chemical reaction and Schottky-barrier formation at V/Si interfaces, Phys. Rev. B**29**, 1540 (1984)

3.218   C.R. Crowell, J.C. Sarace, S.M. Sze: Tungsten-semiconductor Schottky barrier diodes, Transactions of the Metallurgical Soc. AIME **233**, 478 (1965)

3.219   C.Y. Chang, P.L. Chiu, C.H. Ma: Fabrication of Zn-Si Schottky barrier diode by controlling the substrate temperature during evaporation, Solid-St. Electron. **16**, 646 (1973)

4 SiGe Heterojunction Bipolar Transistors

A. GRUHLE

Daimler-Benz AG, Forschung and Technik, Wilhelm-Runge-Str. II, 89081 Ulm, Germany

The world-wide electronics market is estimated to reach 200 billion $ in the year 2000. CMOS digital circuits represent the largest share of this market with their low-power consumption and the possibility of dynamic memories. Silicon Bipolar Junction Transistors (BJTs) are used in about 20% of all integrated circuits, mostly high-speed and analogue applications. The trend of steadily increasing switching speeds and communication rates demands further improvement of the performance of BJTs. In addition, at present the considerable number of BiCMOS circuits, which combine the high-current drive capability of BJTs with the low-power CMOS, demonstrates new applications for bipolar transistors.

The SiGe Heterojunction Bipolar Transistor (HBT) is one device that can meet these demands. III-V devices offer superior performance compared to silicon devices due to better material parameters as mobility and carrier velocity. The possibility of band-gap engineering and the application of quantum effects have yielded new devices as, i.e., the MODFET with its low noise. However, the ten-fold higher substrate-material cost and the incompatibility with existing silicon technology have kept the market share of non-optic III-V devices below 2% of the total electronics market. Compared with III-V devices the advantage of the silicon HBT is not only its superior performance compared to the classical BJT. It is also compatible with existing silicon technology, which enables its integration into a variety of other digital or analogue silicon circuits.

The HBT was actually invented together with the transistor itself by W. *Shockley* in 1948. However, it was not until the early seventies that technology was mature enough to provide the first III-V-HBTs. Their performance has steadily been improved since then, reaching an impressive gate delay of only 1.9 ps [4.1] and an f_{max} of 350 GHz [4.2]. The lack of a lattice – matched semiconductor for a combination with silicon has, for a long time, prevented the implication of HBTs in silicon. Only in the late 80's with advances in the growth of strained SiGe – layers the SiGe HBT became realizable. After only a few years of research these devices now outperform classical silicon BJTs in switching speed, transit frequency and noise level [4.3–5].

In order to understand the advantages of the HBT let us look at the physical limitations of classical BJTs. A fast device requires a thin base for a short transit time. This, in turn will cause a high base sheet resistance which makes the design of practical devices difficult. This trade-off between a thin base and a low base resistance is dictated by the maximum base doping of only several 10^{18} cm^{-3}

Springer Series in Electronics and Photonics, Vol. 32
Silicon-Based Millimeter-Wave Devices, Eds.: Luy et al.
© Springer-Verlag Berlin Heidelberg 1994

which cannot be exceeded in order to maintain a sufficient current gain. Several different devices have been proposed that try to circumvent the mentioned problem: the idea of the metal-base transistor [4.6] is to use an extremely thin and at the same time highly conductive metal or silicide layer as base. However, due to technological problems and the lack of an appropriate material system no useful devices have been demonstrated so far. The same is true for the camel transistor [4.7] which takes advantage of fast ballistic carrier transport accross the base. The BICFET [4.8] has a thin induced channel at a heterointerface as base, however the high sheet resistance prevents its application to practical devices. On the other hand, the HBT represents an almost ideal solution to the speed – resistance trade-off. The heterojunction allows an extremely high base doping, resulting in a very thin base layer which still has a sufficiently low sheet resistance. Applications of SiGe HBTs include digital circuits for a higher operation frequency or, even more important, reduced power dissipation at a given speed. The excellent low-noise properties may be used in high-frequency pre – amplifiers as satellite receivers or mobile telephones with the possibility of integrating analogue and digital circuits on one chip. Finally the high thermal conductivity of silicon compared to GaAs makes SiGe HBTs interesting for high-frequency power amplifiers.

4.1 Operation Principle of Homojunction and Heterojunction Bipolar Transistors

4.1.1 The Bipolar Junction Transistor and Its Physical Limits

For simplicity, we will consider npn transistors only, pnp devices may by treated in a similar way by exchanging p by n, N_A by N_D, etc. The collector current of a BJT, which is the minority carrier current accross the base, can be described by

$$J_C = \frac{q D_n n_i^2}{Q_{B\,\text{eff}}} \left(e^{V_{BE}/V_T} - 1 \right) = J_n \left(e^{V_{BE}/V_T} - 1 \right) \tag{4.1}$$

where D_n is the diffusion constant of the electrons in the base, n_i the intrinsic carrier concentration, V_{BE} is the base-emitter voltage, $V_T = kT/q$ is the thermal voltage, and $Q_{B\,\text{eff}}$ is the effective base Gummel number, the integral of the base dopant inside the neutral base. "Effective" means that influences as band-gap narrowing, degeneracy, etc. are included. J_n is called the *collector saturation current*. If recombination in the neutral base as well as in the EB space charge layer is neglected, the base current J_B is a pure injection of holes into the emitter. It can be described by an expression similar to (4.1), except that D_n should be replaced by the hole-diffusion constant D_p and that an effective emitter Gummel number $Q_{E\,\text{eff}}$ must be used.

The current gain β of a BJT is

$$\beta = \frac{J_C}{J_B} = \frac{D_n Q_{E\,\text{eff}}}{D_p Q_{B\,\text{eff}}} = \frac{D_n N_E W_E}{D_p N_B W_B}. \tag{4.2}$$

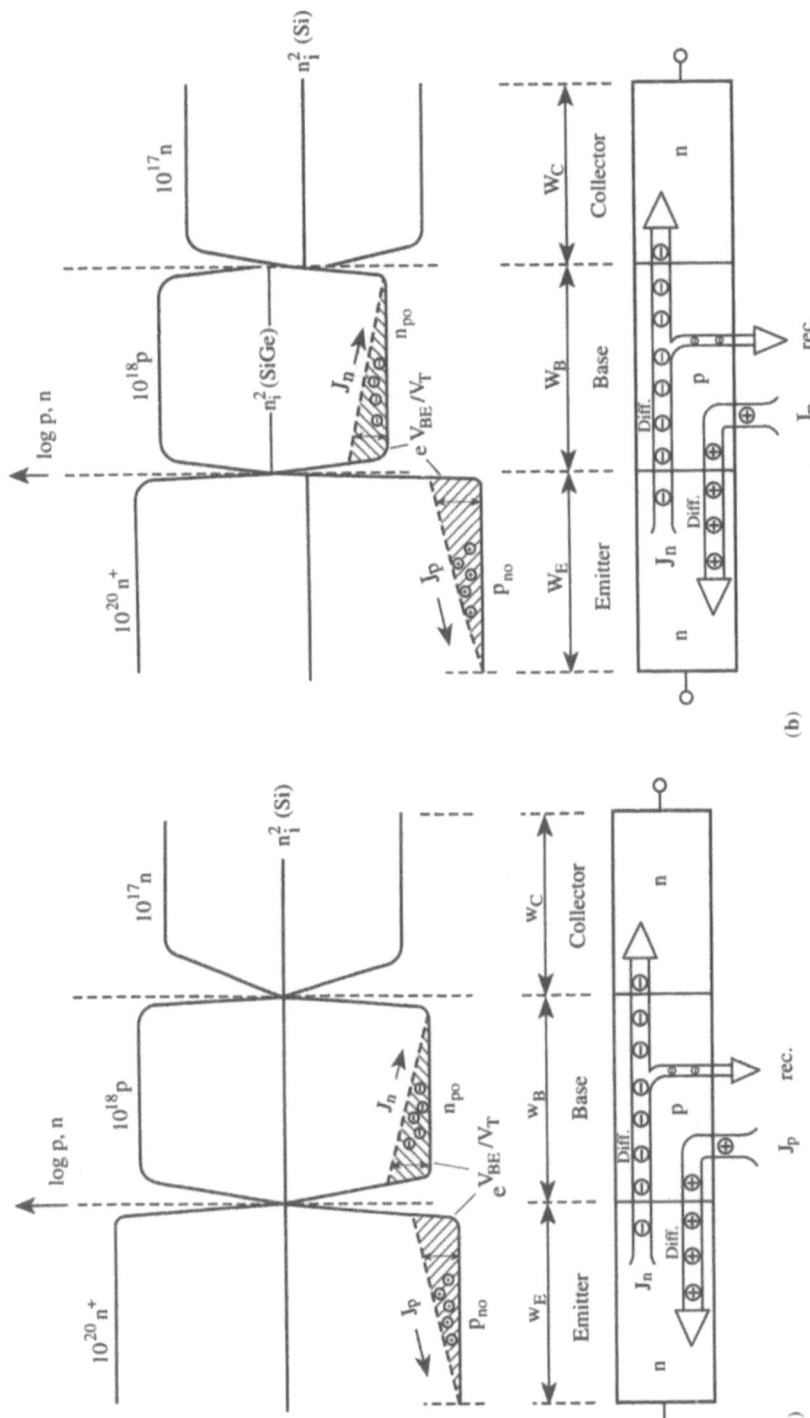

Fig. 4.1. Carrier concentrations and currents in a BJT (**a**) and in an HBT (**b**)

The last expression is a simplification in the case of constant base and emitter dopings N_B and N_E. Base thickness w_B and emitter thickness w_E have the same order of magnitude, i.e., the emitter doping N_E must be much larger than the base doping N_B in order to obtain a sufficient current gain. This is demonstrated in Fig. 4.1a where majority and minority carrier densities of a typical BJT structure are shown. Due to the forward bias V_{BE} the minority carrier concentrations on both sides of the BE space-charge layer are risen by the amount $[\exp(V_{BE}/V_T) - 1]$. Drift and diffusion of these carriers across the distances w_E and w_B represent the base and collector currents, respectively. Note the logarithmic scale of the y-axis. It becomes clear that J_C is much higher than J_B only if $p_{n0} \ll n_{p0}$ or $N_E \gg N_B$.

The base resistance of a BJT should be as low as possible. This will give low noise, a high f_{max} and it will reduce emitter current crowding. On the other hand, fast switching requires a thin base for short carrier transit times. This will consequently lead to high base sheet resistances. Figure 4.2 illustrates this tradeoff between base thickness, which approximately determines f_T, and the base sheet resistance. The diagonal lines indicate the base doping in the case of a uniform dopant distribution. (In practice, devices have a graded impurity profile with a

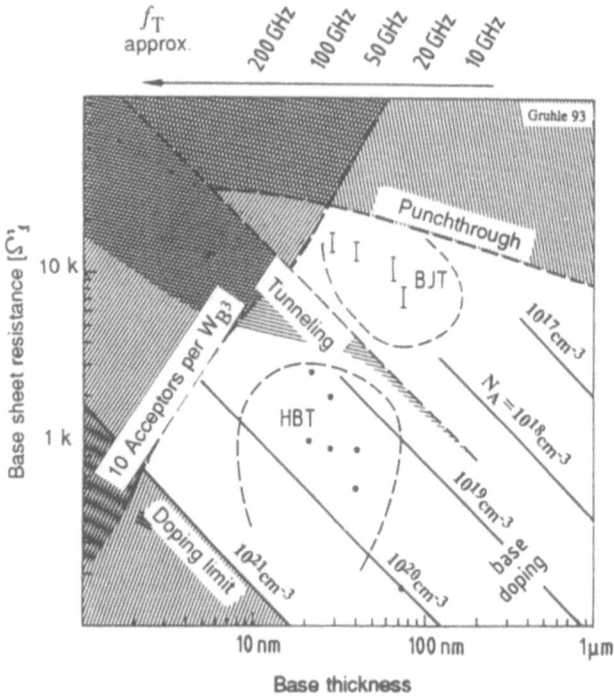

Fig. 4.2. Limits of base thickness and doping for BJTs and HBTs. The diagonal lines are for uniform base doping. The punchthrough curve is for $V_{CB} = 1$ V. (Note that the base sheet resistance is lower than the base pinch resistance, which is usually measured)

higher peak level at the emitter side and a lower concentration at the collector side). The upper line is the punchthrough limit, i.e., when complete depletion of the base occurs due to the applied collector voltage (the curve holds for V_{CB} = 1 V). Base sheet resistances above a value of several 10 kΩ/sq. would degrade device performance anyhow because of the large base resistance, even in the case of submicrometer emitter widths. On the other hand, the base doping cannot be made arbitrarily high: tunneling currents between the much higher-doped emitter and the base will short-circuit the BE junction if the base impurity level exceeds about 5×10^{18} cm^{-3}. This limit is represented by the line labeled "tunneling". Figure 4.2 demonstrates the "dead-end street" the BJT runs into when trying to increase its switching speed above 50 GHz. An additional forbidden area on the left side further narrows the possible BJT range. It describes the situation where less than 10 acceptors are in a cube of w_B^3. This may cause local punchthroughs by statistical deviations of the dopant-atom distribution in the base. The experimental points in Fig. 4.2 are from recently reported BJTs that had f_T values in the order of 50 GHz. Although they appear to be still far away from the tunneling line, their actual peak base doping already touches this limit. (Note that the usually reported base pinch resistance is higher than the sheet resistance shown in Fig. 4.2 because of the partial base depletion at $V_{BE} = 0$).

4.1.2   The Heterojunction Bipolar Transistor

An HBT has a BE-heterojunction where the band gap of the emitter semiconductor E_{gE} is larger than the band gap in the base E_{gB}. In order to fabricate HBTs in silicon, one approach is to use a wide-band-gap emitter material, e.g., Semi-Insulating Polycrystalline Si (SIPOS), amorphous silicon, microcrystalline silicon, β – SiC and GaP [4.9]. However, in most cases the high emitter resistance associated with the wide-gap material degrades device performance. The second approach is to use a narrow-band-gap base material as SiGe. We will concentrate in the following on Si/SiGe/Si HBTs which are correctly spoken, double-heterojunction transistors. However, as will be explained in Sect. 4.2.3 the Collector-Base (CB) heterojunction is of minor importance for the fundamental behaviour of an HBT.

Figure 4.1b shows an extension of Fig. 4.1a where the case of an HBT with a narrow-band-gap SiGe alloy base has been included. The new intrinsic carrier concentration n_i^2 (SiGe) in the base has increased due to the smaller band-gap and consequently the level of the minority carriers, so p $\cdot$ n = n_i^2 remains valid. The base doping N_A is assumed to be constant and p $\approx N_A = 1 \cdot 10^{18}$ cm^{-3}. The difference between the HBT and BJT is now that the concentration of the injected electrons is several orders of magnitude higher (note the logarithmic scale). This means that the collector current of the HBT becomes

$$J_c = \frac{q D_n n_i^2 [\text{SiGe}]}{Q_{B \text{ eff}}} \left(e^{V_{BE}/V_T} - 1 \right) \tag{4.3}$$

where n$_i$ [SiGe] is the intrinsic carrier concentration of the SiGe material which is roughly (Sect. 4.2.2)

$$n_i^2[\text{SiGe}] = n_i^2[\text{Si}]e^{\Delta E_g/kT}. \tag{4.4}$$

A band-gap difference ΔE_g between the SiGe and the Si semiconductors of 200 meV, for example, will increase n_i^2 [SiGe] by a factor of 2300. The collector current will rise by the same amount. Note that the base current is not affected. This means that the current gain of an HBT will be by a factor of $\exp(\Delta E_g/kT)$ higher than in a similarly doped BJT.

Some researchers explain the advantage of a SiGe HBT by the fact that the back-injection of holes into the emitter is suppressed. This is no contradiction when BJT and HBT are compared at constant collector current instead of constant V_{BE} as above.

A different way of describing the behaviour of an HBT is to look at the band diagram of the transistor. From Fig. 4.3 it becomes clear that the electrons injected into the base see a much lower conduction-band barrier than the shift that the holes see in the valence band, when trying to reach the emitter.

The real advantage of the HBT is not to achieve high current gain but to trade it against a high base doping. In the case of an HBT (4.2) becomes

$$\beta = \frac{D_n N_E w_E}{D_p N_B w_B} e^{\Delta E_g/kT} \tag{4.5}$$

where the base doping N_B may now be much higher than the emitter impurity level N_E. The exponential factor will still assure a sufficient current gain. A high base doping allows a reduction of the base sheet resistance and at the same time of the base thickness, resulting in short transit times. Figure 4.2 exhibits the advantages of HBTs over BJTs: The base sheet resistance of SiGe HBTs is roughly an order of magnitude lower despite the reduced base thickness. Note that the tunneling limit does not apply to HBTs because the emitter doping N_E may be lower than N_B. Experimental points are recently fabricated devices from [4.3,10,11].

A more general expression for the collector saturation current of an HBT is [4.12,13]

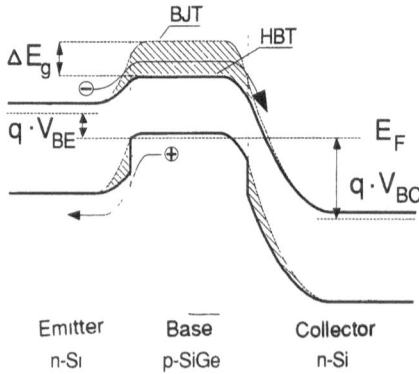

Fig. 4.3. Band diagram of a SiGe HBT

$$J_C = q \left(\int\limits_0^{w_B} \frac{p}{D_n n_i^2} dx \right)^{-1} \left(e^{V_{BE}/V_T} - 1 \right). \tag{4.6}$$

This equation correctly describes the behaviour of HBTs with arbitrary dopant and germanium profiles as, for example, an intentionally introduced grading to achieve an acceleration field in the base (Sect. 4.2.2). Assuming $p \approx N_A$ and a constant minority carrier mobility μ_n in the base, and using the Einstein relationship $\mu_n V_T = D_n$ (valid because the Fermi level is far away from the conduction band) we can write 4.6) as

$$J_C = \frac{q V_T \mu_n}{\int\limits_0^{w_B} \frac{N_A(x)}{n_i^2(x)} dx} \left(e^{V_{BE}/V_T} - 1 \right). \tag{4.6a}$$

Comparison with (4.1) reveals that the base Gummel number $Q_{B\ eff}$ has been replaced by the integral in the denominator. The base dopant concentration $N_A(x)$ at the position x is now weighted by the value of $n_i^2(x)$, which depends on the germanium content there.

The transit frequency f_T of a bipolar transistor is defined as the frequency at which the short-circuit current gain is unity. It may be expressed as [4.9]

$$f_T = \frac{1}{2\pi \tau_{EC}}, \tag{4.7}$$

where the total delay τ_{EC} is the sum of

$$\tau_{EC} = \tau_E + \tau_{E'} + \tau_B + \tau_C + \tau_{C'}. \tag{4.8}$$

The emitter delay τ_E represents the charge of holes injected from the base into the emitter. In the case of a classical BJT it depends strongly on the properties of the polysilicon used for the emitter formation. It contributes a large part to the total delay in BJTs, typically 0,5–2 ps [4.14]. In well-designed HBTs, however, τ_E becomes negligible because of the suppressed hole injection.

The emitter charging time $\tau_{E'}$ is given as ($I_E \approx I_C$ assumed)

$$\tau_{E'} = \frac{V_T}{I_C} C_{BE} \tag{4.9}$$

It exists in both BJTs and HBTs. By plotting τ_{EC} against $1/I_C$, as shown in Fig. 4.4, it may be separated from the other delay components when extrapolating the measured curve to the limit $1/I_C = 0$. If transistors are operated at the highest possible current density before the onset of the Kirk effect (Sect. 4.2.3) the contribution of $\tau_{E'}$ is usually small, i.e. a few tenths of a ps. There is an advantage of HBTs over BJTs at lower collector currents: The lowly doped emitter of an HBT also reduces C_{BE}. Compared with a classical n$^+$ polyemitter the curve in Fig. 4.4 is less steep. This means the HBT has a high f_T even at low currents.

The base transit time τ_B is the time for the removal of the minority carriers inside the neutral base: If accelerating fields (Sect. 4.2.2) are absent, then it holds that

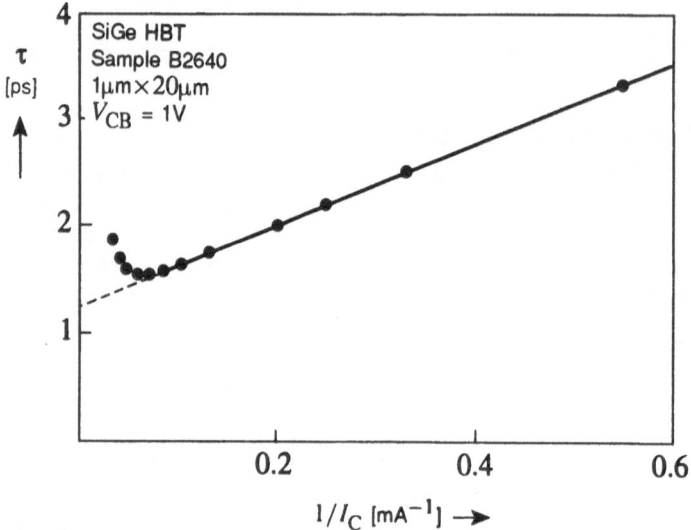

Fig. 4.4. Total transistor delay as a function of the inverse collector current

$$\tau_B = \frac{w_B^2}{2D_n} + \frac{w_B}{v_s}. \tag{4.10}$$

Usually the first term is the dominating part and its quadratic dependence calls for a thin base, a property easily achievable with an HBT, even without sacrificing in the base sheet resistance. This reduction of τ_B is one of the most important advantages of the HBT.

The collector transit time

$$\tau_c = \frac{x_c}{2v_s} \tag{4.11}$$

is the time that the carriers need to pass the CB space-charge layer of thickness x_c with saturated velocity v_s. The collector charging time

$$\tau_c' = C_{BC}\left(R_E + R_C + \frac{V_T}{I_C}\right) \tag{4.12}$$

is caused by the parasitic collector and emitter resistances R_E and R_C. Both components are independent of a possible heterojunction within the device. The current-dependent part $C_{BC}V_T/I_C$ behaves similar to τ_E' and is usually negligible at high current densities.

The maximum oscillation frequency f_{max} is often calculated as (4.15)

$$f_{max} = \sqrt{\frac{f_T}{8\pi C_{BC}R_B}}. \tag{4.13}$$

The low base resistance of HBTs will therefore enable the fabrication of devices with very high f_{max} values, important for all non-digital applications. It seems that by reducing R_B (e.g., by going to submicrometer devices) almost infinite f_{max} values may be achieved. However, (4.13) is only a first approximation and it should therefore be applied with caution.

As a conclusion the advantages of the HBT with regard to high-frequency performance are the elimination of the emitter delay τ_E, a short base transit time τ_B, and a low base resistance that increases f_{max}. Base and emitter of an HBT may therefore be considered as an almost ideal injector of carriers. The device performance is then exclusively determined by the collector design (Sect. 4.2.3).

Logic gates with HBTs will benefit from the improvement of both f_T and R_B. Despite the dificulties in predicting the gate delay of ECL or CML circuits from the device parameters, several expressions can be found in the literature [4.15–18]. They all contain f_T or τ_B, the base resistance R_B, the load resistance R_L, and C_{CB}.

4.1.3 The Si/Ge Material System

Compared with silicon, pure Ge or SiGe alloys have a smaller band gap. For this reason these materials may be used as the base of an HBT. Starting with a silicon substrate, which usually forms the collector, the result will necessarily be a Si/SiGe/Si double-heterojunction structure. However, there is a problem when combining the two semiconductors. Ge has a 4.2% larger lattice constant than silicon. This suggests that roughly every 25th atom in a row will be improperly bonded leading to dislocations, as depicted in Fig. 4.5a. However, below a critical thickness t_c a layer may be elastically deformed and it will adapt the lattice constant of the adjacent crystal (Fig. 4.5b). This is called *pseudomorphic growth* contrary to the growth of a relaxed layer. Figure 4.6 shows the critical thickness of SiGe alloy layers depending on their Ge content. The stable thickness of the calculated equilibrium curve can fortunately by exceeded [4.19], leading to so-called *metastable layers* if the growth temperature is kept low enough, which will inhibit dislocation

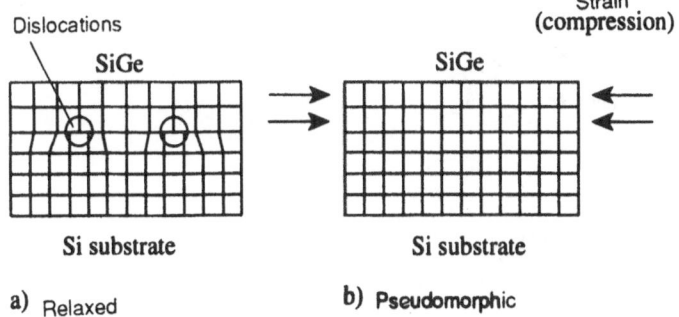

Fig. 4.5. Schematic lattice diagram of SiGe/Si heterostructures, relaxed layer with dislocations (**a**) and pseudomorphic growth (**b**)

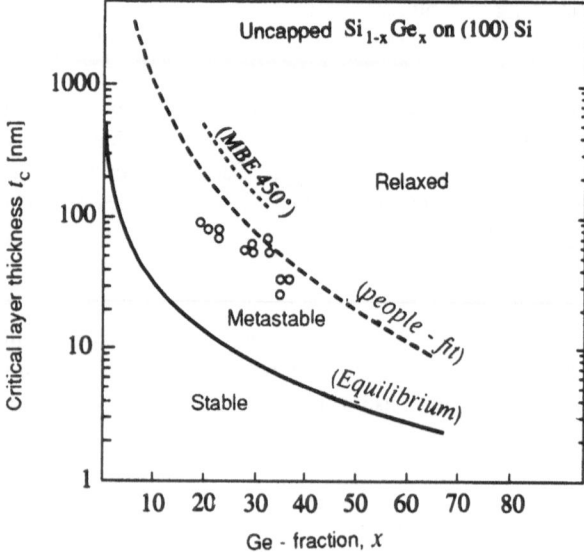

Fig. 4.6. Critical layer thickness of SiGe alloys versus Ge content [4.19]

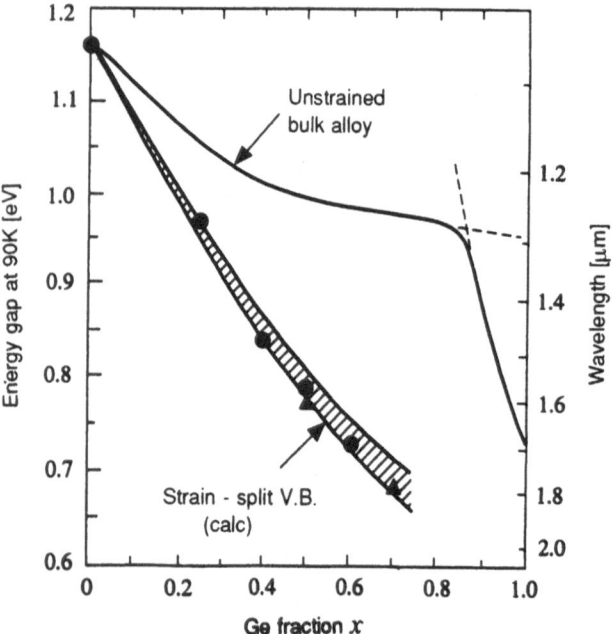

Fig. 4.7. Band gap of SiGe alloys versus Ge content [4.20]

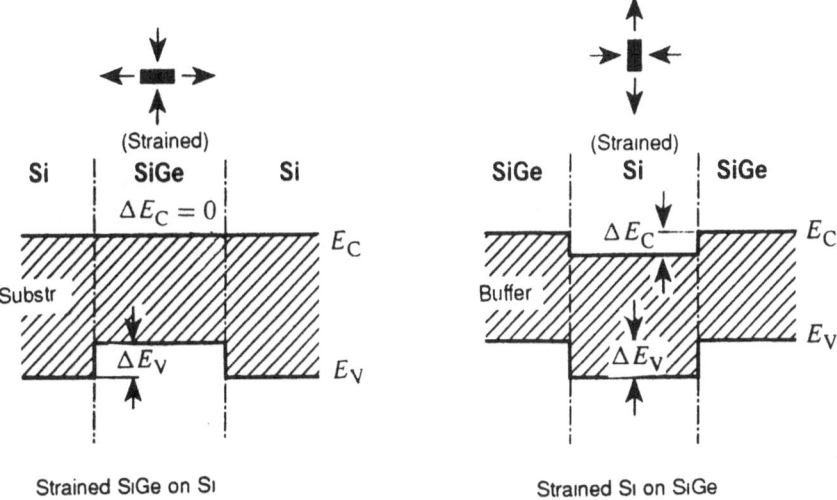

Fig. 4.8. Band alignments of Si/SiGe heterostructures

formation. The allowable thickness of a pure Ge layer is only about 1 nm. It is therefore out of question to use it as the base of a SiGe HBT. SiGe alloys with 10–40% Ge content, however, will allow practical HBT base thicknesses of several 10 nm in the metastable region.

The band gap of SiGe alloys depending on the Ge percentage is plotted in Fig. 4.7 [4.20]. In the case of SiGe HBTs the lower curve for strained SiGe layers on Si applies. The considerable difference to the unstrained alloy curve comes from the influence of the mechanical strain on the band structure. The 6-fold degenerate conduction-band valleys are separated, 4 of them are lowered in energy, 2 are raised. At the same time the degeneracy of the valence band is lifted. (In Fig. 4.7 only the relevant upper valence band is shown.) The result is an overall band gap shrinkage which exceeds the value of the unstrained SiGe alloy. This is favorable for the SiGe HBT which can benefit from the larger difference in band gap.

When forming a heterostructure, the alignments between the two conduction and valence bands are important. Figure 4.8 displays the two cases of strained SiGe on Si and strained Si on SiGe. For SiGe HBTs only the first possibility applies. The fact that the complete band-gap shift occurs in the valence band only favors npn HBTs. No conduction-band step or spikes will disturb the flow of the electrons, a problem known in GaAs HBTs, which can only be solved by graded ("soft") heterojunctions.

4.2   Design of SiGe HBT Layers

In order to exploit the advantages of the HBT, the layer parameters have to be optimized according to the particular application of the transistor, e.g., low noise, high frequency or power. The emitter, base and collector layers and their impact on device performance will be considered separately.

4.2.1 Emitter Design

The polysilicon emitter of BJTs provides a high emitter efficiency and at the same time a reduced hole storage. This is due to the formation of a very shallow monocrystalline emitter region by outdiffusion from the n^+-polysilicon (usually As-doped by ion implantation). Compared with an ohmic metal contact the poly-mono interface has a very low surface recombination velocity for the minority carriers, which reduces the base current. In addition, the residual SiO_2 film between single and polycristalline silicon forms a tunneling barrier for holes leading to a low minority carrier storage inside the polysilicon emitter. The effect is some what similar to the heterojunction in an HBT. However, the disadvantage of the interface oxide is an increased emitter resistance. The device performance of BJTs is therefore governed by the trade-off between the emitter efficiency and its resistance [4.9,21]

In the case of a SiGe HBT the base-emitter heterojunction will increase the emitter efficiency without an additional emitter resistance. Standard polysilicon emitters may therefore be "improved" by the introduction of Ge into the base, because the above-mentioned trade-off is no longer valid. The combination of a SiGe base with a polysilicon emitter is particularly useful if the Ge percentage is limited to low values (e.g., less than 10%) or if a strong Ge grading in the base (see next section) requires a small Ge content at the EB interface. Technological aspects may also play an important role when choosing a polysilicon emitter, as explained in Sect. 4.3.

The most rigorous approach when designing an HBT is to completely drop the polyemitter concept. A single-crystal emitter of an HBT can be designed without considering the backinjection of holes which in the case of a BJT dictates the possible parameter range. Due to the suppressed hole current in an HBT, neither the minority carrier diffusion length nor the emitter Gummel number have restrictions. The transistor current gain can always be tailored by an appropriate germanium content in the base. An important feature of an HBT is that the emitter doping can be reduced even below the base level. This lowers the base-emitter capacitance and increases device speed. In addition, an n^+ emitter cap layer may be added with arbitrarily high doping level and thickness, e.g., for optimum ohmic contacts and thick silicide layers. The classical polysilicon emitter with its poor reproducibility and its need for a high-temperature anneal is no longer necessary.

In order to investigate the impact of doping and thickness of a monocrystalline emitter on its injection efficiency the E-B-part of an HBT with different emitter designs has been modeled with a 1-dimensional finite-element simulator using the drift-diffusion equations. The base was 50 nm thick, it had a $1 \cdot 10^{19}$ cm^{-3} doping level and an ideal minority carrier drain at the collector side. Complete hole recombination at the n^+ side of the emitter was assumed. The band-gap difference was chosen to be 200 meV which corresponds to about 30% Ge. The

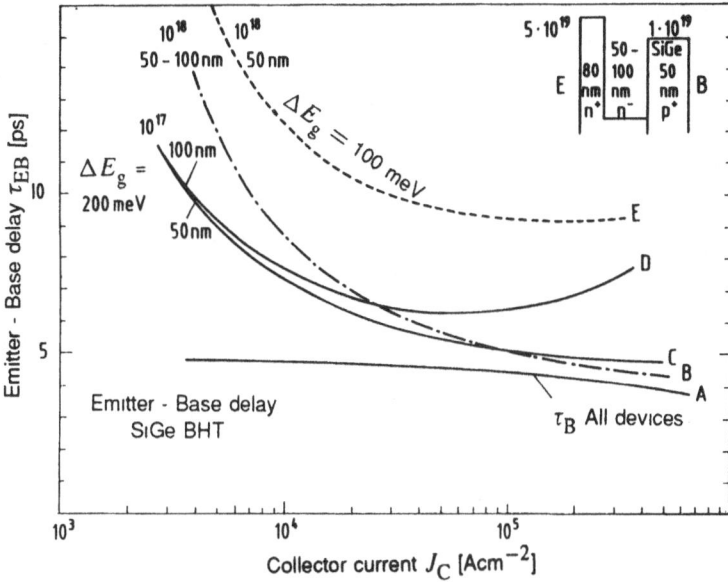

Fig. 4.9. Emitter-Base delay of different emitter designs for SiGe HBTs

total E-B delay $\tau_{EB} = \tau_E + \tau'_E + \tau_B$ was calculated for different emitter designs and plotted in Fig. 4.9 as a function of collector current. One part is the base transit time τ_B shown by the lower curve A. It is equal for all devices because it represents the injected electrons inside the base which is identical in all cases. (The delay time is rather high because of the absence of any accelerating field and the simplifying assumption of equal minority and majority carrier mobility. The non-trivial problem of minority-carrier mobility, which is about 3 times higher than the one of the majority carriers, will be discussed in Sect. 4.2.2. Although the absolute τ_{EB} values may not be exact, Fig. 4.9 nevertheless describes the qualitative behaviour correctly). The increase of all curves at low current densities is caused by the changing of the emitter-base capacitance. Although this capacitance increases with the square root of the base-emitter voltage τ'_E becomes negligible at high levels of the current, which increases exponentially. The emitters with low doping of 10^{17} cm^{-3} (curves C and D) have a smaller capacitance and are favourable at current densities below 10^4 A cm^{-2} compared with curve B which represents a 10^{18} cm^{-3} doped emitter. There is only little dependence on the thickness because the space-charge layer is thinner than 50 nm (large forward bias). However, for a practical device the emitter should be as thin as possible in order to reduce the resistance of the undepleted part. At high current densities curve B is the best choice. The lowly doped emitters (curves C and D) have to be filled up with carriers above the impurity level which causes additional delays, in particular in the case of the 100 nm thick layer. Further simulations indicate that undoped i-layers within the emitter have the same drawback. To demonstrate the speed advantage of the base-emitter heterojunction the device B has been simulated again (curve E) with a much lower Ge concentration. Now the hole injection into the emitter is not completely suppressed and the additional emitter delay shifts the curve upwards. This is the basic problem of the all-silicon "pseudo-heterojunction transistor" [4.22] where a careful design of a low-doped emitter allows a highly doped base maintaining sufficient current gain without the need of a heterojunction. However, simulations show that the emitter delay cannot be pushed below 1 ps [4.22] due to the injected holes.

Unlike in the case of BJTs it may be advantageous for the emitter of an HBT to have a small minority-carrier lifetime, or an interface with complete recombination (i.e., n^+ -poly or silicide) closely spaced from the BE junction. This is because the suppressed hole injection at high Ge contents does not only reduce the emitter charge but will, at the same time, increase transistor current gain. However, practical devices usually need β values below 100, as explained in the next section. Therefore under certain conditions an intentional increase of the base current may be desirable by reducing the hole-diffusion length in the emitter (which, in addition, will even further reduce τ_E).

High-speed HBTs that operate at large current densities should have a highly doped emitter to reduce (electron and hole) charge storage and the emitter series resistance. However, the maximum allowable impurity concentration is limited to about $5 \cdot 10^{18}$ cm^{-3} by the onset of tunneling currents between base and emitter [4.23]. The fact that the base is usually degenerately p^+ -doped and that it has extremely steep doping gradients (see next section) favours these currents. (The same problem exists for BJTs, however there the p- and n-doping levels are opposite). In ion-implanted junctions these tunnel currents are up to 3 orders of magnitude higher than in junctions formed by CVD [4.24], probably because of insufficiently annealed crystal damage. In practical circuits the base-emitter junction may temporarily become reverse-biased and Zener-tunneling currents have to be considered as well. This may restrict the emitter doping to concentrations even below 1.10^{18} cm^{-3} [4.23].

4.2.2 Base Design

a) Influence on Device Current Gain, Switching Speed and Early Voltage

According to (4.3) the Ge content will determine the enhancement of the collector saturation current. If pure hole injection into the emitter is responsible for the base current, the latter will be constant, independently of the base parameters. Figure 4.10 shows the Gummel plots of several similar HBTs with different Ge contents [4.25]. The collector-current curves are shifted upwards with increasing Ge percentage, whereas the base currents remain almost indistinguishable. By introducing enough Ge in the base the transistor current gain may be pushed to very high values. At the same time the emitter delay vanishes as the injected hole charge becomes negligible for a given collector current. Even with moderate Ge fractions of 20–35%, very high current gains have been achieved: the maximum reported room-temperature β was 5000 [4.26]. As expected from (4.5) the effect

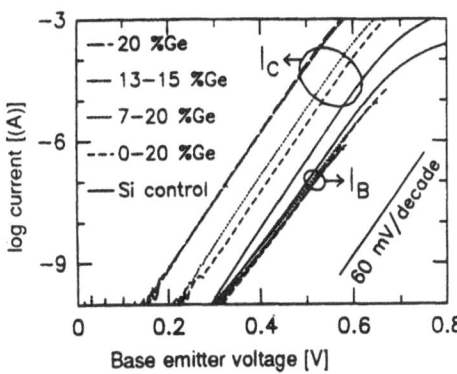

<pre style="white-space: pre-wrap" class="border rounded p-2 bg-muted/40 text-sm overflow-x-auto block">**Fig. 4.10.** Gummel plot of several HBTs
with different Ge contents [4.25]</pre>

of the band-gap difference is much more pronounced at low temperatures. A current gain of 13000 has been demonstrated at 77K [4.10], for comparison at room temperature the same device had a β of only 400. In these devices, however, the current gain was limited by a rather high base current because of surface leakage at the unpassivated mesa side walls. An indication for this is a non-ideal base current that scales with periphery rather than emitter area. A high-temperature oxide passivation, however, yields nearly ideal Gummel plots (as in Fig. 4.10) maintaining a sufficient current gain even in the pA range, which is state-of-the-art for silicon BJTs. The possibility of a high-quality passivation is an advantage of SiGe HBTs compared to III-V devices where the high surface recombination rate ruins the current gain of small devices, a particular problem for narrow emitter fingers. High-current-gain HBTs, however, will only be useful in some special analogue applications. Much more important is the improvement of the current gain – Early voltage product to be described later. In most cases a high current gain is not even desirable as it will always reduce the collector – emitter (avalanche) breakdown voltage V_{CE}. This is because the holes created by impact ionization in the CB space-charge layer represent an additional base current. Practical β values are between 10 and 100, and may be designed by tailoring base doping and Ge content according to (4.5).

Neutral base recombination plays an important role in III-V HBTs with direct-band-gap material due to the short minority-carrier lifetimes. The base current in this case will be increased or even dominated by a certain percentage of the collector current, which represents the recombining electrons. Therefore only part of the injected carriers from the emitter will reach the collector. This ratio is called the *base transport factor* α_T. Figure 4.11 is a nice demonstration of neutral base recombination in a SiGe HBT. The experiment was similar to that one of Fig. 4.10, however a large amount of oxygen incorporation (10^{20} cm^{-3}) during the CVD growth of the SiGe layer significantly reduced the carrier lifetime within the neutral base [4.27]. The collector-current increase of the HBT is accompanied by an almost equal enhancement of the base current, which is roughly 1% of the former. However, Fig. 4.11 represents an exception, usually the minority-carrier lifetimes in Si and SiGe are so large that neutral base recombination does not

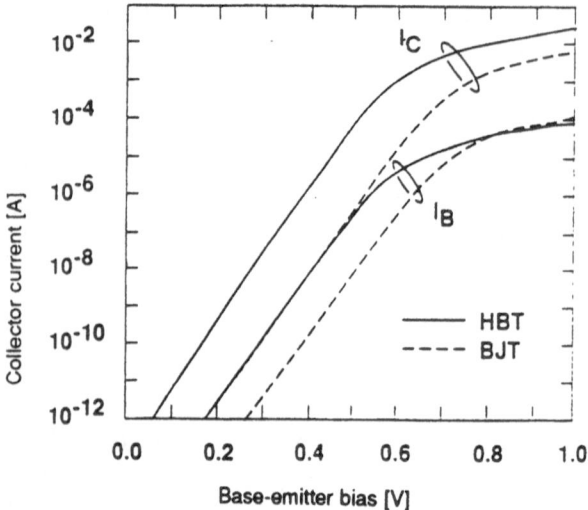

Fig. 4.11. Gummel plot of an HBT with neutral base recombination [4.27]

play a role in BJTs and HBTs: The nearly-ideal devices of Fig. 4.10 were grown in a similar CVD reactor with a load lock to reduce the oxygen content. HBTs grown by MBE have been reported [4.28,29] that show no evidence of neutral base recombination either, despite the results of *Shafi et al.* [4.30] who claimed to have found the effect in their device. Nevertheless, at very high base dopings Auger recombination may become important for neutral base recombination because it increases with the square of N_B.

In order to improve HBT device performance one tends to increase the Ge amount in the base. However, for a given base thickness the maximum allowable Ge content is limited by the onset of strain relaxation, as described in Sect. 4.1.3 and Fig. 4.6. If the HBT fabrication process requires high-temperature treatments (e.g., anneal, oxidation) that exceed the growth temperature, the layers loose their metastability. The trade-off between Ge content and base thickness becomes even more severe due to layer relaxation by the formation of misfit dislocations [4.31]. Fortunately, in high-speed HBTs one will try to make the base as thin as possible anyhow to reduce its transit time. Practical values in the range of 20–30 nm thickness and 20–30% Ge are well below the critical thickness given in Fig. 4.6. In addition, there is a constant tendency to reduce process temperatures, in particular to avoid excessive dopant diffusion (see below). The ultimate performance of advanced SiGe HBTs will therefore not be limited by possible strain relaxation. In fact, the large lattice mismatch between Si and SiGe is no longer a constraint, an often heard argument against SiGe HBTs when comparing them with III-V devices.

An interesting aspect of HBTs is the possibility of grading the Ge content across the base. The varying band gap then produces an electric field E_B for the minority carriers inside the neutral base. A linear increase of the Ge content from the emitter to the collector side will yield a constant accelerating field $E_{acc} = \Delta E_{gr}/q \cdot w_B$ for the injected electrons thus reducing the base transit time τ_B [4.12]

$$\tau_B = \tau_{B0} \frac{2kT}{\Delta E_{gr}} \left[1 - \frac{kT}{\Delta E_{gr}} \left(1 - e^{-\Delta E_g/kT} \right) \right] \qquad (4.14)$$

where $\tau_{B0} = w_B^2/2D_n$ is the diffusion-governed transit time of the quadratic term in (4.10) for a uniform base. A varying Ge level from 10% to 25% will roughly give a ΔE_g of 100 mV and according to (4.14) a τ_B reduction of almost a factor of three. Even the 75 GHz – SiGe HBT [4.32] with a moderate 0–7% Ge grading in its 45 nm base exhibited only 0.75 ps base transit time. Compared with (4.10) this is about half the value for a uniform base. Note, however, that a strong grading will necessarily lead to a low Ge content at the BE junction. This may reduce the current gain β and cause a non-negligible emitter storage time. An additional unwanted effect is a variation of β with collector current because the BE space-charge layer edge "sees" a different Ge content with varying V_{BE}. At small base widths below 50 nm the second term of (4.10) is no longer negligible. As it is not affected by any accelerating field the benefit of a Ge grading in the base may be reduced.

A built-in accelerating field can be found in almost all silicon BJTs as well because of the varying base dopant concentration that decreases from the emitter to the collector. This grading is automatically provided in most fabrication processes because the base is usually diffused or implanted from the wafer surface yielding a decreasing concentration in direction of the substrate which is the collector. On the emitter side, however, the p n junction will never be completely abrupt. Within a certain region the base doping will have an opposite grading that produces a delay field. The overall base transit time is therefore usually expressed as $\tau_B = w_B^2/2D_n\eta$ where η is the effective acceleration factor. Depending on the base dopant distribution the delaying part may be quite important and may ruin the speed advantage of the accelerating side. Some researchers [4.21] even claim that in most transistors η is near 1. (Note that it is difficult to determine η experimentally because large errors may occur when measuring τ_B, D_n and w_B).

The effects of dopant gradients may be even more pronounced in HBTs than they are in BJTs because of the larger base doping concentration range. A peak impurity level of $5 \cdot 10^{19}$ cm^{-3} dropping to $1 \cdot 10^{17}$ cm^{-3} at the collector junction, for example, will cause a conduction band shift of about 80 meV. This is the same accelerating field as a 12% absolute increase of the Ge content. (The 80 meV have been calculated using the apparent band-gap data of pure silicon [4.33]. In SiGe the values seem to differ only slightly, see below). This demonstrates that the high base doping levels in an HBT influence the minority-carrier transit to the same extent as a possible Ge grading does. Unfortunately the delaying field on the emitter side exists as well in HBTs. It may be even larger than in BJTs because of the higher peak base doping and the resulting large dopant slope towards the low-doped emitter. The detrimental effect of this delaying field has been observed experimentally [4.34]. Simulations [4.35] indicate that it contributes 40–80% of the total base transit time. For this reason the EB-junction should be as abrupt as possible to avoid any delaying field. The ideal base dopant concentration for the shortest transit time should have its peak at the EB-junction and decrease exponentially towards the collector, which gives a constant accelerating field. An additional Ge grading may be added for further improvement. Advanced growth methods as MBE or CVD offer the possibility of depositing arbitrary base dopant and Ge profiles. However, diffusion during growth, segregation or autodoping still limit the abruptness of the achievable dopant transients. Improving these restrictions is one of the challenges in order to approach the SiGe HBT with ultimate performance.

Another interesting application of a graded Ge profile in the base is an HBT with a high current gain – Early voltage product. The Early effect is the increase of collector current with collector-base voltage V_{CB} due to the modulation of the CB space charge layer, which changes the base Gummel number. The higher the base doping is compared to the collector level, the smaller will these variations be and the higher the Early voltage. However, this has to be traded against a lowering of the current gain and a slowing down of the device. This is because the peak base doping is on the collector side forming an inevitable delaying field. In an HBT these restrictions are no longer valid. *Prinz* and *Sturm* [4.13] showed that the current gain – Early voltage product of an HBT can be written as

$$\beta V_A = \frac{q^2 D_n}{J_B C_{BC}} n_i^2 (w_B) .$$
(4.15)

The base current J_B and the capacitance C_{BC} are constant because they are exclusively determined by emitter and collector properties, respectively. Therefore, the βV_A product depends only on the intrinsic carrier concentration at the collector edge of the neutral base region. A high Ge content at this position will therefore yield high βV_A products without the need of a high doping level at the collector side. The explanation is that, although the variation of the CB space charge layer with V_{CB} remains unaffected by the Ge, the contribution of the modulated part to the total base Gummel number in the integral of (4.6) is significantly reduced. A graded Ge profile with increasing Ge content towards the collector will therefore not only be helpful for fast transit times but will also provide transistors with high Early voltages. A 100-fold increase in the βV_A-product has been demonstrated with SiGe HBTs [4.13] compared to BJTs with similar f_T-values.

b) Material Aspects of SiGe and Their Impact on HBT Performance

Let us now look in detail at the relationship between the Ge content and the intrinsic carrier concentration n_i^2, the parameter which is most important for HBT performance according to (4.6). In order to verify the theoretical calculations [4.20] of the band-gap difference between SiGe and Si layers in Fig. 4.7 optical absorption experiments [4.36] were performed and more recently admittance spectroscopy measurements [4.37]. Some experimental results are included in Fig. 4.7. They are in good agreement with theory, however, the spread of the experimental data still leaves an error of about 20 meV in the determination of the exact band gap. In the case of an HBT this corresponds to an unacceptable factor of two in the collector current or transistor gain. This uncertainty demands a more exact knowledge of the relationship between Ge percentage and band gap. An interesting method was first reported by *King et al.* [4.27] who used the temperature dependence of the ratio of the collector saturation current of an HBT compared to a BJT with an identical base Gummel number by dividing (4.3) by (4.1):

$$\frac{J_{n0}(T)[\text{HBT}]}{J_{n0}(T)[\text{BJT}]} = c \exp\left(\frac{\Delta E_g}{kT}\right). \tag{4.16}$$

Plotting this ratio versus temperature gives a curve the slope of which corresponds to ΔE_g. However, looking at (4.1 and 3) it is clear that the above relationship only holds if certain conditions are met. Dividing n_i^2 [SiGe] by n_i^2 [Si] does not necessarily yield $\exp(\Delta E_g/kT)$ because n_i^2 also depends on the density of states N_V and N_C which differ in Si and SiGe. A second assumption is that the temperature variation of $D_n(T)$ is equal for both devices, which is very unlikely considering the effects of band splitting due to strain on carrier mobilities.

Even if the exact relationship between the Ge content and the band gap is known, the calculation of n_i^2 will remain difficult. The effect of Band-Gap Narrowing (BGN) has to be considered, in particular in the case of HBTs with high base doping exceeding degeneracy. In addition, at these high impurity levels the Fermi level approaches the valence band and the Boltzman statistics is not longer valid. Considering Fermi-Dirac statistics instead the correct value of n_i^2 is

$$n_i^2 = n_i^2[\text{Boltz}]\frac{2}{\sqrt{\pi}}F_{1/2}\left(\frac{E_V - E_F}{kT}\right)\exp\left(\frac{E_F - E_V}{kT}\right)\exp\left(\frac{\text{BGN}}{kT}\right). \tag{4.17}$$

This is also known as the Apparent Band-Gap Narrowing (ABGN) which allows to compute n_i^2 correctly. A large amount of data exists for pure silicon, both theoretical and experimental [4.33]. The case of strained SiGe layers, however, is much more complicated. The splitting of light and heavy holes increases the value of the BGN [4.38]. However, at the same time the density of states is reduced [4.39] which influences the position of the Fermi level. Calculations of the apparent band-gap narrowing in SiGe [4.40] therefore yield values almost equal to pure silicon, at high Ge contents they may be even lower. Figure 4.12 displays the calculated ABGN for three different Ge contents.

Pruijmboom et al. [4.41] claimed to have found experimental evidence for the reduction of the density of states in strained SiGe layers. Writing (4.16) as

$$\frac{J_{n0}(T)[\text{HBT}]}{J_{n0}(T)[\text{BJT}]} = \frac{N_C N_V[\text{HBT}]}{N_C N_V[\text{BJT}]}\exp\left(\frac{\Delta E_g}{kT}\right) \tag{4.18}$$

and performing temperature measurements he found an $N_C N_V$-ratio of 0.4 which agrees well with the theoretical value from [4.39]. On the other hand, if one applies (4.18) to the data of [4.27] the obtained $N_C N_V$-ratios are all far below 0.1. It must be doubted that (4.18) is useful at all because it does not consider different minority-carrier mobilities in Si and SiGe, neither its temperature dependence, neither possible variations of the ABGN as explained above. At low Ge contents, when the band splitting is on the order of kT, theory predicts a strong temperature

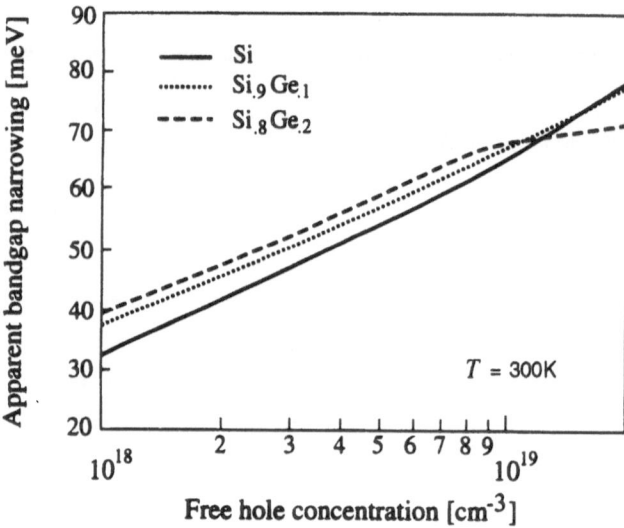

Fig. 4.12. Calculated apparent band gap narrowing of SiGe for three different Ge contents [4.40]

dependence of the density of states [4.39] and the $N_C N_V$-ratio will not even be constant. In addition, any parasitic barrier (see below) will completely change the temperature behaviour of J_{n0} [4.39]. The absence of such barriers must be checked first before using (4.18).

The majority carrier mobility in the SiGe layer determines the base sheet resistance. Due to the strain-induced changes in the band structure the mobility is a tensor with different components parallel and perpendicular to the growth plane. Calculations [4.42] predict that the in-plane mobility, which determines the base sheet resistance, increases with Ge content, despite the additional alloy scattering. HBTs with practical values of 20%–30% Ge will benefit from a mobility increase of around 50%, which helps to reduce the base resistance of the device. It should be noted, however, that the above calculations assume undoped SiGe layers. There is some experimental evidence, too, for enhanced in-plane carrier mobility in strained SiGe layers [4.43].

According to (4.6a) the minority-carrier mobility μ_n is important for analysis and design of BJTs and HBTs. It is difficult to determine its value independent of other parameters. In most experiments BJTs have been used themselves, and $\mu_n n_i^2$ was actually determined. The problems in connection with the correct n_i^2 have already been discussed. There are fortunately methods [4.44,45] that extract the minority-carrier mobility independent of n_i^2, using photogenerated carriers. These results agree well with the recent analytical model of *Klaassen* [4.46]. Figure 4.13 shows the electron mobilities of both majority and minority carriers for pure silicon. Note that the minority mobility largely exceeds the majority value for highly doped p-type silicon and that it has a completely different temperature dependence. As a consequence measurements of the base sheet resistance, which only determine the majority-carrier mobility, are of limited use as an estimate for μ_n in (4.6a).

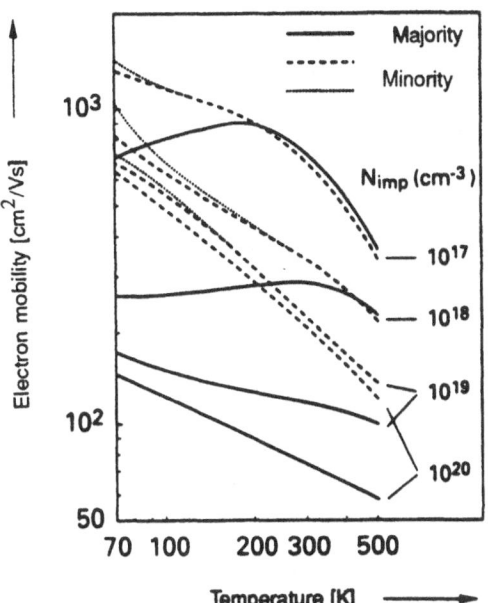

Fig. 4.13. Majority and minority electron mobility in silicon [4.46]

In the case of SiGe HBTs there is no data available at all for μ_n in SiGe. An independent measurement of the minority-carrier lifetime, mobility and diffusion length according to the methods of [4.44,45], will only yield the in-plane parameters because these experiments depend on lateral devices. However, the strain-induced changes of the band structure in SiGe will lead to anisotropic carrier mobilities. In the case of a strained SiGe base layer theory predicts an increase of the electron mobility for transport perpendicular to the heterojunction. At present, however, only the data of Fig. 4.13 for pure silicon is available and may serve as a rough estimate. A large amount of work both experimental and theoretical needs to be done to fully describe the anisotropic carrier transport in SiGe HBTs.

c) Parasitic Conduction-Band Barriers

An important design issue for SiGe HBTs is a possible outdiffusion of the highly concentrated base dopant into the emitter and collector during the fabrication process. Practical devices will therefore not have completely abrupt BE and BC junctions nor will they necessarily coincide with the SiGe interfaces, contrary to the assumptions made when deriving (4.3). The diffusion of Ge in Si is negligible compared with the one of dopants [4.47], so the SiGe interfaces may be considered as fixed. Let us define the base of an HBT as the part which contains Ge, regardless of the position of the pn junctions. If one of these junctions lies outside the SiGe base, for example, as a consequence of an outdiffusion of the acceptors inside the base into the (lower doped) collector, a parasitic barrier will be formed

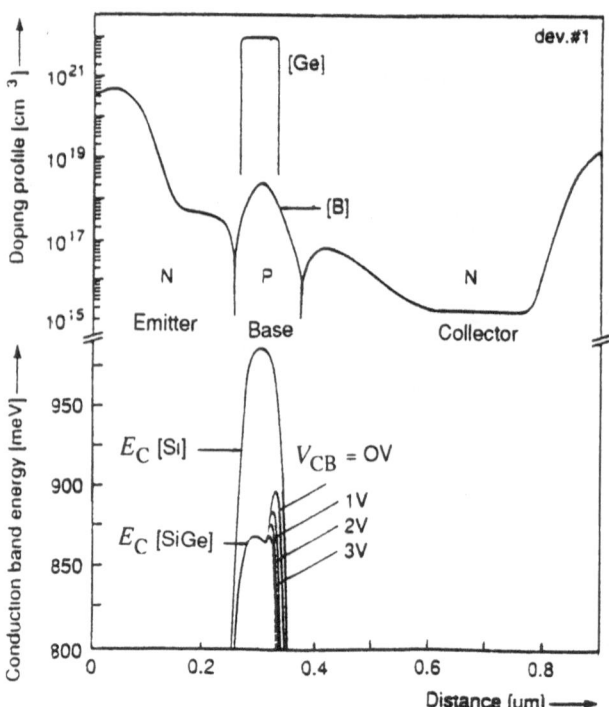

Fig. 4.14. Modulation of a parasitic conduction band barrier by the applied collector base voltage

in the conduction band [4.48,49], as can be seen in Fig. 4.14. This is because inside the p-doped region the valence band is "clamped" to the Fermi level. The parasitic barrier dramatically reduces the current gain of the HBT. Even if the outdiffused part represents only a small percentage of the total base doping its contribution to the integral in (4.6) is large due to the low n_i^2 of silicon. (Note that the boundaries of the integral are between the BE and BC pn junctions, not according to the above definition). The reduction of the collector saturation current occurs regardless of whether the outdiffusion is on the emitter or collector side or on both. A parasitic barrier on the collector edge is strongly modulated by the collector-base voltage, as illustrated in Fig. 4.14. This completely ruins the Early voltage of the HBT [4.13] due to the strong modulation of the integral in (4.6). In addition, device speed is slowed down [4.41] because of the delaying field formed by the barrier for the minority carriers crossing the base. In [4.48] equations have been derived that allow to estimate the collector saturation current and the base transit time in the presence of a parasitic barrier. An additional drawback that will particularly degrade device performance in logic circuits is a collector-emitter offset voltage which will as well result from parasitic conduction band barriers [4.50].

The described effects may also result from conduction-band offsets fortunately negligible in SiGe HBTs. In III-V HBTs, however, specially graded heterojunction designs are necessary. For the same reason the band alignments of a Si/SiGe/Si heterostructure (Fig. 4.8) are completely inadequate for a pnp transistor. Despite this, pnp SiGe HBTs have been fabricated, also using graded heterojunctions [4.51].

Because of the strong effects of parasitic barriers on the performance of HBTs it is important to detect these barriers. A very efficient method is to look at the ideality factor n of the collector current J_c

$$J_c = J_{n0} \left(e^{V_{BE}/nV_T} - 1 \right) \tag{4.19}$$

where J_{n0} is the collector saturation current, V_{BE} the base-emitter voltage, and V_T the thermal voltage. This exponential behaviour can usually be measured quite exactly over several decades in the case of a Gummel plot set-up where leakage currents of the CB junction are suppressed because $V_{CB} = 0$. The collector current is a pure drift and diffusion current and its ideality factor should be exactly $n = 1$. If $n > 1$ this deviation must come from a change of J_{n0} with V_{BE}

$$J_c = J_{n0}(V_{BE}) \left(e^{V_{BE}/V_T} - 1 \right) \tag{4.20}$$

because of the changes in base Gummel number. BJTs have typical ideality factors between 1.001 and 1.010, depending on how near the punch-through condition they are operated. In an ideal SiGe HBT with a base doping above $1 \cdot 10^{19}$ cm^{-3} the variation of the depleted part of the base with V_{BE} is much smaller, particularly because of the low doped emitter where most of the space charge is located. This leads to ideality factors even below 1.001. However, in the presence of a parasitic barrier at the EB junction the strong modulation of the effective-base Gummel number will considerably increase the ideality factor. It may attain values up to $n = 1.1$. As an example Fig. 4.15 compares the Gummel plots of an HBT after two different anneals at 700°C and at 800°C. The larger outdiffusion during the higher-temperature anneal causes a reduction of the collector saturation current

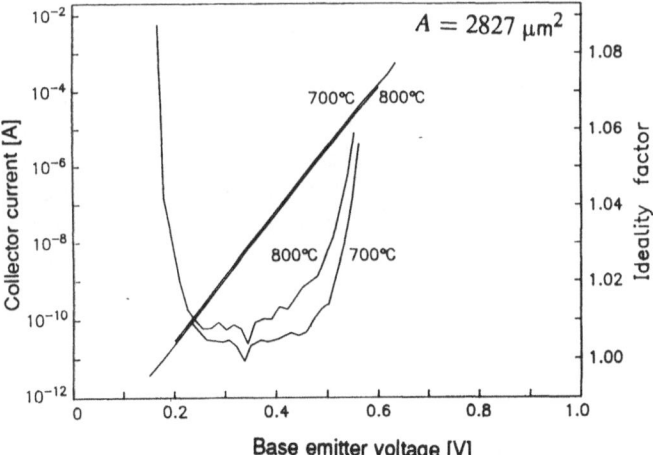

Fig. 4.15. The effect of base dopant outdiffusion on the collector saturation current and its ideality factor

and an increase of the ideality factor. Note the n scale which is between $n = 1.0$ and $n = 1.1$. Despite the required exact temperature control of less than 1°C during the Gummel plot measurement this method has proved to be very sensitive when looking for parasitic barriers [4.52]. Knowing the n-factor in both normal and inverse operation (collector and emitter exchanged) allows a separate examination of the BE and CB junction. As already mentioned, the Early voltage is as well affected by parasitic barriers: It can be shown [4.52] that a simple relationship exists between the Early voltage V_A and the ideality factor $n_{(i)}$ in inverse operation:

$$V_A = V_T \frac{n_{(i)}}{n_{(i)} - 1}. \tag{4.21}$$

The inverse Early voltage and the normal ideality factor are related in the same way. Some practical aspects concerning the measurements of V_A and n have been described in [4.52].

If the highly doped p$^+$-base region is embedded between two undoped spacer layers [4.49] the pn junctions may be kept inside the SiGe region. The resulting HBT tolerates a limited amount of diffusion without forming parasitic barriers. The thickness of the undoped layer on the collector side i_{CB} may be designed much larger than necessary in order to stay on the safe side. The effect on device performance of a thick i_{CB}-layer is small: It will form part of the depleted CB space-charge layer and slightly increase the collector transit time. For example, carriers at saturated velocity will transfer a 10 nm spacer in 0.1 ps.

The design of the emitter side however is crucial. As already explained, the pn junction should be as abrupt as possible to avoid a delaying field. But if diffusion is inevitable during layer growth or the following fabrication process, the pn junction must stay within the SiGe region to avoid the formation of a parasitic barrier. This may be achieved by the introduction of an undoped i_{EB}-spacer layer, which will shift the base-dopant diffusion tail (and with it the pn junction) back into the base. The above process can be described as a shift of the Si/SiGe interface with respect to the junction. The device performance of an HBT depends not only on this position of the Si/SiGe interface with respect to the EB junction but also on the base dopant slope with its delaying field. These two parameters were varied in a simulation of the EB part of an HBT (similar to the one of Fig. 4.9) in order to investigate the impact on device behaviour. The results are shown in Fig. 4.16 where the normalized collector current density, the ideality factor and τ_{EB} are plotted as a function of the Si/SiGe interface position with respect to the pn junction at $x = 0$. The curves (a) represent the case of an ideal abrupt doping profile, which will be considered first. The Si/SiGe heterointerface is always assumed abrupt. At $x = 0$ the Si/SiGe and the pn interfaces coincide. For $x > 0$ the Si/SiGe interface lies within the p base, for $x < 0$ within the n emitter. In the first case a rapid decrease of the collector saturation current can be observed with increasing distance x because of the change in the effective base Gummel number. At the same time the ideality factor rises which reflects the formation of a parasitic conduction-band barrier similar to Fig. 4.14. The n factor reaches a maximum and drops again when the barrier becomes too stable to be modulated by V_{BE}. The

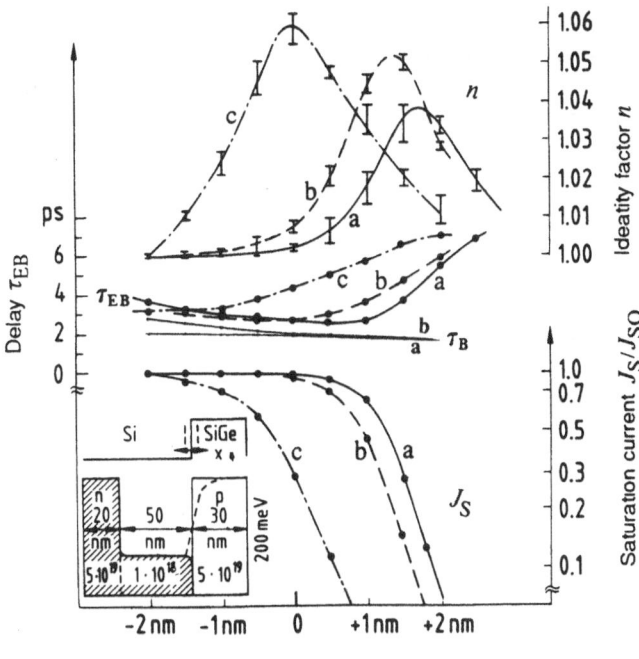

Position of SiGe heterointerface x

Fig. 4.16. The influence of the SiGe/Si heterointerface position (with respect to the EB pn-junction) on the collector saturation current, its ideality factor and the base-emitter delay of a SiGe HBT

τ_{EB} curve increases rapidly for $x > 0$. A more detailed analysis of the different τ contributions shows that this rise comes exclusively from the τ_E part due to hole injection into the emitter. The τ_B part remains almost constant as can be seen from the lower curve. The delay associated with the BE capacitance τ_{EC}-τ_B-τ_E has an almost constant value of 0.3–0.4 ps (not plotted). With other words, in the $x > 0$ regime the emitter storage time of the device, which has become a BJT, is not longer negligible.

Moving the Si/SiGe interface into the emitter ($x < 0$) has little effect on the form of the conduction band, and J_S and n are not affected. However, τ_{EB} slightly increases again caused by the τ_E part, this time due to the injected holes between the pn junction and the SiGe/Si interface.

Curves b (dashed) and c (dashed-dotted) in Fig. 4.16 describe the case where the ideal rectangular-base dopant profile with the concentration N_A has been broadened due to diffusion during growth or post-epitaxial anneals. It has been assumed that this leads to an error function distribution with $N_A/2$ at the original step interface position. Two different diffusion lengths $1 = (4Dt)^{1/2}$ have been considered, $1 = 2.5$ nm (curves b) and $1 = 5$ nm (curves c).

Note that the new position of the pn junction moves 3.7 (7.4 nm) to the left. (This means that $x = 0$ is the pn junction position of the initial abrupt case). The general behaviour of the HBT remains similar, however the curves are shifted to the left and their bending "softens". At $x = 0$ parasitic barriers form when outdiffusion takes place and both n and τ_{EB} increase. For a minimum τ_{EB} the Si/SiGe interface must be moved in the direction of the emitter. Note that at the optimum position a small parasitic barrier remains ($n > 1$), which seems to be beneficial for the device speed. Nevertheless, the minimum τ_{EB} increases with dopant-transition broadening, the difference coming from the τ_B part, which increases particularly at $x < 0$. This is because of the delaying base field from the dopant slope, part of which falls into the neutral base.

As a conclusion it has been shown that the design of the base-emitter junction of a SiGe HBTs is extremely critical and that it has a major influence on the high-frequency performance of the device. An abrupt pn junction is best and minor diffusion gradients of only a few nm will already increase the transistor delay. There is an optimum position of the Si/SiGe interface with respect to the pn junction which depends on the dopants slope and which should be maintained within an accuracy of 0.5 nm.

Jorke [4.53] proposed a completely different base design with a planar doped layer embedded between two intrinsic layers. First calculations indicate device delay times similar to the conventional SiGe HBT structure but no experimental verifications have been reported so far.

4.2.3 Collector Design

The design of the collector layer is governed by the trade-off between a short collector transit time $\tau_c = w_c/2v_s$ across the collector space-charge region w_c and the collector-base breakdown voltage

$$V_{CBO} = \frac{\varepsilon E_{br}^2}{2q N_{DC}} \tag{4.22}$$

where E_{br} is the breakdown field, and N_{DC} the collector doping density. The BC junction can be considered as one-sided abrupt because of the much higher base-doping density. For the same reason the problem of punchtrough plays no role in HBTs. For a given breakdown voltage the maximum collector doping can be calculated from (4.22) and for the thickness w_c the smallest value should be chosen, which is $w_c = 2V_{CBO}/E_{br}$. Otherwise, the undepleted part of the collector layer forms an unwanted series resistance R_c. It will increase the total device delay by an additional $C_{BC}R_C$ term. For this reason the collector resistance should be kept as small as possible. In practical devices V_{CBO} will be lower than the value in (4.22) due to junction curvature. In addition, transistor performance is usually limited by V_{CEO}, which can be much smaller than V_{CBO} depending on the current gain. Taking these two effects into consideration, the device and consequently N_{DC} and w_c should be designed for a higher V_{CBO} than the actual operation voltage. In this context another advantage of the SiGe HBT becomes clear: The fact that the collector is made of the wide-gap semiconductor (not necessarily the case in III-V-HBTs) leads to a small avalanche-multiplication factor and high V_{CEO} compared to a SiGe collector. Combining (4.22) with the expression for τ_c one can estimate the maximum possible f_T of an ideal HBT, assuming $V_{CEO} \approx V_{CBO}$ and $\tau_E = \tau_B = \tau_{RC} = 0$

$$f_{TJ} = \frac{E_{br} v_s}{2\pi V_{CBO}}. \tag{4.23}$$

This is the *Johnson* limit [4.54] which demonstrates the trade-off between operating voltage (and power, see [4.54]) and switching speed. It only depends on the material parameters of silicon. For example, at $V_{CBO} = 10$ V a maximum f_T of 50 GHz may be expected.

In BJTs at high collector current densities the Kirk effect becomes important. When the concentration of the injected electrons exceeds the collector doping, base pushout occurs, that is holes will accumulate in a part of the collector region to compensate the excess electron charge. The result is a drop off in current gain and transit frequency due to the widening of the base. An applied collector voltage V_{CB} will shift the critical current J_K

$$J_K = q v_s \left(N_{DC} + \frac{2\varepsilon(V_{CB} + V_{bi})}{q w_c^2} \right) \tag{4.24}$$

above which the Kirk effect starts [4.21] to higher values because of the additional reverse-bias field in the CB space charge region (V_{bi} is the built-in voltage). A useful application of (4.24) is a thin intrinsic collector region at the base side that will lower the maximum CB junction field for a given V_{CB} (this is favourable in order to reduce the avalanche currents [4.55,56]): Despite the low collector doping the Kirk effect is suppressed as long as w_c, which is now the thickness of the i-layer, is kept small, even at $V_{CB} = 0$ because of the V_{bi} term.

In SiGe HBTs the Kirk effect is somewhat different. The base-collector heterojunction suppresses the injection of holes into the collector. This may be an advantage in integrated circuits where the transistors are operated in saturated-mode switching [4.57] because of the reduced carrier storage. At high current densities, however, when the collector current exceeds the value J_K of (4.24) the accumulation of electrons in the collector region can no longer be compensated by holes from the widening of the base as in the case of a BJT. The result is the formation of a potential barrier that opposes the flow of electrons into the collector and increases the minority carrier storage in the base. The onset of the Kirk effect in SiGe HBTs will therefore be marked by a very rapid drop in β and f_T. A detailed analysis of this accelerated f_T rolloff can be found in [4.58,59].

In AlGaAs/GaAs-HBTs the so-called *ballistic collection transistor* [4.1] takes advantage of the velocity overshoot of the electrons injected into an intrinsic of even p-doped part of the collector to reduce τ_c. There is some experimental evidence for velocity overshoot in silicon as well [4.60]. If this effect has a positive effect on device speed at all it will not depend on the above collector structure, which in the case of GaAs is necessary to prevent the scattering of carriers into the L-valley.

4.3 Fabrication Technologies and Device Performance

The approach of the IBM researchers for their SiGe HBTs was to modify an existing polyemitter bipolar transistor process. A very limited number of fabrication steps had to be changed, and allowed quick development and the use of existing circuit designs. Figure 4.17a depicts the schematic cross section of the non-self-aligned SiGe HBT that had reached an f_T of 75 GHz in 1990 [4.32]. The corresponding SIMS profile can be seen Fig. 4.17b. The base of the device is grown at 500°C by UHV-CVD. The deposition is non-selective, i.e. the base layer extends over the recessed field oxide isolation as polysilicon, which forms part of the external base with a low parasitic capacitance. The 0.3 μm wide emitter is formed in a conventional poly-emitter process. This restricts the possibilities of the HBT design to a

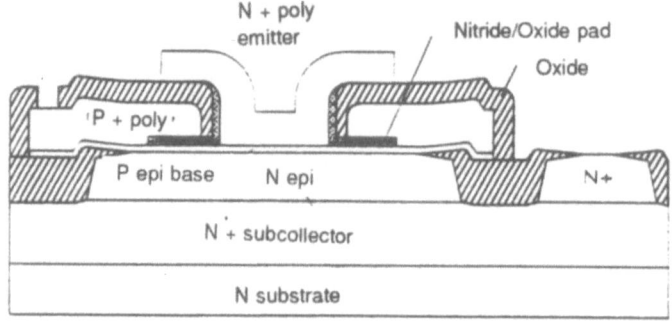

(a)

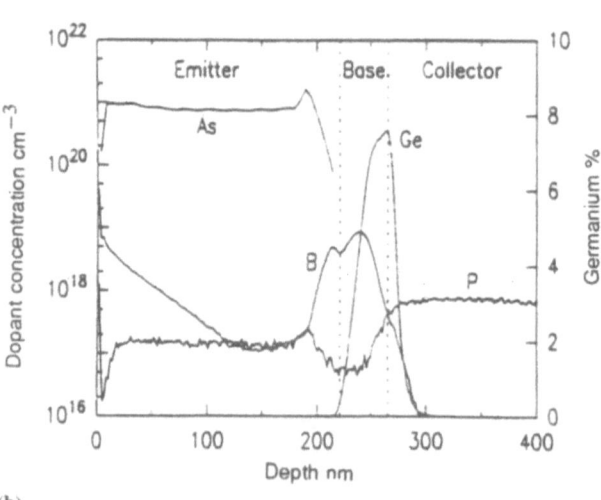

(b)

Fig. 4.17. Cross section of a SiGe HBT with poly-emitter (**a**) and the corresponding doping profile (**b**) [4.32]

typical BJT doping profile where the emitter impurity level exceeds the base doping. The necessary high-temperature poly-emitter forming step limits the allowable Ge content in the base because of the danger of strain relaxation. A triangular-graded profile with a modest 0–7% Ge was employed that adds a drift field to the base and increases device speed. The restrictions in base doping to avoid BE tunneling are the same as in BJT. Therefore the obtained base pinch resistance of 17 kΩ/sq. was rather high and limited the performance of the fabricated ECL ring oscillators [4.61]. However, an improved self-aligned technology has recently yielded ECL delays of only 18.9 ps [4.5], despite a similar base sheet resistance.

A reduction of the intrinsic base resistance calls for a smaller emitter width. A special technology called Selective Epitaxial Emitter Window (SEEW) has been developed that allows linewidths of 0.35 µm using 0.8 µm optical lithography [4.62]

A high base doping in order to obtain a low base sheet resistance is only allowed in connection with a low-doped emitter to avoid tunneling. *Comfort et al.* [4.63] introduced a single-crystal emitter cap grown by low-temperature epitaxy between base and poly-emitter. They were able to put 6×10^{19} cm^{-3} boron into the base and achieved a base sheet resistance of 4 kΩ/sq resulting in a high f_{max} of 40 GHz. However, the necessary As poly-emitter anneal cycle provoked a non-negligible base dopant diffusion. It broadened the base thickness and caused base leakage currents due to the insufficient separation between n$^+$ and p$^+$ regions. In addition, the Ge content had to be kept low, only 9% were used.

The only solution for a better performance of poly-emitter type HBTs is a reduction of the thermal budget. This was achieved using phosphorus-doped polysilicon instead of arsenic [4.64]. The much higher diffusivity of P offers the advantage of a lower thermal anneal cycle. The highest temperature was an RTA at 860°C compared to 970°C in the case of an As-doped poly-emitter. The improvements are considerable: A base thickness below 30 nm, Ge contents up to 25%, and a maximum transit frequency of $f_T = 113$ GHz [4.65]

Sato et al. [4.66] selective epitaxial growth (APCVD) to grow a single-crystalline base with 15% Ge that merges with an overhanging p$^+$-polysilicon external base layer (Fig. 4.18). Although the poly-emitter is in-situ phosphorus doped, a rather high-temperature RTA at 950°C drives the emitter in. Reduction of the extrinsic base resistance is achieved by outdiffusion from a BSG film. Both f_T and f_{max} were 50 GHz, ECL ring oscillators reached 19 ps.

A completely different approach to SiGe HBTs is to grow the entire layer structure in a CVD or MBE system without interruption. This excludes any contamination during fabrication, in contrast to the case of the above-described sequential deposition where a possible contamination will occur exactly at the EB and BC interfaces, which are most sensitive. Growth temperatures both in CVD and MBE systems are much lower than during a classical poly-emitter drive-in process. Without the latter the freedom of design for an optimized HBT largely increases. High Ge contents may be used without the fear of relaxation, and steep dopant slopes enable an ultrathin base. The price which has to be payed when starting with a complete HBT layer structure are mesa-type devices and therefore a non-planar

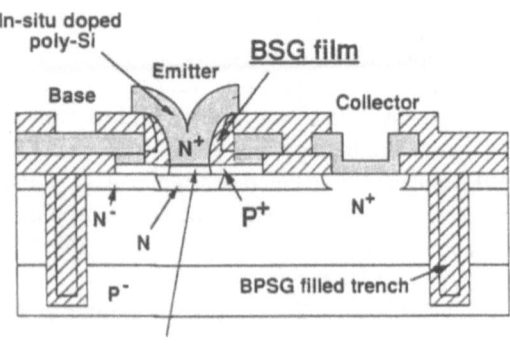

Fig. 4.18. Cross section of a SiGe HBT with selectively grown SiGe base and p-doped poly-emitter [4.66]

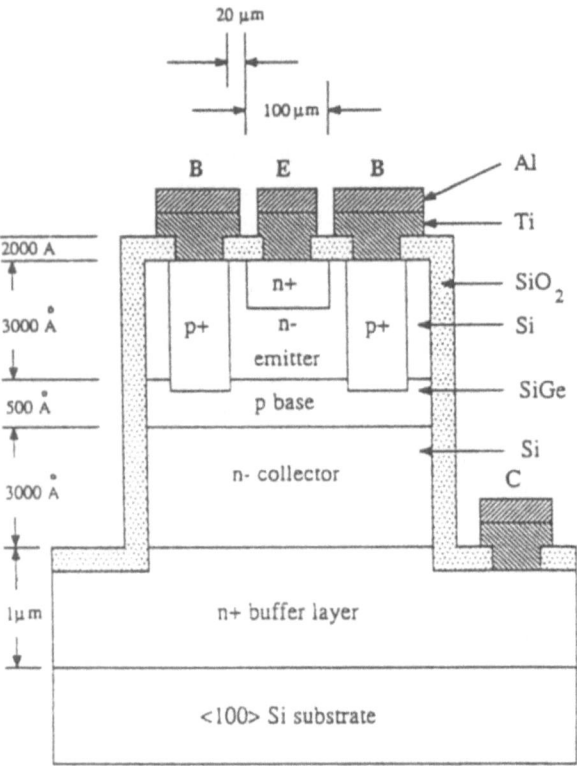

Fig. 4.19. Mesa-type SiGe HBT with ion-implanted base contacts [4.39]

surface. Figure 4.19 exhibits a cross section of the SiGe HBT from [4.37]. Several other groups [4.27,30,41] have used very similar structures. Base and emitter contacts are provided by ion implantation and after the mesa etch the whole device is passivated with a CVD oxide followed by a moderate implantation anneal for 10 min at 800°C (or RTA 825°C [4.30], 10 min. at 850°C [4.41], RTA 850°C

[4.27]). The side walls of the p^+ implantation form a parasitic BE junction. Its effects on the HBT, however, is negligible due to the larger band gap inside the silicon, and the reduced current density compared with the actual BE heterojunction. The smallest devices fabricated with this technology were $1\,\mu m \times 10\,\mu m$ emitter stripes on a mesa of about $7 \times 14\,\mu m$ [4.67] using e-beam lithography. They reached $f_T = 29$ GHz.

A mesa-type HBT usually has a collector area A_c that is larger than the emitter area A_E. This will not affect the behaviour of the transistor unless it is operated in the inverse mode. Then only part of the injected electrons will be collected by the smaller collector. A simple calculation gives the inverse current gain

$$\beta_i = \frac{1}{\dfrac{1+\beta_0}{\beta_0 A_c/A_E} - 1} \tag{4.25}$$

which will in most cases be smaller than 1 (β_0 is the inverse current gain of a device with $A_C = A_E$). If in Fig. 4.19 the base implant is deep enough to reach below the SiGe layer (and provided it extends to the mesa surface) then the active, i.e., the heterojunction emitter and collector areas become almost equal. The inverse HBT will exhibit a current gain of β_0 which usually is much larger than unity, depending on the HBT layers. This has certain advantages for integrated-circuit design [4.15].

A double-mesa process has been reported by *Schreiber* [4.68]. Emitter- and base implants are used to improve contact resistance. The anneal is carried out at 900°C in a furnace. Self-alignment between emitter and base contacts is accomplished by an oxide spacer at the edge of the emitter mesa. A transit frequency of 39 GHz has been reported.

The Daimler-Benz group [4.3,10,11,29,34] utilized a completely "cold" fabrication technology that does not even need the external base implantation and the following anneal. The MBE-grown layers see the highest temperature during their growth, which is about 550°C only. This allows extremely steep dopant gradients because diffusion becomes negligible. High base dopings of up to 10^{20} cm^{-3} may be achieved that have yielded base sheet resistances as low as 200 Ω/sq [4.10]. The danger of temperature-induced strain relaxation does not exist either. Details of the MBE system can be found in Chap. 7. The devices are of the double-mesa type and have a self-alignment between emitter and base. Figure 4.20 shows the device fabrication steps. The process starts with 4-inch substrates with the completely grown HBT layer sequence. A 300 nm thick PtAu emitter metal is defined by liftoff and acts as a mask for the following selective wet chemical etch using a modified KOH solution. It stops at the SiGe layer (a). Figure 4.21 depicts some emitter test structures after the removal of the n^- and n^+ emitter layers by the KOH etchant. The slight overetch provides a self-alignment of the following 150 nm PtAu base metallization with respect to the emitter (Fig. 4.20b). At this point devices on n^+ substrates are ready for static measurements. Substrates with buried layers are dry etched and the TiPtAu collector metallization is evaporated (Fig. 4.20c). A second dry etch forms deep trenches that separate the contact pads

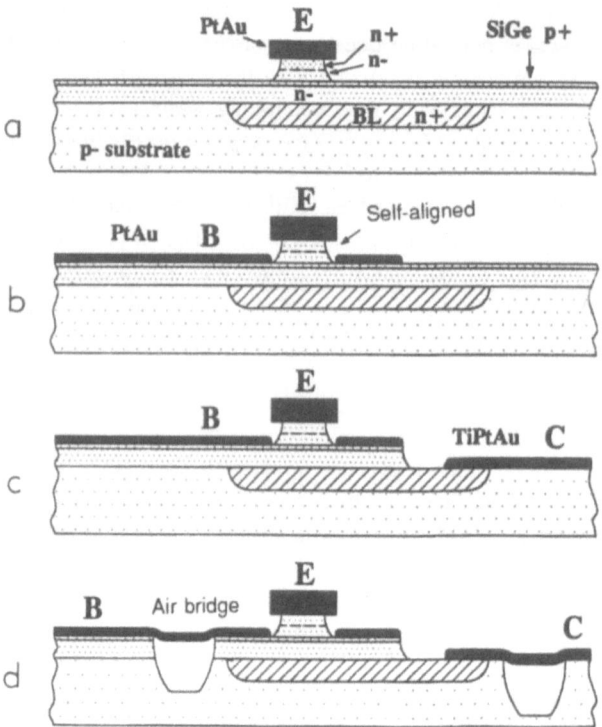

Fig. 4.20. Fabrication process of mesa-type SiGe HBTs with zero thermal budget

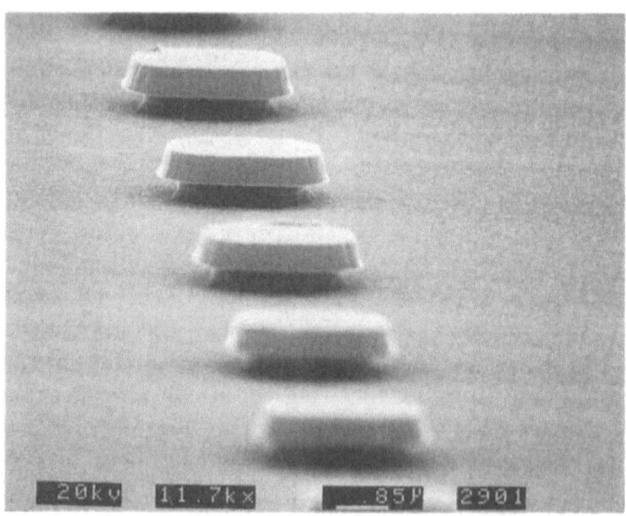

Fig. 4.21. Emitter test structures after removal of the emitter layer by a selective wet chemical etch

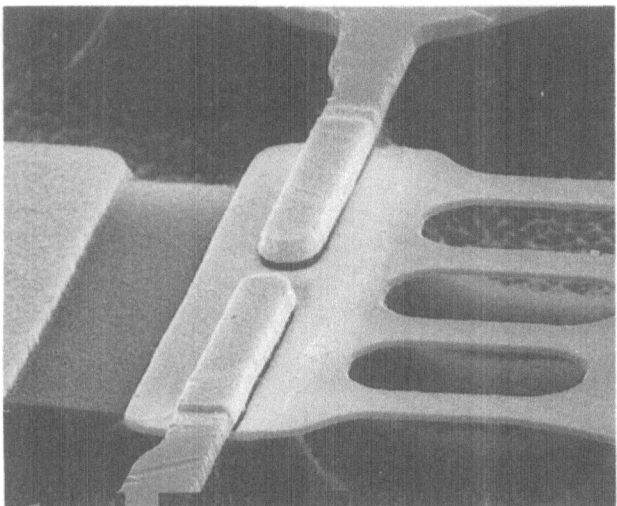

Fig. 4.22. Micrograph of a finished mesa-type SiGe HBT

from the active area leaving air bridges (d). During this step the emitter stripe and the buried layer are protected by photoresist. Figure 4.22 displays a micrograph of a finished SiGe HBT. Below the large emitter-contact pads the complete HBT layer structure remains, the base pad lies on the collector and base layers. The deep trenches and the air bridges, however, separate these large parasitic devices from the actual HBT. The high-resistivity substrate serves as an excellent isolation. Even the pad capacitances are so low ($<$ 10 fF) that no deembedding is necessary for the high-frequency measurements.

The advantage of the described process is not only the zero thermal budget but as well the simplicity and speed of the fabrication sequence (a few days only). The measured device characteristics may therefore immediately be used as a feedback for necessary changes in the layer-growth parameters. A disadvantage is the existence of open surface mesas. The metallization prohibits a high-temperature oxidation for a high-quality passivation of the BE and BC junctions. A sputtered SiO_2 or a low-temperature oxide ($< 300°C$) have little effect. The base currents are therefore not ideal and scale with periphery at low levels ($<$ 1 μA). Despite these leakages, transistors with current gains of more than 500 have been demonstrated with this technology. In addition, for high-frequency applications the transistor will be operated at high current densities where small surface leakage currents play no role. Another disadvantage of the described fabrication process is the highly non-planar surface that makes integration of the HBTs difficult.

The MBE growth allows precise control of composition and thickness of the different HBT layers, which will not be degraded by the fabrication process. The desired optimum layer parameters according to Sect. 4.2 and depending on the particular device application can therefore be grown quite accurately. As an example Fig. 4.23 shows the SIMS profile of a SiGe HBT with a reported f_T of 101 GHz

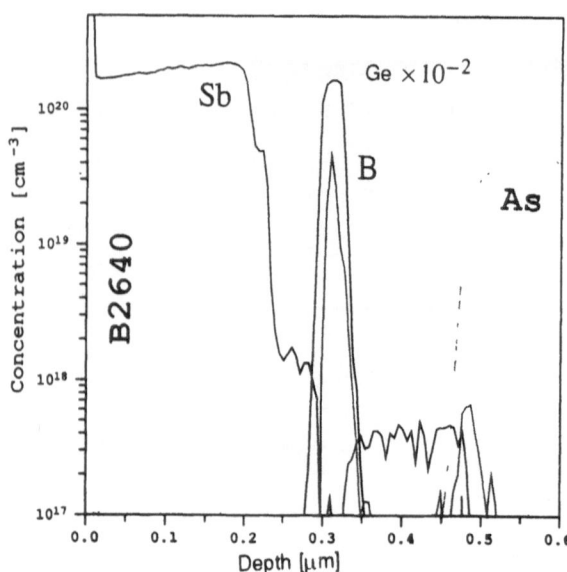

Fig. 4.23. SIMS profile of a SiGe HBT with in f_T of 101 GHz

[4.11]. The complete HBT layers were grown on 4-inch p-substrates with As-implanted and subsequently diffused buried layers with 5–10 Ω/sq. The wafers received an RCA clean and an in situ 900°C flash-off. The 150 nm thick collector was doped to $3 \cdot 10^{17}$ cm^{-3} by Sb doping using secondary implantation. The growth of the SiGe base with 28% Ge by coevaporation at 530°C starts with a 10 nm undoped spacer layer followed by 10 nm doped 3×10^{19} cm^{-3} by evaporation of elementary boron [4.69]. The boron flux is switched off shortly before the germanium flux in order to introduce a 1–2 nm thick emitter spacer layer. The 2×10^{18} cm^{-3} Sb doped emitter layer is 70 nm thick and grown by spontaneous dopant incorporation at relatively low temperatures between 425 and 450°C followed by the 230 nm thick n$^+$ emitter cap grown at 320°C to achieve 1–2×10^{20} cm^{-3} electrically active Sb concentration for a good ohmic contact.

The high-frequency behaviour of a 1 μm × 20 μm large device can be seen in Fig. 4.24. The f_{max} value of 45 GHz is remarkable, considering the fact that the base contact is on one side of the emitter finger only in order to reduce the parasitic CB area. This is the merit of the very low base sheet resistance of only 1.6 kΩ/sq. Devices with double-sided base contacts have reached an f_{max} of 65 GHz [3]. Here again the value of 2 kΩ/sq together with the self-alignment between base and emitter results in a very low base resistance and a high f_{max} according to (4.13). Unlike in the case of BJTs there is no need for submicrometer emitter widths because of the low internal base resistance of HBTs. In fact, transistors with 2 μm emitters had still more than 30 GHz f_{max}.

As explained in Sect. 4.1 a detailed analysis of the f_T against I_C curves yields the different components of the total delay τ_t according to (4.8). The above-described device had a base transit time of about $\tau_B = 0.5$ ps. The unknown minority-carrier diffusion constant makes it difficult to compare this delay with

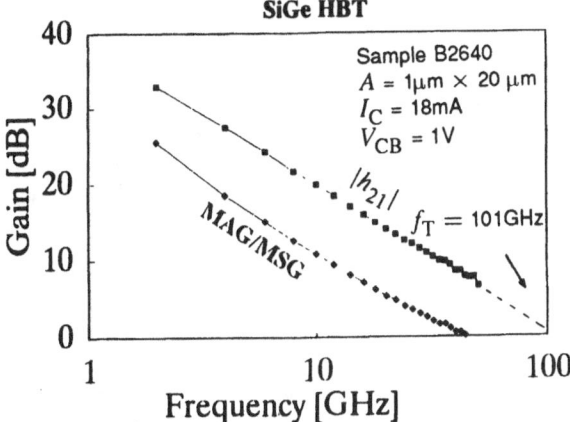

Fig. 4.24. High frequency behaviour of the SiGe HBT of Fig. 23

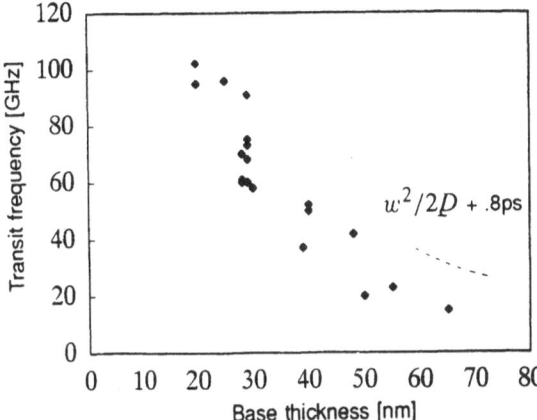

Fig. 4.25. Maximum obtained transit frequency of SiGe HBTs with different base thicknesses

the pure diffusion value of (4.10). No information about accelerating or delaying fields in the base is gained. Figure 4.25 displays a plot of the maximum-obtained transit frequencies of recently fabricated HBT samples against their base thickness. Emitter and collector parameters of all samples were very similar. Although there is considerable spread in the experimental data, the overall tendency follows the dotted line which is $f_T = 1/2\pi\tau_{EC}$ with $\tau_{EC} = w_B^2/2D_n + 0.8$ ps and an assumed $D_n = 5$ cm^2 s^{-1}. About half of the devices shown in Fig. 4.25 had Ge and dopant gradients, however no correlation with f_T has been found. With other words, accelerating and delaying fields seem to cancel each other.

It is interesting to compare the different deposition methods that have been employed for the growth of the SiGe HBT layers. The IBM group has used a UHV-CVD reactor [4.70] for most of their devices. The ultrahigh-vacuum conditions after loading the wafers are necessary to reduce oxygen and moisture partial pressures below 10^{-8} Torr in order not to oxidize the wafers. Epitaxial growth takes place at temperatures between 550°C and 800°C at pressures in the milli-Torr range. The hot-wall reactor allows batch processing, however the large thermal mass requires several hours for cooling. So far only the p-doped base layers have been grown. In the case of a complete HBT layer growth autodoping might occur. A costly multichamber reactor could solve these problems.

Low-Pressure CVD (LPCVD) has widely been used to grow single SiGe layers or complete HBT sequences [4.39,67,71]. Deposition of SiGe was done at 640°C at pressures of a few Torr. The wafers are heated directly by tungsten lamps which enables fast temperature gradients. The epitaxial growth may therefore be controlled by quick changes in temperature, not only by variations of gas composition. This has been called Limited Reaction Processing (LRP) [4.27]. Temperature uniformity across the wafer is a problem which can be circumvented with the help of a susceptor. Its thermal mass, however, slows down the process and outdiffusion from the surface may cause autodoping.

Atmospheric Pressure CVD (APCVD) is very similar except that there is no need for vacuum pumps. The price is a high gas consumption and expensive moisture filters. These are necessary because otherwise the partial pressures of residual O_2 and H_2O in the process gases would be too high when operating at atmospheric pressure. First results of grown SiGe layers [4.72] indicate that APCVD can meet the demands for the growth of complete SiGe HBT layers.

All the above-described CVD growth methods offer the possibility of selective growth, i.e. there is no material deposition on oxide-covered areas. MBE growth, on the other hand, is always non-selective. MBE provides excellent layer thickness and doping control. Complete SiGe HBT layer sequences have yielded devices which are among the fasted reported until now [4.3,30,41].

When comparing the throughput of CVD and MBE it should be kept in mind that the growth rates of the critical HBT layers are very similar in all systems (2–20 nm/min). Except for the hot-wall UHV-CVD all methods are single-wafer processes, i.e., a close stacking of the wafers in a batch is impossible. An issue of increasing importance is the environmental aspect: MBE deposition is the only method that does not use and exhaust toxic gases.

4.4 Applications of SiGe HBTs

The fastest silicon integrated-circuit technology is the Emitter-Coupled Logic (ECL) that uses bipolar transistors in non-saturation mode. A figure of merit is the propagation delay τ_d of a single logic gate measured in a ring oscillator configuration. Minimum τ_d values obtained with silicon BJTs were around 20 ps [4.73] using transistors with an f_T of 40 GHz. (A special circuit with an coupled active pulldown even reached 13 ps [4.74]). A higher transit frequency will only have a small influence on the switching speed because τ_d is a weak function of f_T [4.15,18]. The base resistance R_B, however, plays an important role and for this reason SiGe HBTs with low R_B could significantly reduce τ_D in ECL circuits. First simulations predict values of 10 ps [4.16]. So far reported experimental results of 18.9 ps [4.5] and 19 ps [4.66] from ring oscillators with SiGe HBTs hardly exceeded the performance of BJTs. The reason is that both designs used polysilicon emitter HBTs with high sheet resistances above 10 kΩ/sq. Until now no experimental results exist for ECL circuits using low-resistance base HBTs.

An important advantage of SiGe HBTs is that because of the low-base-resistance SiGe HBTs do not need to be scaled down to submicrometer emitter widths in order to achieve high speed. Note that the above-mentioned 20 ps Si BJT ring oscillators of [4.73] had 0.25 μm-sized emitters. For comparison *Shafi et al.* [4.16] predicted 15 ps gate delay for a 1 μm design using SiGe HBTs.

Other bipolar integrated-circuit families that employ a saturating inverter (TTL, DTL, I^2L) will also benefit from SiGe HBTs. The reduced minority carrier injection into the collector during saturation mode will reduce the storage time. Another speed advantage comes from the reduced EB capacitance of HBTs (unlike in ECL circuits, where it has negligible influence on τ_d).

An interesting application of SiGe heterostructures is the double-base HBT [4.75], which has two doped SiGe base layers separated by a 20 nm thin undoped silicon layer. The transistor current may be controlled independently by the two base electrodes thereby forming a single-transistor NAND gate. Operation at 77 K has been demonstrated experimentally [4.75].

In analogue integrated circuits the possible high current gain of SiGe HBTs may improve circuit design and performance. Exactly balanced current mirrors can be built with only two transistors leading to less components on the chip and therefore less area. The same is true for the output conductance of a current source which depends on the current gain – Early voltage product of a transistor. Extremely high values may be obtained with HBTs without speed sacrifices (Sect. 4.2). Employing HBTs with high β values in input stages will, in addition, reduce input offset currents.

One of the most important applications of SiGe HBTs will be low-noise amplifiers. The high-frequency noise of a transistor is mainly determined by its base resistance R_B, the base transit time τ_B and the BE capacitance C_{BE} according to the theory of *Hawkins* [4.76]. All three parameters can be made very small in an HBT leading to low-noise figures. Figure 4.26 shows the noise-equivalent circuit and Fig. 4.27 the measured noise figures of a SiGe HBT with an emitter size of 0.8 μm × 16 μm and a sheet resistance of only 700 Ω/sq. The obtained values of 0.5 dB at 2 GHz and 0.9 dB at 10 GHz exceed the performance of III/V HBTs. The agreement with *Hawkins'* theory, represented by the line in Fig. 4.27, is excellent. Further improvement is possible as can be seen in Fig. 4.28 which shows the calculated minimum achievable noise figures of an optimized SiGe HBT. At 10 GHz values below 0.6 dB appear feasible, which is state-of-the-art of III-V HEMTs.

The low-frequency noise behaviour of bipolar transistors is governed by recombination currents inside the space-charge layers and at the periphery. The noise level rises above the thermal noise (i.e., high frequency noise) when going to low frequencies. The slope is usually proportional to l/f. The frequency where thermal and l/f-noise are equal is called the *corner frequency*. It is used as a figure of merit. In principle, there should be no fundamental difference between BJTs and

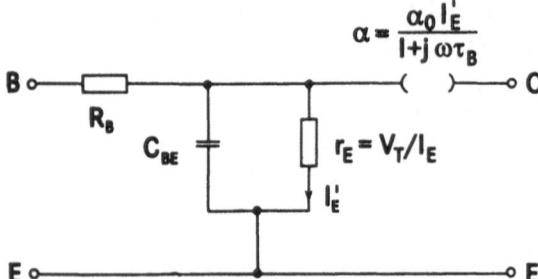

Fig. 4.26. Noise equivalent circuit of a SiGe HBT

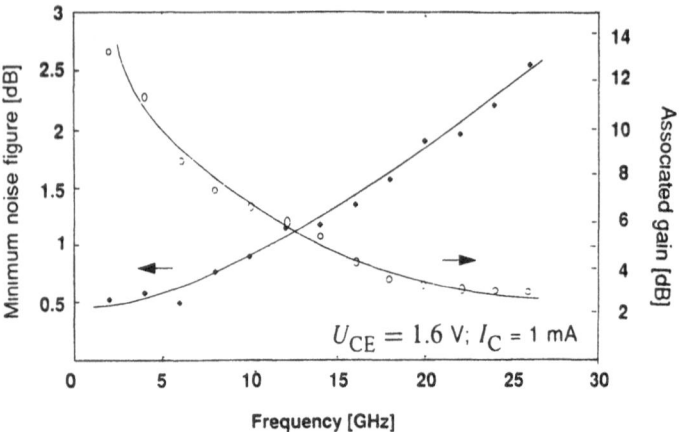

Fig. 4.27. Noise figure and associated gain of a SiGe HBT with a base sheet resistance of 700 Ω/sq. (Si/SiGe: F_{min} and G_a vs. frequency)

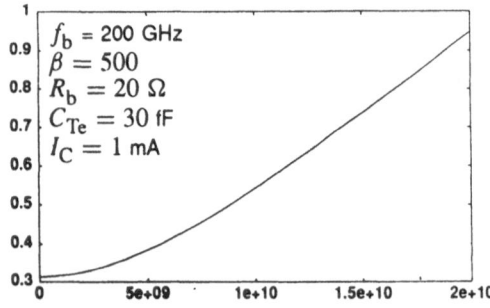

Fig. 4.28. Calculated minimum achievable noise figures for SiGe HBTs

HBTs concerning the $1/f$ noise sources. They rather depend on the individual fabrication technology. For example, the polysilicon-silicon interface in the case of a poly-emitter is known to be a possible noise source. Mesa-type transistor structures with a poor-quality passivation will suffer from surface recombination noise. State-of-the-art corner frequencies of silicon BJTs are between 1 Hz and 10 kHz.

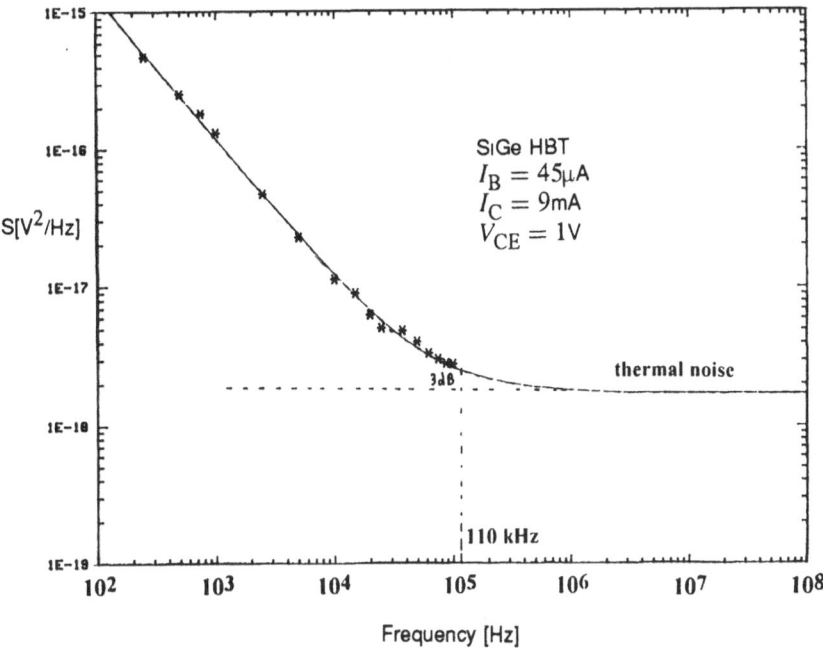

Fig. 4.29. $1/f$ – noise of a SiGe HBT

First measurements on SiGe HBTs [4.77] revealed a corner frequency of 100 kHz, as shown in Fig. 4.29. Considering the fact that these devices were unpassivated mesa structures [4.3] without a high-quality passivation oxide these first results are satisfactory. A comparison with III-V microwave transistors reveals that this figure is among the lowest-reported corner frequencies. Unfortunately, no noise results are available from the poly-emitter SiGe HBTs [4.32,65] which might have even better results due to their high-temperature oxide passivation.

Why is the low-frequency behaviour important for devices that operate at microwave frequencies? The reason is intermodulation between the high-frequency signal and the $1/f$ noise. This will produce phase noise in the case of oscillators which are particularly sensitive because they should deliver extremely stable reference frequencies in both transmitter and receiver applications. A 9.6 GHz hybrid dielectric resonator oscillator with SiGe HBTs has been reported [4.78] with −85 dB noise at 100 kHz off carrier (see also Chap. 5). More recently a hybrid 24 GHz oscillator with −92 dB noise has been integrated on a silicon substrate by the same group. These noise levels are similar to the performance of III-V devices. Again, it should be mentioned that the SiGe HBTs used were unpassivated mesa structures with expected improvement in the case of a high-quality oxide passivation.

As a conclusion, the SiGe HBT offers excellent microwave noise behaviour largely exceeding the performance of BJTs. The technology-dependent $1/f$ noise may reach very low levels due to the possibility of a high-quality oxide passivation, impossible for III-V devices. In addition, the integration of SiGe HBTs into existing circuit technologies will lead to one-chip solutions (e.g., for mobile communication systems) that contain low-noise microwave pre-amplifiers, highly stable oscillators, the signal-processing part and even the power output amplifier. The last point, power applications, has already been addressed in Sect. 4.23. The advantage of SiGe HBTs over III-V devices is the 2 to 3-fold higher thermal conductivity of the substrate that may compensate for the smaller E_{br} of silicon that determines the Johnson limit in (4.23). In fact, unequal heat distribution in GaAs HBTs demands costly technological solutions [4.79]. So far, no experimental results exist for power SiGe HBTs.

Finally, there is some interest in liquid-nitrogen-temperature operation of integrated circuits for computer applications [4.80]. SiGe HBTs offer high current gain at low temperature and the high base doping prevents carrier freeze-out as in the case of BJTs.

4.5 Conclusion

By combining SiGe and Si the heterojunction has been introduced into the silicon system and it is no longer restricted to III-V semiconductors. One of its most important applications is the SiGe HBT, which offers the possibility of a thin and highly doped base leading to fast transit times and low base resistances. After only a few years of research these devices now outperform classical silicon bipolar transistors in gain, noise level and transit frequency, the latter exceeding 100 GHz.

The growth of the thin SiGe base layers has matured and problems as strain relaxation and dopant profile control are mastered both with MBE and CVD growth methods. However, the sensitivity of the pseudomorphic SiGe layers to high temperatures requires new fabrication technologies, differing from the processes presently employed. Several new fabrication schemes have been reported, but all of them so far either restrict the possible range of HBT layer composition (and therefore device performance), or the process prohibits device integration. Further development of fabrication technologies is required in order to fully exploit the advantages of SiGe HBTs.

The application of SiGe HBTs in digital integrated circuits will only slightly increase switching speed, which is mainly determined by the interconnect capacitances. There may be some power advantage if HBTs are used. However, in this case the size of the HBTs must not be larger than the one of today's BJTs. Any increase of circuit performance always has to be weighed against the additional complexity of the fabrication technology.

The most important application of SiGe HBTs will be high frequency, low-noise amplifiers taking advantage of the high f_{max} and low base resistance. The integration of these circuits together with a digital signal processing part is particularly interesting for one-chip mobile communication devices.

References

Section 4.0

4.1   T. Ishibashi, O. Nakajima, J. Nagata, Y. Yamauchi, H. Ito: Ultra-high-speed AlGaAs/GaAs HBTs. IEDM *88*, 826–829 (1988)

4.2   W.J. Ho, N.L. Wang, M.F. Chang, A. Sailer, J.A. Higgins: Self-alignment AlGaAs/GaAs HBT with extrapolated maximum oscillation frequency of 350 GHz. DRC '92, IVA-1 (1992)

4.3   A. Gruhle, H. Kibbel, U. Erben, E. Kasper: 91 GHz SiGe HBTs grown by MBE. Electr. Lett. *29*, 415–416 (1993)

4.4   H. Schumacher, U. Erben, A. Gruhle: Noise characterisation of SiGe HBTs at microwave frequencies. Electr. Lett. *28*, 1167–1168 (1992)

4.5   D.L. Harame, E.F. Crabbé, J.D. Cressler, J.H. Comfort: A high performance epitaxial SiGe-base ECL BICMOS technology. IEDM *92*, 19–22 (1992)

4.6   S.M. Sze, H.K. Gummel: Appraisal of semiconductor-metal-semiconductor transistor. Sol.-St. Electr. 9, 751–769 (1966)

4.7   J.M. Shannon, A. Gill: High current gain in monolithic hot-electron transistors. Electron. Lett. *17*, 620–621 (1981)

4.8   R.C. Taft, J.D. Plummer, S.S. Iyer: Demonstration of a p-channel BICFET in the GeSi/Si system. Electr. Dev. Lett. *10*, 14–16 (1989)

Section 4.1

4.9   P.M. Asbeck: Bipolar transistors in *High-Speed Semiconductor Devices*, ed. by S.M. Sze, (J Wiley, New York 1990)

4.10 A. Gruhle, H. Kibbel, U. König, U. Erben, E. Kasper: MBE grown Si/SiGe HBTs with high β, f_T and f_{max}. EDL-*13*, 206–208 (1992)

4.11 A. Gruhle, H. Kibbel, U. Erben. E. Kasper: Base thickness and high frequency performance of SiGe HBTs. Dev. Res. Conf. IIA-2 (1993)

4.12 H. Kroemer: Two integral relations pertaining to the electron transport through a bipolar transistor with a nonuniform energy gap in the base region. Sol.-St. Electr. *28*, 1101–1103 (1985)

4.13 E.J. Prinz, J.C. Sturm: Analytical modeling of current gain – Early voltage products in Si/SiGe/Si HBTs. IEDM *91*, 853–856 (1991)

4.14 T.F. Meister: Vertical scaling considerations for polysilicon emitter bipolar transistors, in *Ultra Silicon Bipolar Technology*, ed. by L. Treitinger, M. Miura, Springer Ser. Electron. Photon., Vol. 27 (Springer, Berlin, Heidelberg 1988) pp. 57–58

4.15 H. Kroemer: "HBTs and integrated circuits. Proc. IEEE, *70*, 13–25 (1982)

4.16 Z.A. Shafi, P. Ashburn, G.J. Parker: Predicted propagation delay of Si/SiGe heterojunction bipolar ECL circuits. IEEE J. Sol. St. Circ. *25*, 1268–1276 (1990)

4.17 M.Y. Ghannam, R.P. Mertens, R.v. Overstraeten: An analytical model for the determination of the transient response of CML and ECL gates. IEEE Trans. ED- *37*, 191–201 (1990)

4.18 P.K. Tien: Propagation delay in high speed silicon bipolar and GaAs HBT digital circuits. Int'l J. High Speed Electr., *1*, 101–124 (1990)

4.19 R. People, J.C. Bean: Calculation of critical layer thickness versus mismatch for GeSi/Si strained-layer heterostructures. Appl. Phys. Lett. *49*, 229 (1986)

4.20 R. People, J.C. Bean: Band alignments of coherently strained SiGe heterostructures on GeSi substrates, Appl. Phys. Lett. *48*, 538–540 (1986)

Section 4.2

4.21 D.J. Roulston: *Bipolar Semiconductor Devices* (McGraw-Hill, New York 1990), pp. 225–226

4.22 Z.A. Shafi, P. Ashburn: Silicon-based pseudo-heterojunction bipolar transistors. J. Sol. St. Electr. *68*, 72–74 (December 1990)

4.23 J. Stork, R.D. Isaac: Tunneling in base-emitter junctions, IEEE Trans. ED-*30*, 1527–1534 (1989)

4.24 Z. Matutinovic-Krstelj, E.J. Prinz, P.V. Schwartz, J.C. Sturm: Reduction of p^+n^+ junction tunneling current for base current improvement in SiGe HBTs, IEEE EDL. *12*, 163–165 (1991)

4.25 J.C. Sturm, E.J. Prinz, C.W. Magee: Graded-base SiGe HBTs grown by RTCVD with near-ideal electrical characteristics. IEEE EDL-*12*, 303–305 (1991)

4.26 H.-U. Schreiber, B.G. Bosch: SiGe HBTs with current gains up to 5000. IEDM *89*, 643–646 (1989)

4.27 C.A. King, J.L. Hoyt, J.F. Gibbons: Bandgap and transport properties of SiGe by analysis of nearly ideal Si/SiGe/Si HBTs. IEEE Trans. HED-*36*, 2093–2104 (1989)

4.28 A. Pruijmboom, C.E. Timmering, J.v. Rooij-Mulder, D.J. Gravesteijn, W.B. de Boer. Microelectr. Eng. *19*, 427–434 (1992)

4.29 A. Gruhle: High performance Si/SiGe HBTs grown by MBE. J. Vac. Sci. Techn. B*11*, 1186–1189 (1993)

4.30 Z.A. Shafi, C.J. Gibbings, P. Ashburn, I.R.C. Post: The importance of neutral base recombination in compromising the gain of SiGe HBTs. IEEE Trans. ED-*38*, 1973–1976 (1991)

4.31 R. Hull, J.C. Bean: Thermal stability of Si/GeSi/Si heterostructures. Appl. Phys. Lett. *55*, 1900–1902 (1989)

4.32 G.L. Patton, J.H. Comfort, B.S. Meyerson: 75 GHz f_T SiGe HBTs, IEEE EDL-*11*, 171–173 (1990)

4.33 D.B.M. Klaassen, J.W. Slootboom, H.C. de Graaff: Unified apparent BGN in n- and p- type silicon. Solid State Electr. *35*, 125–129 (1992)

4.34 A. Gruhle, H. Kibbel, E. Kasper: The influence of MBE layer design on the high frequency performance of Si/SiGe HBTs. Microelectr. Eng. *19*, 435–437 (1992)

4.35 G. Gao, H. Morkoc: Base transit time for SiGe-base HBTs. Electr. Lett. *27*, 1408–1410 (1991)

4.36 D.V. Lang, R. People, J.C. Bean, A.M. Sergent: Measurement of the band gap of SiGe strained-layer heterostructures. Appl. Phys. Lett. *47*, 1333–1335 (1985)

4.37 K. Nauka, T.I. Kamins, J.E. Turner, C.A. King, J.L. Hoyt: Admittance spectroscopy measurements of band offsets in SiGe heterostructures. Appl. Phys. Lett.-*60*, 195–197 (1992)

4.38 S.C. Jain, D.J. Roulston: A simple expression for BGN in heavily doped Si, Ge, GaAs and SiGe strained layers. Sol.-Stat. Electr. *34*, 453–465 (1991)

4.39 E.J. Prinz, P.M. Garone, P.V. Schwartz, X. Xiao, J.C. Sturm: The effect of BE spacers and strain-dependent densities of states in SiGe HBTs. IEDM *89*, 639–642, (1989)

4.40 J. Poortmans, S.C. Jain, M. Caymax, A. van Ammel: Evidence of the influence of heavy-doping BGN on the collector current of strained SiGe-base HBTs. Micro-electr. Eng. *19*, 443–446 (1992)

4.41 A. Pruijmboom, J.W. Slotboom, D.J. Gravesteijn: Heterojunction bipolar transistors with SiGe base grown by MBE. IEEE Trans. EDL-*12*, 357–359 (1991)

4.42 T. Manku, A. Nathan: Effective mass for strained p-type SiGe, J. Appl. Phys. *69*, 8414–8416 (1991)

4.43 V. Grivickas, V. Netiksis, D. Noreika, M. Petranskas, W. Willander: Ambipolar diffusion is strained $Si_{1-x}Ge_x$ (100) layers grown by molecular beam epitaxy. J. Appl. Phys. *70*, 1471–1474 (1991)

4.44 K. Misiakos, C.H. Wang, A. Neugroschel: Method for simultaneous measurement of diffusivity, lifetime and diffusion length with application to heavily doped silicon. IEEE Trans. EDL-*10*, 111–113 (1989)

4.45 S.E. Swirhun, Y.H. Kwark, R.H. Swanson: Measurement of electron lifetime, mobility and BGN in heavily doped p-type silicon. IEDM *86*, 24–27 (1986)

4.46 D.B.M. Klaassen: A unified mobility for device simulation. Sol. State Electr. *35*, 961–967 (1992)

4.47 F. Schäffler, H.-J. Herzog, H. Jorke, E. Kasper: Influence of thermal annealing on the electron mobility in modulation doped SiGe heterostructures. J. Vac. Sci. Techn. B *9*, 2039–2044 (1991)

4.48 J.W. Slotboom, G. Streutker, A. Pruijmboom, D. Gravesteijn: Parasitic Energy barriers in SiGe HBTs. IEEE Trans. EDL-*12*, 486–488 (1991)

4.49 E.J. Prinz, P.M. Garone, P.V. Schwartz, X. Xiao, J.C. Sturm: The Effects of base dopant outdiffusion and undoped SiGe junction spacer layers in SiGe HBTs. IEEE Trans. EDL-*12*, 42–44 (1991)

4.50 B. Mazhari, G.B. Gao, H. Morkoc: Sol.-St. Electr. *34*, 315–321 (1991)

4.51 D.L. Harame, J. Stork, B.S. Meyerson: 30 GHz polysilicon-emitter and single-crystal-emitter graded SiGe-base PNP transistors IEDM *90*, 33–36 (1990)

4.52 A. Gruhle: The influence of EB-junction design on collector saturation current, ideality factor, Early voltage and device switching speed of SiGe HBTs IEEE Trans. ED-41, 198–203 (1993)

4.53 H. Jorke: Injection current in a planar doped base Si bipolar junction transistor, Sol.-St. Electr. *36*, 975–979 (1993)

4.54 E.O. Johnson: Physical limitations on frequency and power parameters of transistors RCA Rev. *26*, 163–177 (1965)

4.55 D.D. Tang, P. Lu: A reduced field design concept for high performance bipolar transistors IEEE Trans. EDL-*10*, 67–69 (1989)

4.56 P. Lu, J.H. Comfort, D. Tang, B.S. Meyerson, J. Sum: The implementation of a reduced-field profile design for high performance bipolar transistors IEEE Trans. EDL-*11*, 336–338 (1990)

4.57 M. Ugajin, Y. Amemiya: The base collector heterojunction effect in SiGe-base bipolar transistors Sol.-St. Electr. *34*, 393–598 (1991)

4.58 G. Gao, Z. Fan, H. Morkoc: Analysis of cut-off frequency roll-off at high currents in SiGe double-heterojunction bipolar transistors Appl. Lett. *58*, 2951–53 (1991)

4.59 P. Cottrell, Z. Yu: Velocity saturation in the collector of Si/GeSi/Si HBTs, IEEE Trans. EDL *11*, 431–33 (1990)

4.60 G.A. Sai-Halasz, M.R. Wordeman, D.P. Kern, S. Rishton, E. Ganin: High transconductance and velocity overshoot in NMOS devices at the 0.1 μ m - gate-length level. IEEE Trans. EDL-*9*, 464–466 (1988)

Section 4.3

4.61 J.H. Comfort, G.L. Patton, J.D. Cressler, W. Lee, E.F. Crabbé: Profile leverage in a selfaligned epitaxial Si or SiGe base bipolar technology. IEDM (1990) Tech. Digest, 21–4,

4.62 J.N. Burghartz, J.H. Comfort, G.L. Patton, B.S. Meyerson, J.Y. -C. Sun: Selfaligned SiGe base heterojunction bipolar transistor by selective epitaxy emitter window (SEEW) technology IEEE Trans. EDL-*11*, 288–90 (1990)

4.63 J.H. Comfort, E.F. Crabbé, J.D. Cressler, W. Lee, J.Y. -C. Sun: Single crystal emitter cap for epitaxial Si and SiGe base transistors IEDM (1991) Tech. Digest, 857–60,

4.64 E.F. Crabbé, J.H. Comfort, W. Lee, J.D. Cressler, B.S. Meyerson: 73 GHz selfaligned SiGe base bipolar transistors with phosphorus doped polysilicon emitters, IEEE Trans. EDL-*13*, 259–61 (1992)

4.65 E. Crabbé, B. Meyerson, D. Harame, J. Stork, A. Megdavis: 113 GHz f_T graded-base SiGe HBTs Dev. Res. Conf. (1993) IIA-3

4.66 S. Sato T. Hahimoto, T. Tatsumi, H. Kitahata, T. Tashiro: Sub 20 ps ECL circuits with 50 GHz f_{max} self-aligned SiGe HBTs. IEEE IEDM (1992) Tech. Digest, pp. 397–400

4.67   T.I. Kamins, K. Nauka, L.H. Camnitz, J.L. Hoyt, C.A. King: High frequency SiGe HBTs.
       IEEE IEDM-(1989) Tech. Digest. pp. 647–650

4.68   H. Schreiber: High-Speed double mesa Si/SiGe HBT fabricated by selfalignment technology.
       Electr. Lett. 28, 485–487 (1992)

4.69   H. Kibbel, E. Kasper, P. Narozny, H.-U. Schreiber: Boron doping of SiGe base of
       heterobipolar transistors Thin Sol. Films, 184, 163 (1990)

4.70   B.S. Meyerson: Low-temperature silicon epitaxy by ultrahigh vacuum CVD. Appl. Phys. Lett.
       48, 797–799 (1986)

4.71   D. Dutartre, P. Warren: Low Temperature Si and SiGe epitaxy by RTCVD in the system SiH_4,
       B_2H_6 and H_2. AVS Fall Meeting (1992) TC2-TuM4

4.72   T.O. Sedgwick, V.P. Kesan, P.D. Agnello: Characterization of devices fabricated in films
       grown at LT by APCVD. IEDM 91, 451–454 (1989)

Section 4.4

4.73   J.D. Cressler, J. Warnock, P.J. Coane, K.N. Chiong, M.E. Rothwell: A scaled 0.25 μ m
       bipolar technology using full e-beam lithography. IEEE Trans. EDL-13, 262–264 (1992)

4.74   K. Toh, J.D. Warnock, J.D. Cressler, K.A. Jenkins: Sub-15 ps gate delay with new coupled
       active pull-down ECL circuit. IEEE Trans BCTM-33 Vol. 136–138, (1991)

4.75   E.J. Prinz, X. Xiao, P.V. Schwartz, J.C. Sturm: A novel double base HBT for low temperature
       bipolar logic. Proc. DRC (1992) IIA-2

4.76   R.J. Hawkins: Limitations of Nielsen's and related noise equations applied to microwave
       bipolar transistors, and a new expression for the frequency and current dependent noise figure.
       Solid-State Electron 20, 191–196 (1977)

4.77   R. Plana, H. Kibbel, A. Gruhle, L. Escotte, J.P. Roux, J. Graffeuil: Low-frequency noise and
       microwave noise parameters in Si/SiGe HBTs. Proc. ESSDERC (1993) p. 93

4.78   U. Güttich, A. Gruhle, J. -F. Luy: A SiGe HBT dielectric resonator stabilized microstrip
       oscillator at X-band frequencies. Micr. Guid. Wave Lett. 2, 281–283 (1992)

4.79   B. Bayraktaroglu, J. Barrette, R. Fitch, L. Kehias: Thermally stable AlGaAs/GaAs microwave
       power HBTs. Proc. Dev. Res. Conf. (1993) IIIA-5

4.80   J.D. Cressler, E.F. Crabbé, J. Comfort, J. Warnock, K. Jenkins: Profile scaling constraints for
       ion-implanted and epitaxial bipolar technology designed for 77 K operation. IEDM (1991)
       pp. 861–864

5 Silicon Millimeter-Wave Integrated Circuits

J. Buechler

Daimler Benz Research, Wilhelm Runge Str. 11, 89081 Ulm, Germany

The availability of low-cost and reproducible millimeter-wave circuits is of great importance for systems in the future. A promising way to achieve these requirements are integrated circuits based on silicon substrate, the so-called SIMMWICs (Silicon MilliMeter Wave Integrated Circuits).

Due to the short wavelength in the mm-wave region the integration of passive line components, circuits with active elements such as oscillators or mixers and also the integration of small-area planar antenna structures is possible on a single chip. These mm-wave chips are very well suitable for many applications which need to transmit and receive electromagnetic radiation. With a semiconductor material like silicon as the substrate for planar circuits also monolithic integration of active devices is possible. First SIMMWICs were described by *Stabile* and *Rosen* [5.1].

In this chapter hybrid integrated SIMMWICs as well as monolithically integrated SIMMWICs are discussed. After a brief description of the properties of silicon for planar waveguides SIMMWIC sources with IMPact Avalanche Transit–Time (IMPATT) diodes and a Hetero junction Bipolar Transistor (HBT) oscillator are presented. Merging active devices and planar antenna structures yields to single chip SIMMWIC transmitters and receivers described afterwards.

5.1 Silicon as the Substrate Material

5.1.1 Silicon-Substrate Waveguide Parameters

For the behaviour of planar waveguide structures described in Chap. 1 the most important parameters of the substrate material are the permittivity ϵ_r and the loss tangent $\tan \delta$. The measured values for silicon with 10000 Ωcm resistivity are [5.2]

$$\epsilon_r = 11.68(\pm 0.7\%), \quad \tan \delta = 1.3 \times 10^{-3}(\pm 30\%).$$

With this substrate the measured attenuation of 50 Ω microstrip lines is 0.6 dB/cm. For substrate samples with a MBE (Molecular Beam Epitay) layer of 0.1 μm thickness and a doping concentration of 2×10^{16} cm^{-3} the loss tangent increases to 1.8 $\times 10^{-3}$ corresponding to an increase of the microstrip line attenuation of 1.8 dB/cm for silicon substrate with a 1.5 μm MBE layer. After the MBE layer is removed by etching the attenuation is the same as for the unprocessed substrate. Also other properties of silicon do not change throughout various processing steps, as described in

Springer Series in Electronics and Photonics, Vol. 32
Silicon-Based Millimeter-Wave Devices, Eds.: Luy et al.
© Springer-Verlag Berlin Heidelberg 1994

Chap. 7. Thus the prerequisites for monolithic integration of active devices together with passive waveguide structures are fulfilled.

5.1.2 Surface Waves

Printed structures with discontinuities excite surface waves which propagate along the dielectric substrate. Beside the higher-order modes, which may be excited in microstrip circuits depending on the linewidth to wavelength ratio, these surface-wave modes are of great importance for the design of planar circuits. With a given permittivity of the substrate, special care has to be taken to the substrate height to avoid high surface-wave losses particularly if radiating elements are included in the circuit. For finite-sized substrates the surface waves are diffracted from the substrate edges, possibly causing undesirable effects on side-lobe level, polarization, or main beam shape [5.3].

Dependent on the substrate thickness to wavelength ratio two types of surface-wave modes may be possible, Transverse Magnetic (TM) and Transverse Electric (TE). The cut-off frequencies of the N-th TM and TE surface-wave modes for a grounded substrate are given by

$$f_{cTM} = \frac{c_0 N}{2h\sqrt{\epsilon_r - 1}}, \quad f_{cTE} = \frac{c_0(2N - 1)}{4h\sqrt{\epsilon_r - 1}}, \quad N = 0, 1, 2, \dots \quad (5.1a,b)$$

where c_0 is the velocity of light, and h is the substrate thickness. For a parallel-plate waveguide the cut-off frequencies of the N-th TM and TE modes are given by

$$f_c = \frac{c_0 N}{2h\sqrt{\epsilon_r}}, \quad N = 0, 1, 2, \dots \quad (5.2)$$

The fundamental TM mode has a zero cut-off frequency and therefore is always present in the substrate, independent from the substrate thickness and permittivity. In Fig. 5.1 the mode spectrum of a grounded silicon substrate and a parallel-plate waveguide depending on the substrate thickness is depicted. As the substrate becomes thicker, more surface-wave modes can exist and the coupling to the lower-order modes can become stronger.

The loss is proportional to $1/h_e$ where h_e for the parallel-plate waveguide is given by

$$h_e = 2h \quad \text{TEM mode,} \quad (5.3a)$$
$$h_e = h \quad \text{all other modes} \quad (5.3b)$$

and for the grounded substrate due to surface-wave modes by

$$h_{eTE} = h + \frac{1}{p} \quad (5.4a)$$

$$h_{eTM} = h + \frac{1}{p}\frac{1}{1 + p^2 - q^2/\epsilon_r} \quad (5.4b)$$

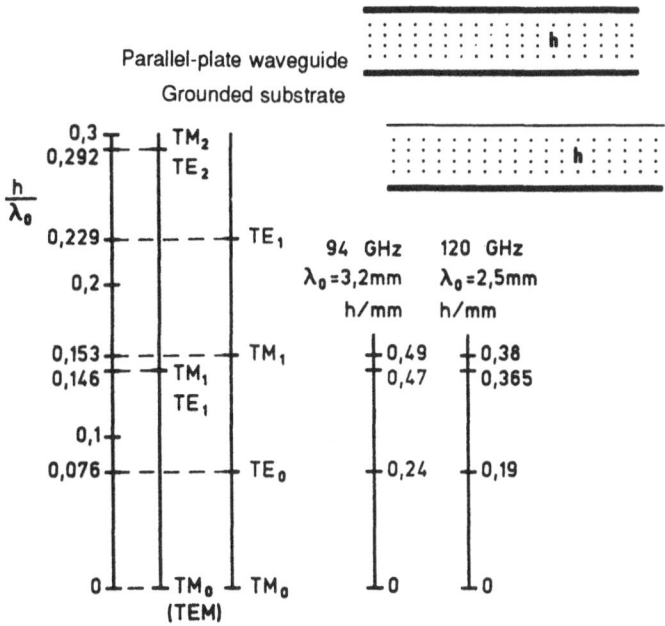

Fig. 5.1. Mode spectrum and corresponding substrate height for silicon with $\epsilon_r = 11.7$

with p and q following from the eigenvalue equations for the TE modes

$$\tan[hq - (N + 1/2)\pi] = p/q, \quad N = 0, 1, 2, \ldots \tag{5.5a}$$

and

$$\tan[hq - N\pi] = \epsilon_r p/q, \quad N = 0, 1, 2, \ldots \tag{5.5b}$$

for the TM modes. The minimum of h_e and thus the maximum of surface-wave loss occurs at substrate thickness to free space wavelength ratios of 0.09 for the TM_0 mode, 0.12 for the TE_0 mode, 0.26 for the TM_1 mode and 0.29 for the TE_1 mode. The surface-wave loss is low for thin grounded substrate. On the other hand, for the parallel-plate waveguide the surface-wave loss is low for thick substrates.

The level of the surface-wave modes excited by a special radiating element, an infinitesimal slot in the ground plane of a grounded substrate as well as a parallel-plate waveguide [5.4] is shown in Fig. 5.2 for a silicon substrate, where λ_d is the substrate wavelength and P_0 is the total power radiated by the slot into free space. For the grounded substrate a low substrate thickness helps to reduce the magnitude of the TM_0 mode to an acceptable low level. For the parallel-plate waveguide the minimum power appears at the cut-off height of the first-order mode.

The properties concerning the power coupled into surface waves of a dipole on a grounded substrate are qualitatively similar to those depicted in Fig. 5.2a [5.4], but the dipole radiates in only one direction whereas the slot radiates into the air and into the substrate.

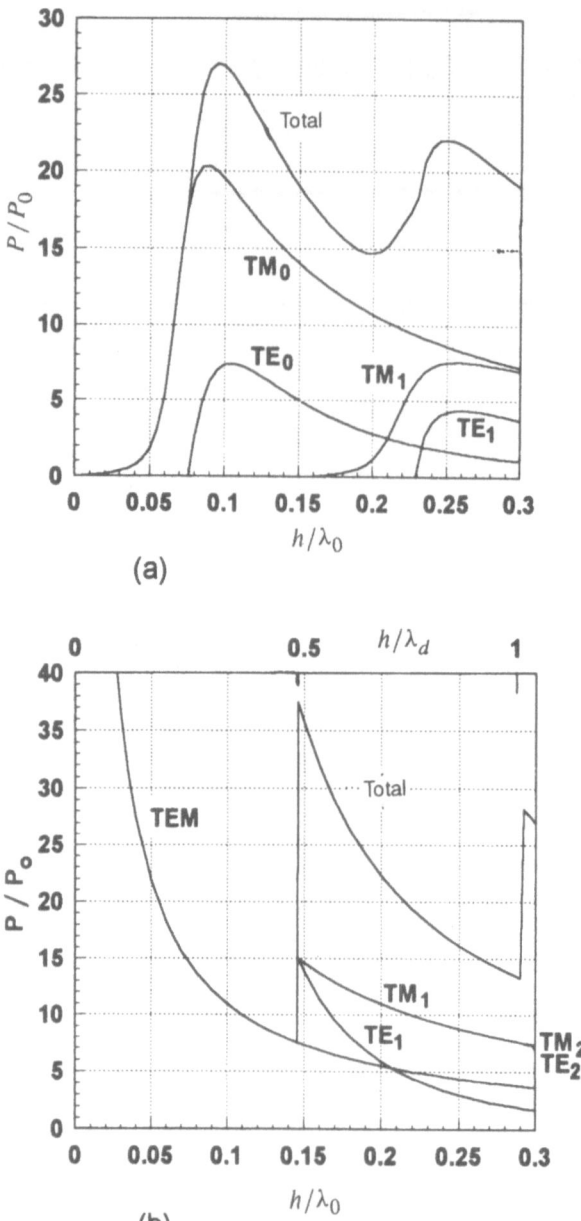

Fig. 5.2. Normalized surface-wave power versus silicon substrate thickness: (**a**) Slot on grounded substrate and (**b**) slot on parallel-plate waveguide

5.2   Millimeter-Wave Sources for SIMMWICs

To lower the cost and to reduce size and weight, and to achieve a better reproducible circuit performance the integration of power sources even in the millimeter-wave range is pushed ahead. For the generation of millimeter-wave power IMPATT diodes are up to now the most powerful solid-state devices. Therefore, in spite of much research effort on three-terminal devices IMPATT diodes can not be replaced by those devices if powerful oscillator operation in the millimeter-wave range is needed. Because of the better noise behaviour also integrated transistor sources at lower frequencies are subject of investigation in SIMMWIC technology.

5.2.1   IMPATT Oscillator

(a) The Radial-Line Resonator

The low impedance level of the IMPATT diode described in Chap. 2 implies a low load impedance according to the oscillation condition

$$R_i(\omega) + R_D(\omega, A) = 0, \quad X_i(\omega) + X_D(\omega, A) = 0 \tag{5.6}$$

where $R_i(\omega)$ is the load resistance and thus the input resistance of the resonator dependent on the frequency, and $R_D(\omega, A)$ is the diode resistance dependent on the frequency and the oscillation amplitude. $X_i(\omega)$ and $X_D(\omega, A)$ are the reactances of the load and the diode, respectively. A suitable resonant structure in planar technology to achieve a low impedance level is the ring resonator, which is illustrated in Fig. 5.3 schematically.

The permittivity of the substrate of this microstrip configuration is

$$\epsilon = \epsilon' - j\epsilon'' = \epsilon'(1 - j \tan \delta), \quad \epsilon' = \epsilon_0 \epsilon_r. \tag{5.7}$$

From the solution of the wave equation for this structure in consideration of $h < \lambda/2$ and $r > 0$ follow the TM_{nm0} modes with the z-component of the Hertzian

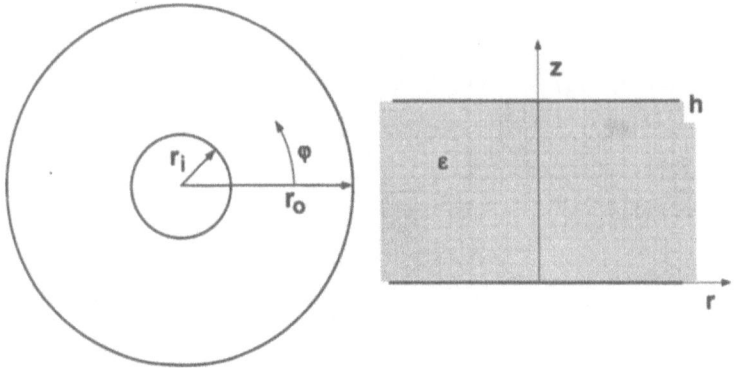

Fig. 5.3. Top view and cross section of a ring resonator

potential [5.5]

$$\Pi_z(r, \varphi) = Z_n(kr) \cos n\varphi, \quad n = 1, 2, 3, \ldots \tag{5.8}$$

with the wave number k and the cylindrical function

$$Z_n(kr) = AH_n^{(2)}(kr) + BH_n^{(1)}(kr). \tag{5.9}$$

The Hankel function $H^{(2)}(kr)$ represents the waves traveling in the $+z$-direction, and $H^{(1)}(kr)$ those traveling in the $-z$-direction. This solution corresponds to the radial-line solution. The fundamental mode of the radial-line resonator is the TM_{010} mode. The existing fields are [5.6]

$$E_z(r) = k^2[AH_0^{(2)}(kr) + BH_0^{(1)}(kr)] \tag{5.10}$$

$$H_\varphi(r) = j\omega\epsilon k[AH_1^{(2)}(kr) + BH_1^{(1)}(kr)] \tag{5.11}$$

with the wave number

$$k = \left[\omega\epsilon'\left(\omega\mu_0 - \frac{R_A}{h}\tan\delta\right) - j\omega\epsilon'\left(\frac{R_A}{h} + \omega\mu_0\tan\delta\right)\right]^{1/2}. \tag{5.12}$$

R_A is the surface resistance given by

$$R_A = \sqrt{\frac{\omega\mu_0}{2\sigma}} \tag{5.13}$$

where σ is the conductivity of the metallization. With the electric and the magnetic field at the outer radius $E_z(r_0)$ and $H_\varphi(r_0)$, the field components inside the radial line are

$$E_z(r) = \frac{\pi kr_0}{2}\left[A_0 E_z(r_0) - jZ_F B_0 H_\varphi(r_0)\right], \tag{5.14a}$$

$$H_\varphi(r) = \frac{\pi kr_0}{2}\left[B_1 H_\varphi(r_0) + j\frac{1}{Z_F}A_1 E_z(r_0)\right] \tag{5.14b}$$

with the wave impedance

$$Z_F = \sqrt{\frac{\mu}{\epsilon}} = \frac{k}{\omega\epsilon} \tag{5.15}$$

and

$$\begin{aligned}
A_p(r) &= N_p(kr)J_1(kr_0) - J_p(kr)N_1(kr_0), \\
B_p(r) &= J_p(kr)N_0(kr_0) - N_p(kr)J_0(kr_0),
\end{aligned} \tag{5.16a, b}$$

where J_n and N_n are the Bessel functions of the first and second kind. The input impedance of the radial line at radius r then is

$$Z_i(r) = \frac{-hE_z(r)}{2\pi r H_\varphi(r)} = Z_{RL}(r)\frac{A_0 Z(r_0) + jB_0 Z_{RL}(r_0)}{B_1 Z_{RL}(r_0) - jA_1 Z(r_0)} \tag{5.17}$$

with the characteristic impedance of the radial line

$$Z_{RL}(r) = Z_F\frac{h}{2\pi r} \tag{5.18}$$

and the load impedance $Z(r_o)$. For small r_i the input impedance becomes high and strongly reactive. The load impedance consists of a fringing capacitance given by [5.7]

$$C_f = \frac{r_o^2 \pi \epsilon'}{h} \left\{ \frac{2h}{r_o \pi \epsilon_r} \left[\ln\left(\frac{r_o}{2h}\right) + 1.41\epsilon_r + 1.77 \right. \right.$$
$$\left. \left. + \frac{h}{r_o}(0.268\epsilon_r + 1.65) \right] \right\} \tag{5.19}$$

and the radiation resistance by [5.6]

$$R_R = \frac{2Z_{F0}}{(k_0 r_o)^2 \pi} \left(\int_0^{\pi/2} J_1^2 \left(k_0 r_o \sin v\right) \sin v \, dv \right)^{-1} \tag{5.20}$$

where Z_{F0} and k_0 are the intrinsic impedance and the wave number of free space. In Fig. 5.4 the radiation resistance is shown.

The loss of the radial line fundamental mode due to the substrate material is

$$P_D = \frac{\omega \epsilon''}{\epsilon'} W \tag{5.21}$$

with the energy stored in the resonator [5.6]

$$W = \frac{\epsilon' \pi^3 (kr_o)^2}{4h} |U(r_o)|^2 \left(\frac{2r_o^2}{(\pi k r_o)^2} \left(1 + \frac{Z_{RL}^2(r_o)}{|Z(r_o)|^2} \right) \right.$$
$$- \frac{r_i^2}{2} \left\{ A_0^2(r_i) + A_1^2(r_i) + \frac{Z_{RL}^2(r_o)}{|Z(r_o)|^2} \left[B_0^2(r_i) + B_1^2(r_i) \right] \right. \tag{5.22}$$
$$\left. \left. + 2\frac{X(r_o)Z_{RL}(r_o)}{|Z(r_o)|^2} \left[A_0(r_i)B_0(r_i) + A_1(r_i)B_1(r_i) \right] \right\} \right).$$

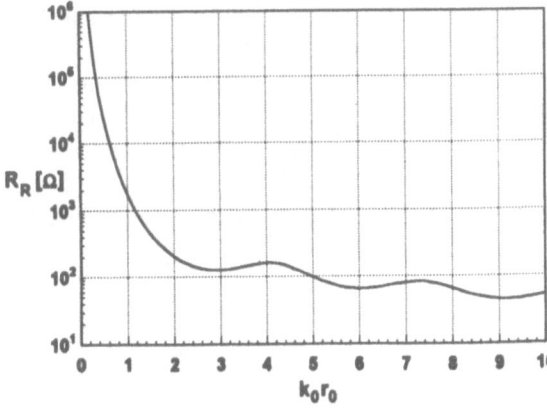

Fig. 5.4. Radiation resistance of the fundamental mode of the radial line

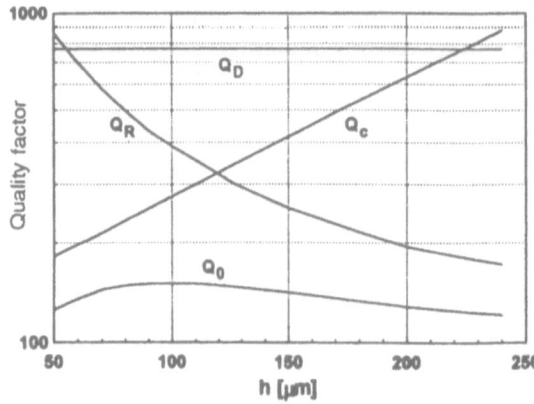

Fig. 5.5. The quality factors for a radial line versus substrate height $\epsilon_r = 11.68$, $\tan\delta = 1.3 \times 10^{-3}$, $\sigma = 45.2 \times 10^6$ AV^{-1}m^{-1}, $r_i = 0.2$ mm, $r_0 = 0.7$ mm, $f = 75$ GHz

The loss due to the metallization is

$$
\begin{aligned}
P_C = {} & \frac{k^2}{8}|I(r_0)|^2 R_A \left(\frac{2r_0^2}{(\pi k r_0)^2} \left[1 + \frac{|Z(r_0)|^2}{Z_{RL}^2(r_0)} - \frac{X(r_0)}{Z_{RL}(r_0)} \frac{\pi k r_0}{2} B_2(r_0) \right] \right. \\
& - \frac{r_i^2}{2} \left\{ B_1^2(r_i) - B_0(r_i)B_2(r_i) + \frac{|Z(r_0)|^2}{Z_{RL}^2(r_0)} \right. \\
& \left[A_1^2(r_i) - A_0(r_i)A_2(r_i) \right] + \frac{X(r_0)}{Z_{RL}(r_0)} \\
& \left. \left. \left[2A_1(r_i)B_1(r_i) - A_0(r_i)B_2(r_i) - A_2(r_i)B_0(r_i) \right] \right\} \right).
\end{aligned}
\tag{5.23}
$$

Knowing the losses the unloaded quality factor of the radial line can be calculated, i.e.,

$$
Q_0 = \left(\frac{1}{Q_D} + \frac{1}{Q_C} + \frac{1}{Q_R} \right)^{-1}
\tag{5.24}
$$

with

$$
Q_D = 1/\tan\delta, \quad Q_C = \frac{h}{\sqrt{2}}\sqrt{\omega\mu\sigma}, \quad Q_R = \frac{2\omega W R_R}{h^2 |E_z(r_0)|^2}.
\tag{5.25a, b, c}
$$

The quality factor referring to the substrate loss Q_D is independent of frequency and substrate height. The quality factor referring to the loss in the metallization Q_C increases with the substrate thickness and with $f^{1/2}$. The quality factor referring to the radiation loss decreases with increasing substrate thickness. The quality factors for a special geometry dependent on the substrate height are displayed in Fig. 5.5.

(b) 73 GHz Oscillator

The oscillator is formed by a ring resonator with a discrete silicon IMPATT diode as the active element. The resonator can be considered as a lossy radial line. For stable

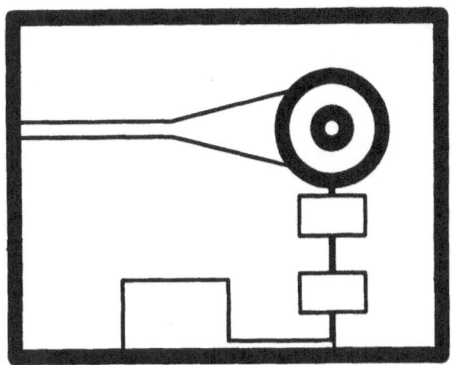

Fig. 5.6. Layout of the oscillator, $r_I = .212$ mm, $r_o = 0.7$ mm, $h = 0.11$ mm

oscillator operation the IMPATT diode has to be terminated in the neighbourhood of a series resonance of the resonator. For optimum oscillator operation the real part of the load resistance has to be minimized. By this the oscillator output power is increased and bias oscillations can be avoided. A low characteristic impedance of the radial line and a high quality factor of the resonator should therefore be realized which can be achieved by a careful design of the planar resonator. Mounting the IMPATT diode in the center of the radial-line resonator minimizes the radiation losses. Reducing the substrate thickness lowers the characteristic impedance of the radial line and also reduces the radiation losses. Figure 5.6 shows the layout of the oscillator.

The oscillator output signal is coupled and emitted via a capacitive gap into a tapered microstrip transmission line. A bias network with a low-pass filter structure is connected to the resonator. The radial-line resonator is terminated by the radiation resistance, the fringing-field capacitance and the external load formed by the gap capacitance and the tapered microstrip line.

Following the reaction concept [5.8] the coupling impedance of the resonator and the transmission line can be calculated. The resulting mutual admittance is [5.9]

$$Y_{SP} = \frac{r_o r_p}{4\pi\omega\mu} \int_0^\Phi \int_{-\varphi_H}^{+\varphi_H} \frac{\cos(\varphi - \varphi')}{r_{SP}^3} (T_R - jT_1) \, d\varphi' d\varphi \tag{5.26}$$

with

$$T_R = R_1 \cos(kr_{SP}) - R_R \sin(kr_{SP}), \quad T_I = R_R \cos(kr_{SP}) + R_I \sin(kr_{SP}),$$

$$\varphi_H = \arccos\left(\frac{r_o}{r_p}\right)$$

$$R_R = 1 + k^2 a^2 + \frac{r_{SP}^2}{r_p^2} + \frac{a}{r_p} + \frac{3a^2}{r_{SP}^2}, \quad R_I = k\left(r_{SP} + \frac{ar_{SP}}{r_p} - \frac{3a^2}{r_{SP}}\right), \tag{5.27}$$

$$a = r_p - r_o \cos(\varphi - \varphi').$$

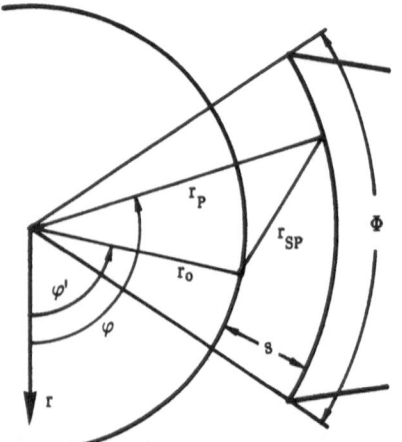

Fig. 5.7. Geometry of the resonator with a coupled line

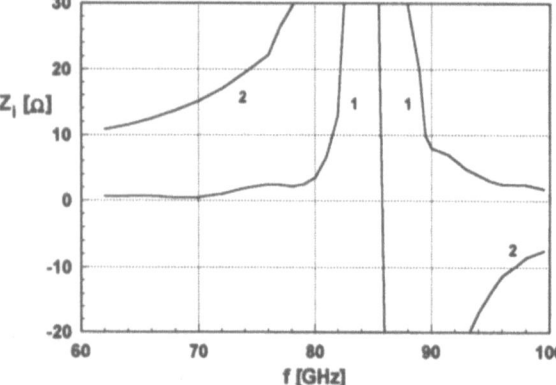

Fig. 5.8. Input impedance of the resonator versus frequency $(1 : \Re\{Z_1\}, 2 : Im\{Z_1\})$ $\epsilon_r = 11.68$, $\tan\delta = 1.3 \times 10^3$, $\sigma = 45.2 \times 10^6$ AV^{-1}m^{-1}, $r_t = 0.2$ mm, $r_o = 0.7$ mm

Figure 5.7 depicts the geometry under consideration in detail. The input impedance of the radial-line resonator can be calculated. The resulting input resistance and reactance are shown in Fig. 5.8, where a bond resistance of 0.05 Ω and a bond inductance of 7 pH are taken into account. For frequencies below 80 GHz the input resistance is rather low and the reactance is inductive. Thus the resonator impedance is suitable to match the IMPATT impedance.

The planar structure is realized on a silicon substrate of 110 μm thickness, which is mounted on a copper block. The chip size is 6×4.5 mm^2. Figure 5.9 shows the cross section and a photo of the oscillator. The substrate is covered with a thin thermal oxide (100 n m) and with an evaporated titanium/gold metallization. The resonator structure is defined by photolithography and selectively electroplated for a thickness of 3 μm. As the active device a quasi-Read double-drift IMPATT diode made from Si-MBE material is used. The diode is thermocompression bonded in the center hole of 425 μm diameter directly on the copper carrier. Electrical connection is achieved by cross ribbons which are also thermocompression bonded.

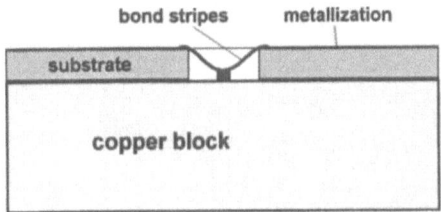
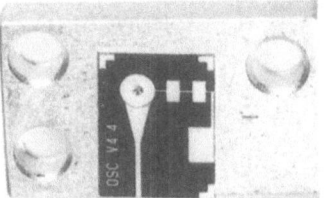

Fig. 5.9. Cross section and photo of the oscillator

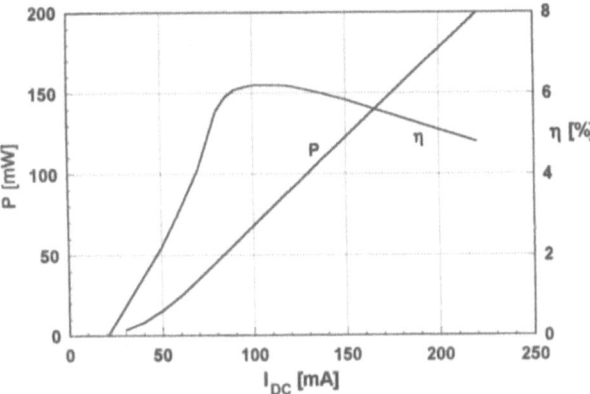

Fig. 5.10. PI characteristic of the oscillator

The maximum RF output power is 200 mW at 73 GHz with an efficiency of 4.5%. Figure 5.10 displays the RF power and efficiency dependence on the dc bias current. The maximum efficiency is 6.1%. The noise-to-carrier ratio is −73 dBc at 200 kHz off the carrier. This is measured at a reduced output power of 4 mW where the IMPATT diode is biased with 33 mA at 17 V.

5.2.2 Varactor-Tuned Oscillator

The oscillator circuit for cw operation at a fixed frequency described in the previous subsection is extended by a radial-line sector coupled to the resonator via a varactor diode to obtain frequency variation by voltage control [5.10]. Figure 5.11 exhibits a photo of the planar circuit.

A hyperabrupt doping profile is used in the varactor diode. A typical CV curve of a small-area diode is shown in Fig. 5.12. A quadratic dependence of the capacitance on the voltage is observed in accordance with the predictions from the doping profile.

The oscillator is produced on a 10000 Ωcm silicon substrate of 100 μm thickness. The chip size is 6×4.5 mm^2. As the active device a Si-MBE made quasi read double-drift IMPATT diode with a diameter of 46 μm is used. Figure 5.13 shows the RF results which were achieved with a 38 μm varactor diode.

Fig. 5.11. Photo of the planar Voltage Controlled Oscillator (VCO)

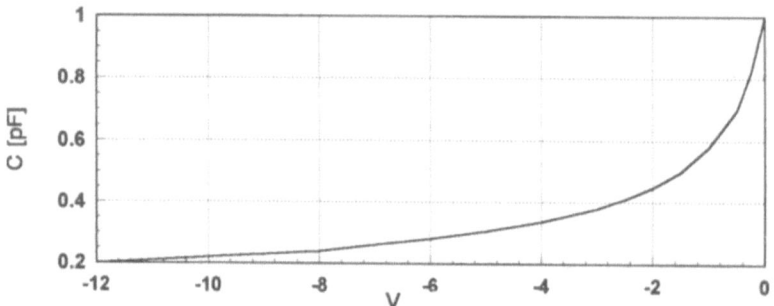

Fig. 5.12. Varactor-diode capacitance versus reverse voltage (Diode diameter: 38 μm)

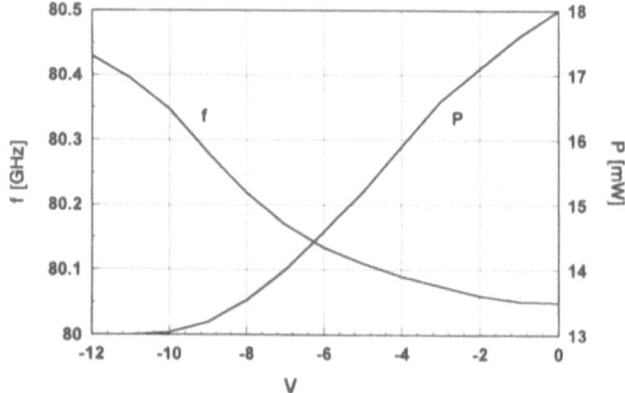

Fig. 5.13. Oscillation frequency and output power of the VCO

Without any varactor tuning the oscillation frequency is 80.05 GHz with an IMPATT diode current density of 10 kA/cm^2. The corresponding output power is 18 mW in cw operation. With constant bias current of the IMPATT diode an increase of the absolute value of the varactor voltage up to 12 V results in an increase of the oscillation frequency by 425 MHz up to 80.43 GHz, while the output power decreases to 13 mW. As expected from theory, the tuning range is smaller when larger-area varactor diodes are used. For example, with a varactor-diode diameter of 160 μm corresponding to a capacitance variation of 5 pF to 30 pF the tuning range is 150 MHz.

5.2.3 HBT Oscillator

For a three-terminal active device the Heterojunction Bipolar Transistor (HBT) in Si-SiGe technology is available (Chaps. 4, 7) with f_T values up to the 100 GHz region. Because of the excellent FM-noise behaviour of Dielectric Resonator Oscillators (DRO) with a Si-SiGe HBT these are promising components for silicon-based integrated circuits. The first reported Si-SiGe HBT-DRO used a 2 μm ×20 μm sized transistor with f_{max} and f_T values of 38 GHz and was produced on a 0.25 mm thick alumina substrate [5.11]. The experimental results were an RF output power of 10 mW at 9.6 GHz and a phase noise of −85 dBc (1 Hz) at 100 kHz off carrier.

Using SIMMWIC technology and an HBT with a f_T value of 50 GHz now a DRO on a silicon substrate of 150 μm thickness has been realized [5.12]. For the oscillator topology a series-feedback configuration and the HBT in common emitter operation is chosen. The RF output port is at the collector side of the transistor, whereas the dielectric resonator is placed on the base side. Figure 5.14 shows the realized oscillator circuit.

The experimental results are an RF output power of 7 mW at 23.3 GHz and a phase noise of −91.6 dB related to the carrier (dBc) in 1 Hz bandwidth at 100 kHz off carrier. The efficiency is 11.4%.

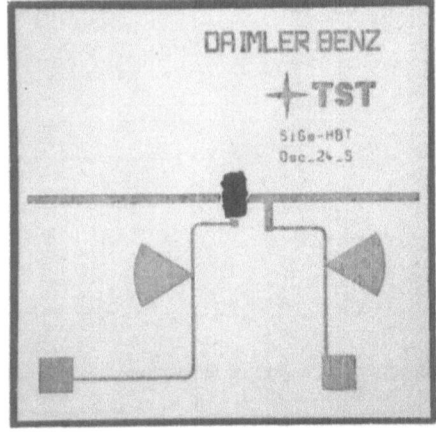

Fig. 5.14. Photo of the DRO chip

5.3 SIMMWIC Transmitter

Combining active devices with passive networks and antenna structures leads to complete single-chip transmitter circuits. Along with a reduction of the chip size this results in an avoidance of the coupling loss between resonator and antenna.

5.3.1 Thermal Limitation of Monolithic IMPATT Diodes

The generated heat in active elements may lead to a reduction of lifetime and to a limitation of the efficiency. Especially in those elements which work near their thermal limit this might become a severe problem if no heat sink can be attached to the element directly like in monolithic coplanar structures.

From a theoretical investigation on IMPATT diodes monolithically integrated on a silicon substrate (Chap. 2) follows the maximum current density

$$J = \frac{1}{2} \frac{\sqrt{U_b^2 + 4\Delta T \left(\frac{R_{sc}}{R_{th}} + \beta U_b^2 \right)} - U_b}{\pi r^2 \left(R_{sc} + R_{th} \beta U_b^2 \right)}, \quad \Delta T = P R_{th} = \text{const.} \quad (5.28)$$

where U_b is the breakdown voltage, R_{sc} is the space-charge resistance, and β is the temperature coefficient. The maximum current density and the dc current dependent on the diode diameter together with experimental data of different diode types are displayed in Fig. 5.15 [5.13].

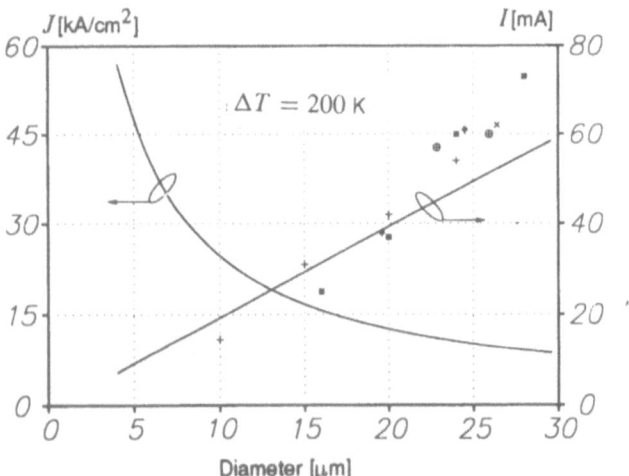

Fig. 5.15. Maximum current density and maximum current for a junction temperature rise of 200 K as a function of the diode diameter

5.3.2 Coplanar Slot-Line Transmitter

The transmitter circuit consists of a coplanar structure with shorted slot lines. In Fig. 5.16 a photo of the SIMMWIC slot line transmitter chip is shown [5.14]. The electrical length of the slot is one wavelength and the slot width is 80 μm. An IMPATT diode is monolithically integrated in the center of the slot. The narrow slots placed by $\lambda/4$ from the end of the resonator allow the biasing of the IMPATT diode.

Due to the high permittivity of silicon more radiation is emitted through the substrate than directly into the air. Therefore, the chips are mounted upside down into a test fixture. The experimental results in cw mode are shown in Fig. 5.17 and in pulsed mode in Fig. 5.18, where ring-shaped IMPATT diodes with different diameters are used. In both cases no additional heat sink is applied. The maximum radiated power in cw mode is 4.4 mW at 89 GHz with a S/N of −80 dBc at 200 kHz off carrier. In pulsed mode the maximum radiated power is 50 mW at about 92 GHz [5.15].

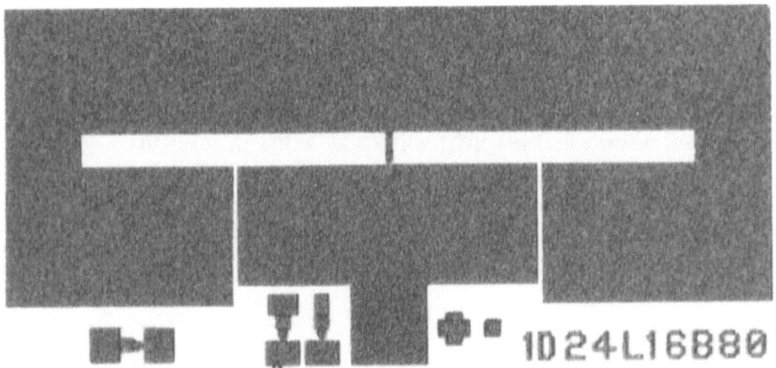

Fig. 5.16. Photo of the slot transmitter

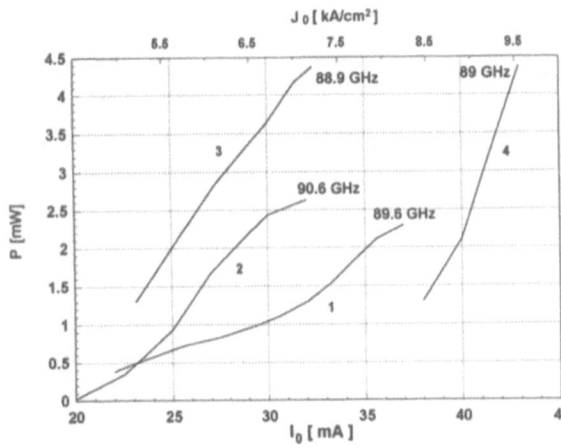

Fig. 5.17. Radiated cw power versus dc current, radiation through the substrate, substrate thickness: 100 μm, slot length: 1.6 mm 1–3: double low-high-low (DLHL) IMPATT structure of different wafers, $\phi = 24$ μm 4: quasi Read double-drift (QRDD) IMPATT structure, $\phi = 24$ μm

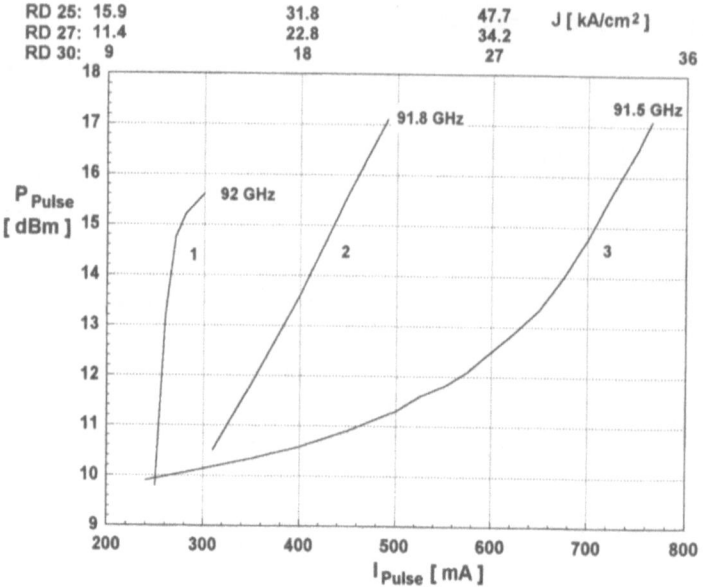

Fig. 5.18. Radiated pulse power versus pulse current (50 ns pulses, 0.5% duty cycle), radiation through the substrate, substrate thickness: 100 μm 1: RD 25, 2: RD 27, 3: RD 30

Furthermore, the slot transmitters are operated up side up with a grounded substrate. In this mode the maximum power radiated directly into the air is 1.3 mW at 75 GHz.

5.4 SIMMWIC Receiver

For the detection of millimeter-wave radiation suitable devices are Schottky barrier diodes or Planar Doped Barrier (PDB) diodes. Planar integrated subharmonic mixers are realized up to D-band frequencies using silicon PDBs grown by molecular beam epitaxy [5.16], and integrated harmonic mixers for W-band frequencies applying silicon Schottkyy diodes are fabricated [5.17]. The integration of those diodes together with planar antenna structures leads to single-chip receivers.

5.4.1 Microstrip Receiver

Differing in the antenna structure and the Schotty-barrier diode type, three types of rectennas (rectifying antenna) based on SIMMWIC microstrip technology are described in this section. A photo of the first receiver is shown in Fig. 5.19 [5.18]. The circuit topology of this rectenna consists of an antenna structure coupled via a microstrip line to a monolithic Schottky-barrier diode which is terminated by a λ/4 line. Filter structures for biasing the Schottky diode and for coupling out the

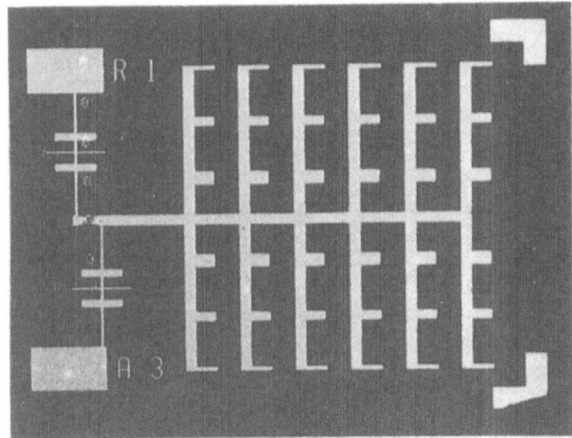

Fig. 5.19. Photo of the receiver

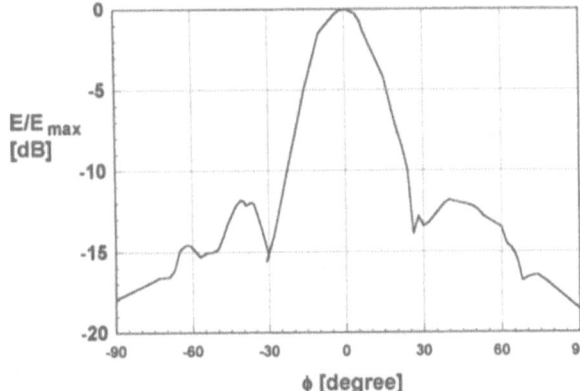

Fig. 5.20. H-plane radiation pattern at 93 GHz

detector signal are connected with the micostrip line. The antenna structure contains 36 radiating elements on an area of 5.4×5.6 mm^2. To achieve high directivity together with high side-lobe attenuation the weights of the radiating elements are designed after Dolph-Chebyshev. In order to achieve uniform phase distribution the distance between the elements is calculated taking into account the discontinuities.

The structure is made on silicon substrate of 190 μm height with a resistivity of 10000 Ωcm. The chip size is 7×9.5 mm^2. The measured radiation pattern of the antenna is exhibited in Fig. 5.20. The half-power beam width is 23° and the side-lobe attenuation is 12 dB. Using the receiver to detect radiation at 93 GHz the Schottky diode is biased at 270 mV. The measured sensitivity which is defined as the detector voltage over RF power at the locus of the antenna is 65 mV/mW cm^2.

Other types of microstrip rectennas (rectifying antennas) with eleven radiating elements on one branch and twenty elements on two branches, respectively, are shown in Fig. 5.21 [5.19]. For the one-branch rectenna a zero bias monolithic p-Schottky barrier diode for the detecting element is used which is terminated by a radial stub. The antenna area is 1×8 mm^2 and the chip size is 2.5×9.5 mm^2.

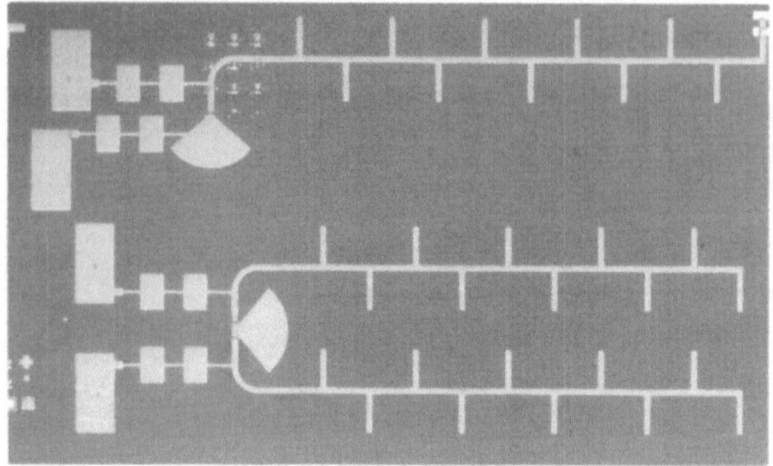

Fig. 5.21. Photo of the rectennas

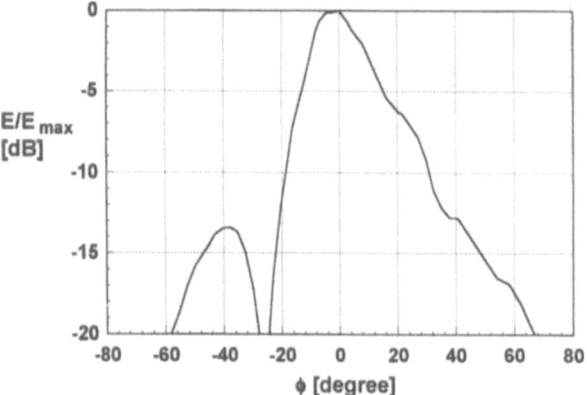

Fig. 5.22. H-plane radiation pattern of the rectenna at 92.2 GHz

The radiation pattern of this antenna is depicted in Fig. 5.22. The half power beam width is 20° at 92.2 GHz and the sensitivity is 83.3 mV/mW cm^2 at zero bias.

Two p-type Schottky-barrier diodes in series are integrated monolithically in the two branch rectenna. The antenna area of this type is 2.6×6 mm^2 and the chip size is 3.5×9 mm^2. The sensitivity is 153 mV/mW cm^2 at 94.1 GHz, while the diodes are biased at 568 mV [5.19]. The corresponding bandwidth is 0.8%.

5.4.2 Coplanar Slot-Line Receiver

As the counterpart to the slot transmitter a coplanar slot receiver is made with SIMMWIC technology. The circuit layout of the rectenna circuit is shown in

Fig. 5.23. The length of the slot is one-half wavelength and the width is 80 μm. At the ends of the slot, two-narrow slots are provided for extracting the rectified signal and for biasing the Schottky-barrier diode which is monolithically integrated in the slot. The total chip size is 2.16×1.0 mm^2.

Measurements were made with chips of 525 μm thickness and a grounded substrate. The resulting sensitivity and the corresponding junction resistance are shown in Fig. 5.24. The maximum sensitivity is 85 mV/mW cm^2. Assuming an antenna gain of the slot of 6 dB this corresponds to a detector sensitivity of 1200 mV/mW [5.20].

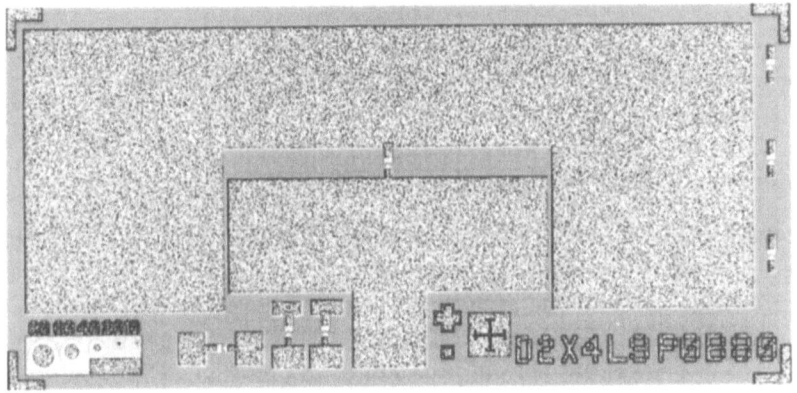

Fig. 5.23. Photo of the coplanar slot rectenna

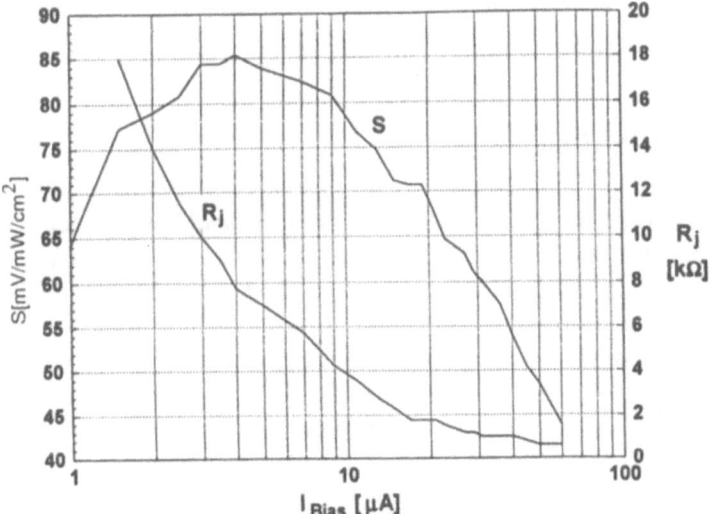

Fig. 5.24. Sensitivity and junction resistance versus bias current at 63.6 GHz

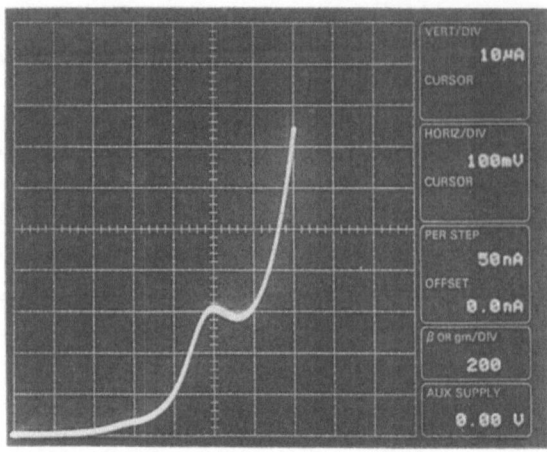

Fig. 5.25. IV characteristic at 77 K

5.4.3   Resonant Tunneling Rectenna

Beside the Schottky diode also Si-SiGe resonant tunnelling structures are suitable
to detect millimeter-wave radiation. Using the passive planar structure described
in Sec. 5.4.1 and a monolithic integrated resonant tunneling device a rectenna is
realized applying the SIMMWIC technology [5.21]. The double-barrier Si-SiGe
layer sequence is grown by MBE on a patterned high-resistivity silicon substrate.
Figure 5.25 shows the IV characteristic at 77 K. The rectenna works biased into
the resonance but also at zero bias.

Biased at 225 mV into the resonant region at 77 K the measured sensitivity is
20 mV/mW cm^2 at 91 GHz. A higher sensitivity of 85 mV/mV cm^2 is obtained if
the device is biased at 460 mV into the second resonant state. This is attributed to
the higher peak-to-valley ratio in the second resonance.

5.5   SIMMWIC Switch

The first SIMMWIC switch has been fabricated by *Stabile and Rosen* [5.22]. This
circuit is a monolithic Single Pole Single Throw (SPST) switch with a minimum
isolation of 21.6 dB and a maximum insertion loss of 2.9 dB across a 20% band
around 36.75 GHz. It incorporates a PIN diode in shunt with a transmission line
and a low-pass dc bias filter on a high-resistivity silicon substrate.

The layout of a new Single Pole Double Throw (SPDT) monolithic SIMMWIC
switch is depicted in Fig. 5.26. It consists of two symmetrical cells for the output
branches connected to a common input port. PIN diodes are used for the switching
elements. Radial stubs are used for RF shorting, and the dc short is achieved by via
holes. The minimum isolation of this switch is 25 dB and the maximum insertion
loss is 2.5 dB in the frequency range from 67 GHz to 80 GHz.

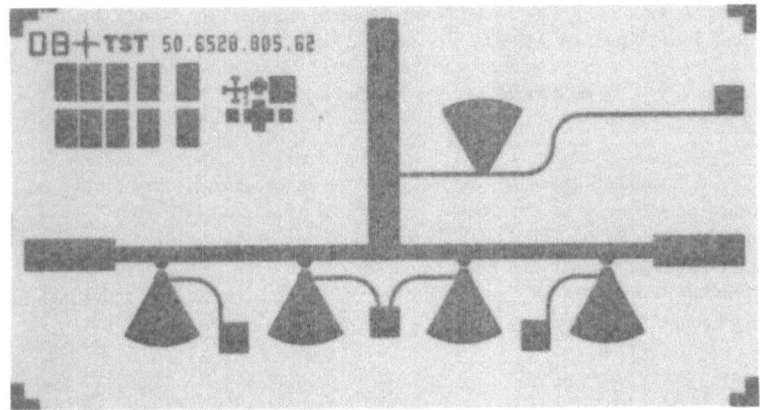

Fig. 5.26. Photo of the SPDT switch

References

Section 5.0

5.1   P.J. Stabile, A. Rosen: A silicon technology for millimeter-wave monolithic circuits. RCA
        Rev. **45**, 587–605 (1984)

Section 5.1

5.2   J. Buechler., E. Kasper, P. Russer, K.M. Strohm: Silicon high-resistivity-substrate millime-
        ter-wave technology. IEEE Trans. MTT-**34**, 1516–1521 (1986)
5.3   D.M. Pozar: Considerations for millimeter wave printed antennas, IEEE Trans. AP-**31**,
        740–747 (1983)
5.4   N.G. Alexopoulos, P.B. Katehi, D.B. Rutlage: Substrate optimization for integrated circuits
        antennas. IEEE Trans. MTT-**31**, 550–557 (1983)

Section 5.2

5.5   K. Simonyi: *Theoretische Elektrotechnik* (VEB Deutscher Verlag der Wissenschaften, Berlin
        1979)
5.6   J. Buechler: Integrierte Millimeterwellenschaltungen auf Silizium. Dissertation Technical
        University of Munich (1989)
5.7   W.C. Chew, J.A. Kong: Effects of fringing fields on the capacitance of circular microstrip
        disk. IEEE Trans. MTT-**28**, 98–104 (1980)
5.8   R.F. Harrington: *Time Harmonic Electromagnetic Fields* (McGraw Hill, New York 1961)
5.9   J. Buechler, E. Kasper, J.F. Luy, P. Russer, K.M. Strohm: Planar W-band oscillator. 18th
        EuMC, Conf. Prof., (1988) pp. 364–369
5.10   J. Buechler, J.F. Luy, K.M. Strohm: Planar W-band oscillator. Proc. IEEE MTT Symp.
        (1989), pp 1205–1206
5.11   U. Güttich, A. Gruhle, J.F. Luy: A Si-SiGe HBT dieletric resonator stabilized microstrip
        oscillator at X-band frequencies. IEEE Microwave and Guided Wave Lett. **2**, 281–283
        (1992)

5.12   U. Güttich A. Gruhle, J.F. Luy: Dielectrically stabilized oscillators for X- and K-band. MIOP
        93, Conf. Proc., (1993) pp. 146-150

Section 5.3

5.13   J.F. Luy, J. Schmidt, J. Buechler: Thermal properties of circular geometry MMICs, to be
        published
5.14   J. Buechler, K.M. Strohm, J.F. Luy, T. Goeller, S. Sattler, P. Russer: Coplanar monolithic
        silicon IMPATT transmitter. 21st EuMC, Conf. Proc. (1991), pp. 352-357
5.15   K.M. Strohm, J. Buechler, J.F. Luy, F. Schäffler: A silicon technology for active high fre-
        quency circuits. Microelectron. Engi. **19**. 717-720 (1992)

Section 5.4

5.16   U. Güttich, K.M. Strohm, F. Schäffler: D-band subharmonic mixer with silicon planar doped
        barrier diodes. IEEE Trans. MTT-**39**, 366-368 (1991)
5.17   P. Nüchter, W. Menzel: A MM-wave frequency divider, Proc. IEEE MTT Symp., (1992)
        pp. 695-697
5.18   J. Buechler, E. Kasper, J.F. Luy, P. Russer, K.M. Strohm: Silicon millimeter-wave circuits
        for receivers and transmitters. Proc. IEEE MTT Symp. (1988) pp. 67-70
5.19   K.M. Strohm, J. Buechler, J.F. Luy: 90 GHz SIMMWIC rectennas, 22nd EuMC, Conf. Proc.
        (1992), pp. 608-613
5.20   K.M. Strohm, J. Buechler, J.F. Luy: A monolithic millimeter wave integrated silicon slot
        line detector. Asia Pacific Microwave Conference, Taiwan (1993), pp. 34-37
5.21   J.F. Luy, K.M. Strohm, J. Buechler: A 91 GHz Si/SiGe resonant tunneling detector, Arch.
        elktr. Uebertrg. (AEÜ) **46**, 370-373 (1992)

Section 5.5

5.22   H.E. Sasse, A. Klaassen, K.M. Strohm, J.F. Luy: Monolithic integrated 76 GHz Si PIN diode
        switch, to be published

6 Self-Mixing Oscillators

M. CLAASSEN

Lehrstuhl für Allgemeine Elektrotechnik und Angewandte Elektronik,
Techn. Univ. München, Arcisstr. 21, 80290 München, Germany

A self-mixing oscillator is a circuitry acting as an RF local oscillator, into which a receiving signal can be conducted to be mixed with the local oscillation due to the nonlinear behaviour of the active device that generates the RF power. This is, in principle, the same as a self-oscillating mixer besides the fact that the produced local-oscillator power may also be extracted from the self-mixing oscillator, e.g., for transmission in Doppler-radar applications.

Self-mixing oscillator circuits are already long known from early vacuum-tube superheterodyne radio-receiver techniques, where the receiving signal was applied to a second control grid of the local-oscillator tube for mixing. For microwave and millimetre-wave oscillators, self-mixing operation of two-terminal devices has been verified for Gunn devices [6.1,2], IMPATT- [6.3], BARITT- [6.3–5], and TUNNETT-diodes [6.6] with, however, quite different mixing properties.

The use of self-mixing oscillators enables, e.g., very simple designs for Doppler-radar systems consisting only of an RF oscillator which is directly connected to an antenna for transmitting the output power. The same antenna may then also be used for receiving the Doppler-shifted signal as reflected from a moving object. The received signal is thus guided back into the oscillator where it superimposes onto the oscillation and is by self-mixing down-converted to the difference frequency for signal evaluation in the bias circuitry.

As the RF part of such a system is quite simple, especially if with proper design even the resonator structure itself may act as antenna, such Doppler-radar systems are well suited for monolithic integration [6.7]. For effective self-mixing operation, however, circuitry and active devices incorporated in the oscillator have to be properly composed. This will be treated in more detail in this chapter.

6.1 Principle of Operation

The basic idea of self-mixing oscillators is that the always inherent nonlinearity of the active device in a local oscillator, which is necessary for limiting the oscillation amplitude, should also be capable for mixing an injected signal with the oscillation. It relies on the fact that the injected receiver signal acts as an additional current- (or voltage-) source in the RF circuit and thus causes a disturbance to the oscillator RF point of operation. This leads to frequency and amplitude distortions from the

Springer Series in Electronics and Photonics, Vol. 32
Silicon-Based Millimeter-Wave Devices, Eds.: Luy et al.
© Springer-Verlag Berlin Heidelberg 1994

free-running oscillation condition. Frequency distortions are not very effective for down-conversion. Amplitude distortions, however, produce bias deviations due to internal rectification in the active device, which is a consequence of its nonlinear current-voltage behaviour. Thus, the down-converted signal can be detected in the bias circuit at the beat frequency.

Two oscillator properties are, therefore, of major importance for the down-conversion; the amplitude response of the oscillator to an injected signal and the shift of bias voltage (or current) at the active device due to oscillation-amplitude variations. The oscillator AM-response is generally strong if the amplitude self-regulation due to amplitude saturation of the amplification mechanism (e.g., of the negative active-device conductance) is weak. This is the case especially near the linear small-signal operation at low oscillator output-power and leads there to efficient signal amplification, i.e., the amplitude-modulation swing may become much larger than the injected-signal amplitude. The shift of the bias current-voltage characteristic with oscillation amplitude due to device-internal rectification is, on the other hand, a typical nonlinear effect. It thus increases with oscillation amplitude and depends strongly on the specific carrier-transport properties of the device. Both phenomena together result, nevertheless, in increasing conversion gain (defined as the ratio of down-converted output power to RF signal input) with decreasing oscillator power.

These principles will be demonstrated in Sect. 6.2. with a linearized disturbance theory of the oscillator as commonly used for noise and injection-locking investigations [6.8–10]. It is derived from a bias-, amplitude-, and frequency-dependent device admittance and dynamic bias current-voltage characteristic, and is applied to a simplified analytic device model. In Sect. 6.3., a conversion-matrix formulation of self-mixing is presented relating currents and voltages at signal-, image-, and down-converted frequency. Noise properties of self-mixing oscillators are treated in Sect. 6.4. Section 6.5 deals with applications to device simulations and Sect. 6.6 with measuring techniques as well as experimental results.

6.2 Linear Disturbance Theory

For calculation and discussion of self-mixing oscillator properties, the down-conversion efficiency is derived from the bias-current-, ac voltage-, and frequency-dependent admittance and bias voltage of the oscillating device with small disturbances from the free-running condition. This conversion can exhibit gain due to signal amplification in the RF circuit, in contrast to conventional, e.g., Schottky-diode mixers with typically 5 to 8 dB conversion loss. The term "conversion gain" will, hence forth, be used for the ratio of downconverted low-frequency output signal to RF input. If it gets smaller than unity (negative in logarithmic scale), it expresses conversion loss.

6.2.1 Model for the Self-Mixing Oscillator

At the frequency of oscillation, the self-mixing oscillator can be represented by an equivalent circuit as used for describing injection-locked oscillators. *Kurokawa* [6.9] applied for this purpose a circuit with an active device being connected to the load impedance as seen from the device (i.e., behind the transformation by the resonator circuitry) in series with an RF voltage source representing the injected signal. Large-signal device simulations, however, mainly yield device admittances, especially for IMPATT diodes [6.11]. Therefore, an RF equivalent circuit, as shown on the lhs of Fig. 6.1, is preferred. Here, the active device is represented by its RF admittance $Y_D(I_0, U_1, \omega)$, the complex value of which depends on the bias current I_0, the RF-voltage amplitude U_1, and, the circular frequency of oscillation ω. It is connected to the frequency-dependent load admittance $Y_L(\omega)$ as transformed by the resonator to the device terminals. The negative real part of the device admittance enables an RF-voltage oscillation $u(t) = U_1 e^{j\omega t}$ to establish across the common terminals under matched conditions. In this case, the input signal to the self-mixing oscillator can more conveniently be represented by a parallel current source, i.e.,

$$i_s(t) = \hat{i}_s e^{j\omega}s^t. \tag{6.1}$$

If the resonator circuit is assumed to perform a lossless passive transformation of the load to the device, the signal-current amplitude $\hat{i}_s$ can be derived from the signal power P_s as

$$\hat{i}_s = 2\sqrt{2P_s G_L}, \tag{6.2}$$

with the load conductance G_L being the real part of Y_L. Resonator losses can, in this case, be introduced by a loss conductance G_1 in parallel to the device reducing its negative conductance, and may be included in the device admittance Y_D. The frequency ω_s of the input signal is characterized by its offset from the frequency of oscillation, $\Delta\omega = \omega_s - \omega$, that can, e.g., arise from the Doppler shift of the transmitted oscillator signal due to reflection at a moving object.

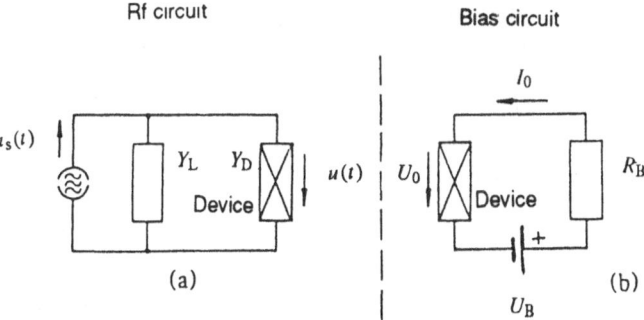

Fig. 6.1. Equivalent circuit of self-mixing oscillator at RF (**a**) and at low frequencies (**b**)

Without signal input, the amplitude and frequency of oscillation establish such that for a given bias current the total circuit admittance $Y_t = Y_D + Y_L$ vanishes. This is the condition of the free-running oscillator labeled with an additional subscript zero. The input signal causes deviations from this operation point (characterized by δ) that vary with time

$$I_0 = I_{00} + \delta I_0(t),$$
$$U_1 = U_{10} + \delta U_1(t), \tag{6.3}$$
$$\omega = \omega_0 + \delta\omega(t),$$

such that Kirchhoff's law of currents is satisfied in the RF circuit on the lhs of Fig. 6.1 at each moment, i.e.,

$$[Y_D(I_0, U_1, \omega) + Y_L(\omega)]U_1 e^{j\omega t} - \hat{i}_s e^{j(\omega + \Delta\omega)t} = 0. \tag{6.4}$$

This combination of time- and frequency-dependent quantities is admissible only if the admittance changes in time, due to deviations from the free-running condition, are slow as compared to the oscillation period; i.e. for $\Delta\omega \ll \omega$, a typical condition for most applications. In addition, the input signal is supposed to be sufficiently small that a linear dependence of (6.4) on the disturbances may be applied, namely

$$\frac{\partial Y_D}{\partial I_0}\delta I_0 + \frac{\partial Y_D}{\partial U_1}\delta U_1 + \frac{\partial(Y_D + Y_L)}{\partial \omega}\delta\omega - \frac{\hat{i}_s}{U_1}e^{j\Delta\omega t} = 0. \tag{6.5}$$

The last term on the lhs of (6.5) acts as a small, time-varying admittance, induced by the input signal, which changes as $e^{j\Delta\omega t}$. It is the source for amplitude and frequency disturbances in the RF circuit.

For the self-mixing oscillator, the RF equivalent circuit has to be supplemented by a corresponding bias circuit as, e.g., introduced by *Thaler* et al. [6.10]. It is depicted on the rhs of Fig. 6.1 showing the active device connected via the bias resistor R_B to the constant-voltage source U_B of the bias supply. Due to device-internal nonlinearities leading to partial rectification, the bias voltage U_0 of the oscillating device will not only depend on the bias current I_0 but also on the RF-voltage amplitude U_1. Thus, variations of the RF amplitude δU_1 induce changes in the bias voltage U_0, which lead to a current disturbance δI_0 in the bias circuit:

$$\frac{\partial U_0}{\partial I_0}\delta I_0 + \frac{\partial U_0}{\partial U_1}\delta U_1 + R_B\delta I_0 = 0. \tag{6.6}$$

The current change δI_0 represents the down-converted signal. It can be detected across the bias resistor R_B acting as the output impedance of the mixer.

Theoretically, the bias voltage U_0 of the oscillating device could also depend, to some extent, on the oscillation frequency ω. As the relative frequency change of a disturbed oscillator, however, is always much smaller than its relative amplitude variation, the effect of a corresponding frequency demodulation may be neglected.

To evaluate the disturbances from (6.5–6), the complex nature of (6.5) has to be regarded. For this reason, the partial derivatives of the admittances are expressed by their absolute values and an angle in the complex plane describing the direction

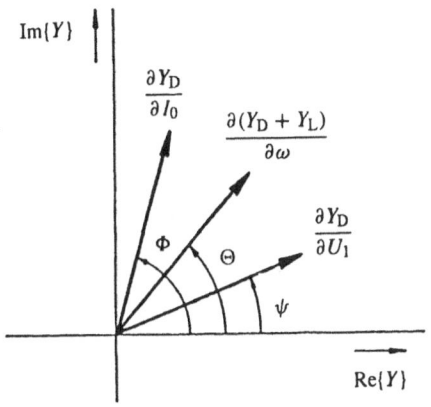

Fig. 6.2. Definition of angles ψ, Θ, and Φ of admittance derivatives (6.7) in the complex admittance plane

of corresponding device and circuit lines, as illustrated in Fig. 6.2.

$$\frac{\partial Y_D}{\partial U_1} = \left| \frac{\partial Y_D}{\partial U_1} \right| e^{j\psi},$$

$$\frac{\partial Y_D}{\partial I_0} = \left| \frac{\partial Y_D}{\partial I_0} \right| e^{j\Phi}, \tag{6.7}$$

$$\frac{\partial (Y_D + Y_L)}{\partial \omega} = \left| \frac{\partial (Y_D + Y_L)}{\partial \omega} \right| e^{j\Theta}.$$

Eliminating the frequency disturbance $\delta\omega$, (6.5) gives the RF amplitude modulation δU_1 due to the input signal. For signal frequencies near the carrier (i.e., neglecting imaginary parts of $\delta\omega$, Sect. 6.3), this can be expressed as

$$\delta U_1 = \alpha \sqrt{\frac{2P_s}{G_L}} \sin(\Theta - \Delta\omega t) + S_{AM}\delta I_0. \tag{6.8}$$

Here, α is an RF matching factor defined by

$$\alpha = \frac{2G_L}{U_{10}|\partial Y_D/\partial U_1| \sin(\Theta - \psi)}, \tag{6.9}$$

which becomes unity for optimum RF power matching[1], and S_{AM} is the amplitude-modulation (AM) sensitivity describing up-conversion of bias-current disturbances to the oscillation amplitude, i.e.,

$$S_{AM} = -\frac{|\partial Y_D/\partial I_0| \sin(\Theta - \Phi)}{|\partial Y_D/\partial U_1| \sin(\Theta - \psi)}. \tag{6.10}$$

For the self-mixing oscillator, (6.8) has to be combined with (6.6) which can be written as

$$\delta I_0 = D_{AM}\delta U_1. \tag{6.11}$$

---

[1] In the case of series- or parallel-resonant behaviour of the RF circuit, the RF matching factor α is equivalent to $2/s$ as used, e.g., by *Kurokawa* [6.9], where $s = (U_1 \partial G_D/\partial U_1)/G_L$ is the amplitude-saturation factor of the device conductance G_D.

Here, D_{AM} is the amplitude-demodulation sensitivity

$$D_{AM} = -\frac{\partial U_0/\partial U_1}{\partial U_0/\partial I_0 + R_B} \qquad (6.12)$$

expressing the down-conversion of the amplitude disturbances δU_1 to the bias current due to device-internal rectification. A combination of (6.8–11) yields either the total amplitude disturbance δU_1 containing down- and up-conversion

$$\delta U_1 = \frac{\alpha\sqrt{2P_s/G_L}}{1 - S_{AM}D_{AM}} \sin(\Theta - \Delta\omega t) \qquad (6.13)$$

or the bias-current disturbance δI_0, which represents the down-converted signal,

$$\delta I_0 = \frac{\alpha\sqrt{2P_s/G_L}\,\partial U_0/\partial U_1}{R_B + \partial U_0/\partial I_0 + S_{AM}\partial U_0/\partial U_1} \sin(\Delta\omega t - \Theta). \qquad (6.14)$$

6.2.2 Conversion-Gain Factors

The down-converted signal of frequency $\Delta\omega$ given in (6.14) leads to an output power P_d at the bias resistor R_B that can be related to signal power P_s to yield the total conversion gain:

$$g_c = \frac{P_d}{P_s} = \frac{\alpha^2(\partial U_0/\partial U_1)^2 R_B}{G_L(R_B + R_{dyn})^2}. \qquad (6.15)$$

where R_{dyn} is a shorthand for

$$R_{dyn} = \partial U_0/\partial I_0 + S_{AM}\partial U_0/\partial U_1. \qquad (6.16)$$

It is the inverse slope of the dynamic current-voltage characteristic of the oscillating device. It contains the contribution $\partial U_0/\partial I_0$ directly due to current, and another one due to the fact that current changes lead to RF amplitude variations, which are down-converted to the bias voltage.

For the purpose of discussion of the different mechanisms influencing down-conversion, the rhs of (6.15) is subdivided into three factors: the open-bias-circuit AM gain g_{AM0}, a matched demodulation sensitivity $g_{dem,m}$, and a bias-matching factor g_{BM}, i.e.,

$$g_c = g_{AM0}g_{dem,m}g_{BM}. \qquad (6.17)$$

The AM gain defined as $\frac{1}{2}\delta\hat{U}_1^2 G_L/P_s$ [6.12] ($\delta\hat{U}_1$ being the amplitude of the AM deviation from (6.8)) expresses the amplitude response of the oscillator to an injected signal of power P_s. It results, for the case of an open bias circuit (i.e., for a large bias resistance R_B giving the bias current: $\delta I_0 = 0$), in

$$g_{AM0} = \alpha^2. \qquad (6.18)$$

As RF matching factor α, g_{AM0} is unity if the RF circuit of the oscillator is matched to maximum output power, but it may become much larger when the amplitude saturation – expressed by $|\partial Y_D/\partial U_1|$ in (6.9) – is low. This is typical for low-amplitude operation of the oscillator near the (linear) small-signal condition. It is the reason for the generally high sensitivity of self-mixing oscillators at a low output power [6.4–6].

In the remaining factors of g_c on the rhs of (6.15), R_B may be optimized to achieve maximum conversion gain. This results in the matched demodulation sensitivity

$$g_{\text{dem,m}} = \frac{(\partial U_0/\partial U_1)^2}{4G_L R_{\text{dyn}}}, \tag{6.19}$$

which gives the down-converted output power P_d of the mixer related to the AM signal of the oscillator $\frac{1}{2}\delta\hat{U}_1^2 G_L$ in the case of the bias resistor R_B matched to R_{dyn}. It contains internal rectification $(\partial U_0/\partial U_1)$ as well as the impedances in RF and bias circuits. Demodulation decreases generally at lower amplitudes, since rectification, typically a phenomenon of nonlinearity, vanishes in the small-signal case. The increase of g_{AM0}, however, mostly overcompensates it [6.12].

The matching influence of the bias resistance R_B on g_c in (6.15) can finally be included in a bias-matching factor

$$g_{\text{BM}} = \frac{4R_B/R_{\text{dyn}}}{(R_B/R_{\text{dyn}} + 1)^2} \tag{6.20}$$

that becomes unity for $R_B = R_{\text{dyn}}$. This is the optimum bias-matching condition with respect to maximum output power of the down-converted signal.

For determination of the mixing-conversion sensitivity, expressed by the Minimum Detectable Signal (MDS), which depends also on the low-frequency noise of the device (Sect. 6.4.), it is more convenient to introduce the open-circuit voltage amplitude $\hat{U}_{d,0}$ of the down-converted signal. This can be found from $\sqrt{2P_d R_B}$ in the limit of large R_B. With (6.15) it follows then

$$\frac{\hat{U}_{d,0}^2}{P_s} = \frac{2\alpha^2(\partial U_0/\partial U_1)^2}{G_L} \tag{6.21}$$

which has to be compared to the open-circuit, low-frequency noise voltage to find the conversion sensitivity.

6.2.3 Simplified Device Model

For demonstration of the typical behaviour of and the influence of matching on self-mixing oscillators, a simplified analytical device model is introduced. It exhibits quadratic amplitude saturation for the negative device conductance $-G_D$

$$-G_D = G_0 - AU_1^2 \tag{6.22}$$

and quadratic RF rectification

$$U_0 = U_{00} - BU_1^2. \tag{6.23}$$

Here, the parameters G_0, A, U_{00}, and B will, in general, depend on the bias current I_0. Quadratic amplitude saturation and RF rectification are for all devices at least valid at small amplitudes U_1, i.e. for low output power. They represent, therefore, a good description near the onset of oscillations. This is a range of operation, where self-mixing oscillators exhibit high down-conversion sensitivity, as will be shown below. The above approximation may qualitatively, and for estimations even quantitatively, be used beyond their strict applicability. More accurate device simulations are presented in Sect. 6.5.

Device and circuit matching losses can be taken into account by an equivalent parallel loss conductance G_1 such that the remaining load conductance G_L for power output is

$$G_L = -G_D - G_1 = \eta G_0 - AU_1^2. \tag{6.24}$$

where η is a matching efficiency that describes the relative influence of losses and is defined as $\eta = 1 - G_1/G_0$. In addition, series- or parallel-resonant behaviours of the oscillator circuit are assumed. This is equivalent to $\text{Re}\{\partial(Y_D + Y_L)/\partial\omega\} = 0$. In this case, the RF matching factor α (6.9) reduces to

$$\alpha = \frac{2G_L}{U_1 \partial G_D/\partial U_1} = \frac{\eta G_0 - AU_1^2}{AU_1^2}. \tag{6.25}$$

The RF output power of the oscillator $P_0 = \frac{1}{2}U_1^2 G_L$ exhibits a maximum with respect to tuning of G_L of

$$P_{\text{max},I} = \frac{\eta^2 G_0^2}{8A}. \tag{6.26}$$

This is the maximum power that can be achieved by matching at a given bias current. The ratio $P_0/P_{\text{max},I}$ is a measure for the oscillation amplitude[2] U_1,

$$U_1^2 = \frac{\eta G_0}{2A}\left(1 - \sqrt{1 - P_0/P_{\text{max},I}}\right). \tag{6.27}$$

At high bias currents, the quadratic amplitude saturation is, in general, not valid up to the output-power maximum. At low powers, the amplitude U_1 can still be expressed as in (6.27), if $P_{\text{max},I}$ is replaced by some effective value that would be obtained by extrapolation of (6.22) to high amplitudes.

With (6.27), the RF matching factor α becomes

$$\alpha = \frac{1 + \sqrt{1 - P_0/P_{\text{max},I}}}{1 - \sqrt{1 - P_0/P_{\text{max},I}}}. \tag{6.28}$$

It is displayed in Fig. 6.3a versus $P_0/P_{\text{max},I}$ (dashed line) and can, at not too high power, be approximated well by $\alpha \approx 4P_{\text{max},I}/P_0 - 2$, i.e., it increases at low power as $1/P_0$ and, therefore, the AM gain g_{AMO} (6.18) increases as $1/P_0^2$.

The demodulation due to RF rectification, expressed by $\partial U_0/\partial U_1$ in (6.12,15, 21) results with (6.23) in

$$\partial U_0/\partial U_1 = -2BU_1, \tag{6.29}$$

---

[2] There is another solution for U_1^2 with the opposite sign infront of the square-root in (6.27). It is, however, not treated here since the self-mixing oscillator degrades at amplitudes above $P_{\text{max},I}$ with respect to conversion gain as well as noise.

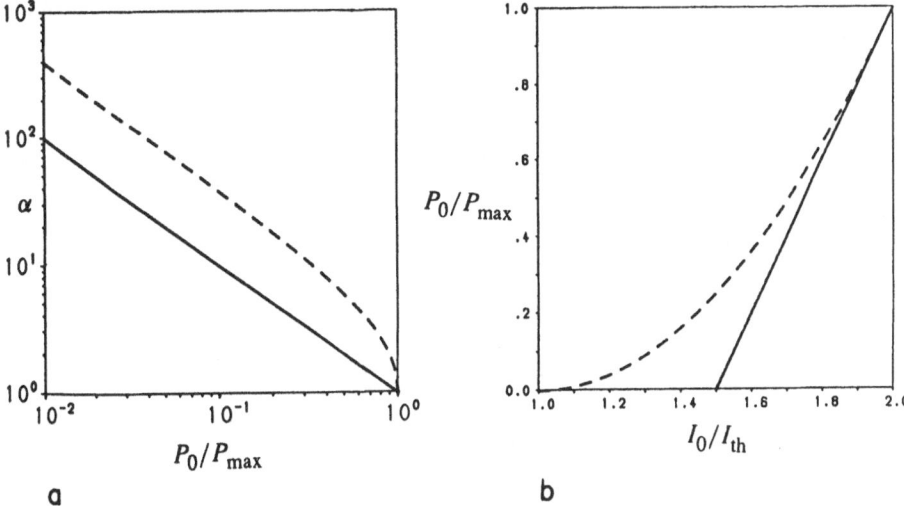

Fig. 6.3. (a) RF-matching factor α versus output power P_0 related to the maximum output power P_{max} for constant bias current and detuning of the RF circuit (dashed line) and for RF matching at one current with current reduction at fixed tuning (solid line) (b) P_0/P_{max} (P_{max} being the maximum power at $I_0 = 2I_{th}$) versus bias current I_0 related to oscillation threshold current I_{th} for RF matching at each current (dashed line) and for an RF circuit matched at $I_0 = 2I_{th}$ (solid line)

where the back-bias factor B can be found from the shift $\Delta U_{0,max,I}$ of the dynamic bias voltage $U_0(U_1)$, when the device is tuned (at constant current) from a non-oscillating condition to maximum output power:

$$\Delta U_{0,max,I} = -B\eta G_0/2A. \tag{6.30}$$

This value has been obtained from (6.23) with U_1^2 for the output-power maximum derived from (6.27) with $P_0 = P_{max,I}$. Insertion into (6.21) gives for the open-circuit voltage-amplitude

$$\frac{\hat{U}_{d,0}^2}{P_s} = \alpha^2 \frac{32A^2 \Delta U_{0,max,I}^2 U_1^2}{\eta^2 G_0^2 G_L}, \tag{6.31}$$

which shows that demodulation, expressed by the factors on the rhs of α^2, increases with amplitude[3] U_1 and tends to vanish in the small-signal case near $U_1 = 0$. If, however, the AM gain is taken into account with the factor α inserted from (6.25), which behaves inversely with U_1, and using (6.24,26), the total mixer conversion-sensitivity becomes

$$\frac{\hat{U}_{d,0}^2}{P_s} = 4\alpha \frac{\Delta U_{0,max,I}^2}{P_{max,I}}, \tag{6.32}$$

---

[3] In (6.31), also G_L is a function of amplitude, as U_1 grows with decreasing G_L to satisfy (6.24), and vice versa. This enhances the amplitude behaviour of (6.31).

It thus behaves, as a function of amplitude U_1 or output power P_0, as matching factor α (Fig. 6.3a, dashed line) and increases at low output power, since the high AM gain overcompensates the reduced demodulation sensitivity. In this regime, α can be expressed by $\alpha \approx 4P_{\mathrm{max},I}/P_0$, which results in

$$\frac{\hat{U}_{d,0}^2}{P_s} \approx \frac{16\Delta U_{0,\mathrm{max},I}^2}{P_0}. \tag{6.33}$$

In this approximation, the down-converted signal $\hat{U}_{d,0}^2$ depends only on the transmitter-receiver system loss P_0/P_s, i.e. the ratio of transmitted to received power, and not on the transmitting power itself.

All device and circuit properties which are essential for down-conversion are included in (6.33) in the shift $\Delta U_{0,\mathrm{max},I}$ of the device bias-voltage from the non-oscillating state to the maximum output power, as given by (6.30). This quantity contains the device ability for RF rectification expressed in (6.23) by the back-bias factor B, the amplitude saturation factor A of the negative device admittance from (6.22), and the matching efficiency η. For a device to be suitable for self-mixing oscillations, it is, therefore, not sufficient that it exhibits a high rectification sensitivity, but it should also show low amplitude saturation and a negative small-signal conductance G_0 that well exceeds all losses expressed by G_1. In particular, the high conversion sensitivities reported in the literature [6.4,5] for a low oscillator output-power level are not achieved when reducing the output power by additional losses or with poor devices but only by detuning the oscillator from maximum power. In any case, the abilities of a device or circuit to become a self-mixing oscillator can, however, easily be tested by the shift of the dynamic bias voltage during tuning.

For experimental verification and in many self-mixing oscillator applications, it is unconvenient to operate the device at constant bias current and to change the matching by proper detuning of the RF circuit, as assumed above. The device is rather matched to the output-power maximum at some bias current I_m well above oscillation threshold (mostly at the highest possible operating current). Bias is then reduced with fixed RF tuning, i.e. with a constant RF load. As a consequence, the device is mismatched with respect to the possible maximum output power at this current, resulting in high conversion sensitivity. This procedure can be reproduced more easily than some defined RF detuning.

In this case, the dependence of the model parameters in (6.22,23) on bias current is essential. The negative small-signal device conductance G_0 (without losses) is in the following assumed to increase proportional to bias current I_0, whereas A and B remain constant. This behaviour is appropriate to most transit-time devices such as IMPATT-, BARITT-, or TUNNETT-diodes at not too high current, since the induced RF conduction-current of the drifting carriers is at least in the sharp-pulse approximation [6.13] proportional to the dc current, and the phase relations remain fairly constant. G_0 can then be expressed by the loss conduction G_1 and the oscillation-threshold current I_{th}, i.e. the minimum current at which the oscillator can be tuned to oscillations (with $G_L \approx 0$).

$$G_0 = G_1 I_0/I_{\mathrm{th}} \tag{6.34}$$

This allows to express the maximum possible output power $P_{max,I}$ for matching at each current I_0 from (6.26) as

$$P_{max,I} = \frac{(I_0/I_{th} - 1)^2 G_1^2}{8A}, \tag{6.35}$$

which is shown versus bias current I_0 in Fig. 6.3b, related to the maximum output power P_m at $I_0 = 2I_{th}$ (dashed line).

For fixed tuning to maximum output power at $I_0 = I_m$, the load conductance G_L remains constant.

$$G_L = \frac{1}{2} G_1 (I_m/I_{th} - 1) \tag{6.36}$$

This produces a change of oscillation amplitude U_1 with bias current according to (6.24) (with G_0 from (6.34) and $\eta = 1 - I_{th}/I_0$):

$$AU_1^2 = G_1[I_0 - (I_m + I_{th})/2]/I_{th} \tag{6.37}$$

Equation (6.37) shows that oscillations do in this case not start at $I_0 = I_{th}$ but at the higher value of $I_0 = (I_m + I_{th})/2$. The corresponding output power $P_0 = \frac{1}{2} U_1^2 G_L$ is

$$P_0 = \frac{(2I_0 - I_m - I_{th})(I_m - I_{th})G_1^2}{8AI_{th}^2} \tag{6.38}$$

or, related to the maximum output power P_m at the matching current $I_0 = I_m$

$$\frac{P_0}{P_m} = \frac{2I_0 - I_m - I_{th}}{I_m - I_{th}}. \tag{6.39}$$

This is also depicted in Fig. 6.3b as a function of I_0 for $I_m = 2I_{th}$ (solid line) and shows that especially near the onset of oscillations at $I_0 \approx (I_{th} + I_m)/2$, output power P_0 is much lower than the corresponding $P_{max,I}$ (dashed line), i.e. there exists strong mismatch.

The RF matching factor α can, in this case, be derived from (6.25) with (6.37):

$$\alpha = \frac{I_m - I_{th}}{2I_0 - I_m - I_{th}} = \frac{P_m}{P_0}. \tag{6.40}$$

This ratio is depicted in Fig. 6.3a by the solid line.

AM demodulation can again be treated as above. It is, however, expressed here by the bias-voltage shift $\Delta U_{0,m}$ between oscillating and non-oscillating device at the matching point for $I_0 = I_m$,

$$\Delta U_{0,m} = -\frac{BG_1}{2A}(I_m/I_{th} - 1) \tag{6.41}$$

which reduces (6.21) with (6.29) to

$$\frac{\hat{U}_{d,0}^2}{P_s} = \alpha^2 \frac{4\Delta U_{0,m}^2 P_0}{P_m^2} \tag{6.42}$$

and with (6.40) to

$$\frac{\hat{U}_{d,0}^2}{P_s} = 4\frac{\Delta U_{0,m}^2}{P_0}. \tag{6.43}$$

Comparing (6.43) with the result obtained from (6.33) for device operation at constant current (with $I_0 = I_m$) and detuning the RF circuit to lower output power, shows at the same RF power P_0 an about 6-dB lower down-converted signal for the case of fixed matching at $I_0 = I_m$ and then reducing the current. Since at lower current the low-frequency noise is less, the sensitivity of the self-mixing oscillator may, in both cases, be comparable. The latter matching procedure is, however, easier to accomplish, and the reduced current can, in addition, favour longterm stability, dc power consumption, etc..

6.3 Matrix Formulation of Conversion Gain

In a common treatment of passive mixers (e.g., Schottky-diode mixers), conversion matrices are applied to relate currents and voltages at all sum and difference frequencies of interest [6.14]. Hereby, the conversion-matrix elements are usually derived from a Fourier analysis of a time-varying device admittance when driven by the local-oscillator signal. A similar procedure has also been applied to self-oscillating BARITT-diode mixers [6.15,16] yielding however, only poor agreement with experiment. Here, signals at the image frequency had been neglected which is not appropriate to typical self-mixing oscillator applications with low-frequency setoff such that both signal and image frequency lie inside the bandwidth of the oscillator's resonance circuit.

6.3.1 Conversion Matrix

In this section, a general conversion matrix of self-mixing oscillators is derived [6.12] from the amplitude-, bias-current-, and frequency-dependent admittances and the dynamic bias current-voltage characteristic of the active device as applied in Sect. 6.2. It relates signal currents and voltages at the mixer-input frequency ω_s, (i.e., the original frequency of the incoming signal) at the image frequency $\omega_i = 2\omega_0 - \omega_s$, and at the down-converted frequency $\omega_d = |\Delta\omega|$, where $\Delta\omega = \omega_s - \omega_0$ is the frequency difference between signal and free-running oscillation equivalent to the frequency setoff introduced in Sect. 6.2, i.e.,

$$\begin{vmatrix} \tilde{I}_s \\ \tilde{U}_d \\ \tilde{I}_i^* \end{vmatrix} = \begin{vmatrix} M_{ss} & M_{sd} & M_{si} \\ M_{ds} & M_{dd} & M_{di} \\ M_{is}^* & M_{id}^* & M_{ii}^* \end{vmatrix} \cdot \begin{vmatrix} \tilde{U}_s \\ \tilde{I}_d \\ \tilde{U}_i^* \end{vmatrix}. \tag{6.44}$$

Here, $\tilde{I}_s$, $\tilde{I}_d$, $\tilde{I}_i$ and $\tilde{U}_s$, $\tilde{U}_d$, $\tilde{U}_i$ are the complex phasors of device current and voltage at the frequencies ω_s, ω_d, and ω_i, respectively; M_{jk} denotes the conversion-matrix elements; and an asterisk means complex conjugation. After determination of the matrix elements, the conversion matrix will be used in the next subsection to derive the conversion gain, leading to similar but more general results, as found in Sect. 6.2.

Complex phasor notation is applied for the total device RF current $\tilde{I}_1 = \tilde{I}_{10} + \delta \tilde{I}_1$ and RF voltage $\tilde{U}_1 = \tilde{U}_{10} + \delta \tilde{U}_1$. These can generally be related by the device admittance

$$\tilde{I}_1 = Y_D(I_0, U_1, \omega) \cdot \tilde{U}_1, \tag{6.45}$$

where the RF voltage amplitude U_1 is the absolute value of $\tilde{U}_1$. One of the phasors may arbitrarily be assumed real as a reference, here the stationary value of the RF voltage $\tilde{U}_{10} = U_{10}$. A linear extension of (6.45) with respect to current, voltage, and frequency deviations caused by the input signal (as defined in (6.3)) yields

$$\delta \tilde{I}_1 = U_{10}[(\partial Y_D/\partial I_0)\delta I_0 + (\partial Y_D/\partial U_1)\delta U_1 + (\partial Y_D/\partial \omega)\delta \omega] + Y_{D0}\delta \tilde{U}_1. \tag{6.46}$$

where $Y_{D0} = Y_D(I_{00}, U_{10}, \omega_0)$ is the undisturbed device admittance at oscillation frequency ω_0. The deviations δU_0 and δI_0 of the bias voltage and current are related due to the dynamic bias current-voltage characteristic $U_0 = U_0(I_0, U_1)$ by

$$\delta U_0 = (\partial U_0/\partial I_0)\delta I_0 + (\partial U_0/\partial U_1)\delta U_1. \tag{6.47}$$

RF current- and voltage disturbances consist of contributions from both sidebands at ω_s and ω_i

$$\delta \tilde{I}_1 = \tilde{I}_s e^{j\Delta \omega t} + \tilde{I}_i e^{-j\Delta \omega t}, \tag{6.48}$$
$$\delta \tilde{U}_1 = \tilde{U}_s e^{j\Delta \omega t} + \tilde{U}_i e^{-j\Delta \omega t},$$

whereas disturbances of bias current and voltage are to be expressed as

$$\delta I_0 = \frac{1}{2}\left(\tilde{I}_d e^{j\Delta \omega t} + \tilde{I}_d^* e^{-j\Delta \omega t}\right), \tag{6.49}$$
$$\delta U_0 = \frac{1}{2}\left(\tilde{U}_d e^{j\Delta \omega t} + \tilde{U}_d^* e^{-j\Delta \omega t}\right).$$

The RF voltage amplitude deviation δU_1 is the real part of $\delta \tilde{U}_1$

$$\delta U_1 = \frac{1}{2}\left[(\tilde{U}_s + \tilde{U}_i^*)e^{j\Delta \omega t} + (\tilde{U}_s^* + \tilde{U}_i)e^{-j\Delta \omega t}\right] \tag{6.50}$$

and the frequency deviation $\delta \omega$ can be deduced from the time derivative of U_1 as

$$\delta \omega = -j(1/U_{10})(d\tilde{U}_1/dt)$$
$$= (\Delta \omega/U_{10})(\tilde{U}_s e^{j\Delta \omega t} - \tilde{U}_i e^{-j\Delta \omega t}). \tag{6.51}$$

This includes not only a real part due to phase changes but also an imaginary part resulting from variations of the RF amplitude. The imaginary part of $\delta \omega$ has been neglected in Sect. 6.2, which is justified at a low frequency offset $|\Delta \omega|$. It leads, however, to a reduction of the conversion gain at large offset, as will be shown below.

Inserting (6.48–51) into (6.46,47), separating terms with $e^{j\Delta\omega t}$ and $e^{-j\Delta\omega t}$, and comparison with (6.44) results in the conversion-matrix elements:

$$
\begin{aligned}
M_{ss} &= Y_{D0} + (U_{10}/2)(\partial Y_D/\partial U_1) + \Delta\omega(\partial Y_D/\partial\omega), \\
M_{sd} &= M_{id} = (U_{10}/2)(\partial Y_D/\partial I_0), \\
M_{si} &= M_{is} = (U_{10}/2)(\partial Y_D/\partial U_1), \\
M_{ds} &= M_{di} = \partial U_0/\partial U_1, \\
M_{dd} &= \partial U_0/\partial I_0, \\
M_{ii} &= Y_{D0} + (U_{10}/2)(\partial Y_D/\partial U_1) - \Delta\omega(\partial Y_D/\partial\omega).
\end{aligned}
\tag{6.52}
$$

6.3.2 Conversion Gain

The conversion matrix (6.52) can be applied to derive the conversion gain of a self-mixing oscillator in conjunction with the oscillator RF and bias circuitry. At signal frequency ω_s, an RF equivalent circuit, as shown on the lhs of Fig. 6.1 will be used, at image frequency ω_i the same but without signal source i_s, and at downconverted frequency ω_d as on the rhs of Fig. 6.1 (without U_B). This results in

$$
\begin{aligned}
\tilde{U}_s &= (\hat{i}_s - \tilde{I}_s)/Y_L(\omega_s), \\
\tilde{U}_i &= -\tilde{I}_i/Y_L(\omega_i), \\
\tilde{I}_d &= -\tilde{U}_d/R_B.
\end{aligned}
\tag{6.53}
$$

In combination with (6.44), (6.53) determines all current and voltage phasors. From these, the output signal and, related to the input, the conversion gain of the self-mixing oscillator can be derived. In this formulation the result is, however, rather cumbersome. The derivation is, therefore, divided into several steps.

As in Sect. 6.2, the open-bias-circuit AM gain $g_{AM0} = \frac{1}{2}\delta\hat{U}_1^2 G_L/P_s$ can be calculated for infinite R_B (i.e., for $\tilde{I}_d = 0$), with $\delta\hat{U}_1 = |\tilde{U}_s + \tilde{U}_i^*|$, see (6.50). When $\tilde{I}_d = 0$, (6.44) can be reduced to the four-element matrix equation

$$
\begin{vmatrix} \tilde{I}_s \\ \tilde{I}_i^* \end{vmatrix} = \begin{vmatrix} M_{ss} & M_{si} \\ M_{is}^* & M_{ii}^* \end{vmatrix} \cdot \begin{vmatrix} \tilde{U}_s \\ \tilde{U}_i^* \end{vmatrix}.
\tag{6.54}
$$

Insertion of $\tilde{I}_s$ and $\tilde{I}_i$ from (6.53) yields $\tilde{U}_s$ and $\tilde{U}_i^*$. These lead to

$$
g_{AM0} = \frac{4G_L^2|Y_L^*(\omega_i) + M_{ii}^* - M_{is}^*|^2}{|(Y_L(\omega_s) + M_{ss})(Y_L^*(\omega_i) + M_{ii}^*) - M_{si}M_{is}^*|^2}.
\tag{6.55}
$$

In the following, a linearized frequency dependence of the load admittance $Y_L(\omega) = -Y_{D0} + (\omega - \omega_0)(dY_L/d\omega)$ is used such that

$$
\begin{aligned}
Y_L(\omega_s) &= -Y_{D0} + \Delta\omega(dY_L/d\omega), \\
Y_L(\omega_i) &= -Y_{D0} - \Delta\omega(dY_L/d\omega).
\end{aligned}
\tag{6.56}
$$

Together with the conversion-matrix elements (6.52) as well as first and last line of (6.7), the open-circuit AM gain g_{AM0} (6.55) can finally be expressed as

$$g_{AM0} = \frac{1}{(1/\alpha)^2 + [(\Delta\omega/\omega_0)Q_L]^2}. \tag{6.57}$$

Here, α is the RF matching factor given in (6.9), and Q_L is a quality factor of the oscillator circuit defined as

$$Q_L = \frac{\omega_0|\partial(Y_D + Y_L)/\partial\omega|}{2G_L}. \tag{6.58}$$

At low frequency setoff $|\Delta\omega| \ll \omega_0/\alpha Q_L$, (6.57) gives the same result as (6.18) in Sect. 6.2; at higher values of $|\Delta\omega|$, however, g_{AM0} is reduced due to the fact that the oscillation amplitude cannot follow the phase changes of the input signal because of the energy stored in the reactances of the resonator circuitry.

For finite values of R_B (i.e., for $\tilde{I}_d \neq 0$), the last line of (6.53) can be used for the transformation of (6.44) to a four-element matrix equation as given in (6.54) with the transformed matrix elements

$$M'_{jk} = M_{jk} - \frac{M_{jd}M_{dk}}{M_{dd} + R_B}. \tag{6.59}$$

This is equivalent to a transformation of $\partial Y_D/\partial U_1$ to

$$\frac{\partial Y'_D}{\partial U_1} = \frac{\partial Y_D}{\partial U_1} - \frac{(\partial Y_D/\partial I_0)\partial U_0/\partial U_1}{\partial U_0/\partial I_0 + R_B} \tag{6.60}$$

in the matrix elements (6.52). With this, amplitude-modulation gain g_{AM} can be expressed as in (6.57) with the transformed matching factor

$$\alpha' = \alpha\frac{\partial U_0/\partial I_0 + R_B}{R_{dyn} + R_B} \tag{6.61}$$

where R_{dyn} is the inverse slope of the dynamic bias current-voltage characteristic as given in (6.16).

The demodulation sensitivity g_{dem} is expressed as

$$g_{dem} = \frac{\frac{1}{2}|\tilde{I}_d^2|R_B}{\frac{1}{2}\delta\hat{U}_1^2 G_L}. \tag{6.62}$$

$|\tilde{I}_d|/\delta\hat{U}_1$ can be found from the second line of (6.44) together with (6.50) and last line of (6.53) giving

$$g_{dem} = \frac{(\partial U_0/\partial U_1)^2}{G_L}\frac{R_B}{(\partial U_0/\partial I_0 + R_B)^2}. \tag{6.63}$$

The demodulation factor contains essentially the device-internal rectification expressed by $\partial U_0/\partial U_1$ and the power matching in the bias circuit due to R_B.

The total conversion gain of the self-mixing oscillator $g_c = \frac{1}{2}|\tilde{I}_d|^2 R_B/P_s$ is the product of g_{AM} [(6.57) with α replaced by α' from (6.61)] and g_{dem} (6.63). In the case of the low-frequency setoff such that the second term in the denominator on the rhs of (6.57) may be neglected, this gives the same result as (6.15) in Sect. 6.2.

6.4 Noise in Self-Mixing Oscillators

The noise properties of self-mixing oscillators which limit its receiver sensitivity are usually expressed by the Minimum-Detectable Signal (MDS) [6.3–6] defined as that mixer-input intensity which produces an output signal of equal power as the mixer-output noise. There are two sources of noise that contribute to it.

One source is the RF noise of the oscillator circuit in the neighbourhood of the oscillation frequency resulting mostly from carrier-transport fluctuations in the active device. This can be represented by an equivalent-noise current source $i_{n,RF}$ that may be included in the RF circuit on the lhs of Fig. 6.1, in addition to the signal-current source i_s. It consists of contributions from both sidebands at $\omega_s = \omega_0 + \Delta\omega$ and $\omega_i = \omega_0 - \Delta\omega$, both of them produce amplitude distortions that can be demodulated and result in down-converted noise at the mixer output.

Another noise source exists in the bias circuit, which exhibits current and/or voltage fluctuations due to the low-frequency carrier-transport properties of the active device including contacts. These are directly detected at the mixer output, they can however, also be up-converted to amplitude distortions due to the modulation sensitivity and then down-converted to the mixer output by internal rectification.

RF and low-frequency noise processes may be assumed to be independent, i.e. uncorrelated, such that the resulting noise powers at the output port are to be added. In most practical cases, however, one of the mechanisms predominates sufficiently that the other may be neglected.

6.4.1 RF Noise

To demonstrate the effect of RF noise, the random-mean-square (rms) value of the RF-noise current $i_{n,RF}$ is described by a noise measure M:

$$\langle |i_{n,RF}|^2 \rangle = 4kTMBG_L, \tag{6.64}$$

where k is Boltzmann's constant, T is the absolute temperature, and B is the bandwidth of the measuring set-up. At a low oscillation amplitude, M is equivalent to the amplifier-noise measure if the negative device admittance is optimally matched to the load G_L. Under these conditions, noise currents at ω_s and ω_i are uncorrelated such that their effects may be added. In the large-signal case, a corresponding effective AM noise measure may also be introduced. This will generally be larger than the small-signal value of M.

The down-conversion of an RF-current amplitude $\hat{i}_s$ to a current amplitude $\delta\hat{I}_0$ in the bias circuit can directly be deduced from (6.14) by substitution of P_s according to (6.2). The same ratio is valid for the rms values. If the equal-noise contributions from both sidebands are quadratically added, this yields for the effective down-converted noise current in the bias circuit $\delta I_{0,n}$

$$\frac{\langle|\delta I_{0,\mathrm{n}}|^2\rangle}{2\langle|i_{\mathrm{n,RF}}|^2\rangle} = \frac{\delta \hat{I}_0^2}{i_{\mathrm{s}}^2} = \frac{\alpha^2(\partial U_0/\partial U_1)^2}{4G_{\mathrm{L}}^2(R_{\mathrm{B}} + R_{\mathrm{dyn}})^2} \tag{6.65}$$

and leads to the noise power P_{n} at the output resistance R_{B}, namely

$$P_{\mathrm{n}} = \langle|\delta I_{0,\mathrm{n}}|^2\rangle R_{\mathrm{B}} = \frac{2kTMB\alpha^2(\partial U_0/\partial U_1)^2 R_{\mathrm{B}}}{G_{\mathrm{L}}(R_{\mathrm{B}} + R_{\mathrm{dyn}})^2}. \tag{6.66}$$

The self-mixing oscillator sensitivity $\mathrm{MDS_{RF}}$ due to RF noise, defined as the mixer input signal P_{s} which gives a down-converted signal P_{d} of equal magnitude as the noise power P_{n}, results from (6.15) with P_{d} substituted by P_{n} (6.66)

$$\mathrm{MDS_{RF}} = P_{\mathrm{n}}/g_{\mathrm{c}} = 2\,kTMB. \tag{6.67}$$

This shows that with an oscillator exhibiting about 20 dB noise measure, a minimum-detectable signal down to -150 dBm should be achievable at $T = 300$ K in 1-Hz bandwidth, which is independent of the output power P_0 if the RF-noise source predominates.

6.4.2 Low Frequency Noise

At not too low output power, the contribution of bias-noise sources is, however, generally much stronger and determines the minimum-detectable signal especially at low frequency setoff ω_{d}. This can be incorporated into the oscillator model by introducing an open-circuit noise-voltage source $u_{\mathrm{n,0}}$ into the bias circuit, shown on the rhs of Fig. 6.1 (e.g., instead of the bias-supply voltage U_{B}). The device acts here with its dynamic resistance R_{dyn} as the internal impedance of the noise-voltage source including the mechanism of up-conversion of current fluctuations due to the AM sensitivity S_{AM} leading to amplitude distortions that are down-converted by internal rectification $\partial U_0/\partial U_1$. The resulting noise power P_{n} at the mixer output-resistance R_{B} thus becomes

$$P_{\mathrm{n}} = \frac{\langle|u_{\mathrm{n,0}}|^2\rangle R_{\mathrm{B}}}{(R_{\mathrm{B}} + R_{\mathrm{dyn}})^2}. \tag{6.68}$$

The Minimum-Detectable-Signal $(\mathrm{MDS_0})$ is then

$$\mathrm{MDS_0} = \frac{P_{\mathrm{n}}}{g_{\mathrm{c}}} = \frac{\langle|u_{\mathrm{n,0}}|^2\rangle G_{\mathrm{L}}}{\alpha^2(\partial U_0/\partial U_1)^2} \tag{6.69}$$

independent of the bias resistance R_{B} since both g_{c} and P_{n} depend in the same way on R_{B}. With (6.21), $\mathrm{MDS_0}$ can also be expressed as

$$\mathrm{MDS_0} = 2\frac{\langle|u_{\mathrm{n,0}}|^2\rangle}{\hat{U}_{\mathrm{d,0}}^2/P_{\mathrm{s}}}, \tag{6.70}$$

i.e., it behaves essentially inversely to the quantity $\hat{U}_{\mathrm{d,0}}^2/P_{\mathrm{s}}$ discussed in Sect. 6.2.3. It thus decreases according to (6.33) or (6.43) at low output power proportional to P_0. The maximum admissable system loss, i.e. the ratio of transmitting output power to minimum-detectable signal reaches there a constant value of

$$\frac{P_0}{\text{MDS}_0} = \frac{8\Delta U_{0,\max,I}^2}{\langle |u_{n,0}|^2 \rangle} \tag{6.71}$$

in the case of constant current with detuning the RF circuit, or

$$\frac{P_0}{\text{MDS}_0} = \frac{2\Delta U_{0,m}^2}{\langle |u_{n,0}|^2 \rangle} \tag{6.72}$$

for RF matching at $I_0 = I_m$ and then reducing the current.

The total Minimum-Detectable Signal (MDS) is the sum of RF and bias-noise contributions:

$$\text{MDS} = \text{MDS}_{\text{RF}} + \text{MDS}_0. \tag{6.73}$$

At very low output power, the RF noise source may dominate the mixer sensitivity. For most practical operation conditions, however, the sensitivity is bias-noise-limited with $\text{MDS} = \text{MDS}_0$.

6.5 Numerical Simulations

As an example, conversion-gain calculations are performed for a self-mixing IMPATT oscillator containing a quasi-Read double-drift diode that operates in the upper V-band. Here, the device properties are not derived from a simplified analytic model as applied in Sect. 6.2.3, but from an accurate large-signal partial-differential numerical computer simulation of the carrier-transport equations in time domain. The simulation program [6.11] is based on a drift-diffusion model [6.17] with material parameters for silicon adopted from *Grant* [6.18] and from *Jacoboni et al.* [6.19].

The device structure consists of a $p^{++}pn^{+}nn^{++}$-series of layers, as described in Sect. 2 with 3×10^{19} cm^{-3} and 1.1×10^{17} cm^{-3} doping concentration for the p^{++}- and the p-layer and 1×10^{18} cm^{-3}, 6×10^{16} cm^{-3}, and 3×10^{19} cm^{-3} for the n^{+}-, the n-, and the n^{++}- layer, respectively. The widths of the inner active p, n^{+}, and n layer are 420 nm, 30 nm, and 270 nm, respectively, that of the p^{++}- and n^{++}-contact regions 200 nm each.

For large-signal simulation of IMPATT devices, it is most appropriate to use impressed bias current to achieve stable operation conditions on the dynamic current-voltage characteristic together with sinusoidal RF voltage drive which avoids excitation of the avalanche resonance. In each operation point, characterized by the bias current I_0 as well as the amplitude U_1 and circular frequency ω of the RF oscillation, the device admittance $Y_D(I_0, U_1, \omega)$ can be derived from the resulting RF current by Fourier analysis, whereas the mean value of the device voltage represents the dynamic bias voltage $U_0(I_0, U_1, \omega)$. The functional dependences of admittance and bias voltage on the operation conditions determine the self-mixing properties of the device as formulated in Sects. 6.2.1,2.

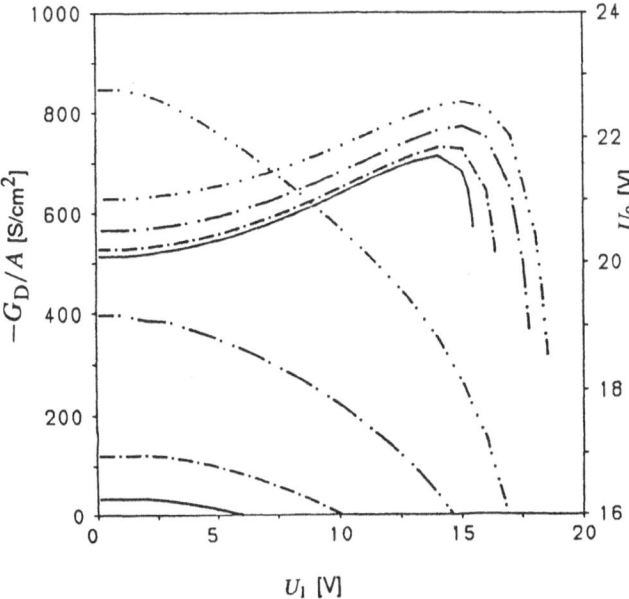

Fig. 6.4. Calculated negative device conductance per unit area $-G_D/A$ (monotonically descending curves, left scale) and dynamic bias voltage U_0 (curves with maximum, right scale) versus oscillation amplitude U_1 of a quasi-Read IMPATT device. (Parameter: bias-current density I_0/A —— 1×10^3 A/cm^2, — · — · 2×10^3 A/cm^2, — ·· — 5×10^3 A/cm^2, — ··· — 1×10^4 A/cm^2)

Figure 6.4 displays the calculated negative conductance per unit area $-G_D/A$ of the quasi-Read IMPATT-device (monotonically descending curves, left scale, A being the active cross-section area) and the dynamic bias voltage U_0 (curves with maximum, right scale), each plotted versus oscillation amplitude U_1 for several values of the bias-current density, as achieved from device simulations at an oscillation frequency of 76 GHz and a temperature of 500 K in the active region regarding self-heating due to dc power dissipation. Both, negative-device conductance and dynamic-bias voltage exhibit similar behaviour as assumed in the simplified analytic model of Sect. 6.2.3.: The negative device conductance $-G_D$ is at low amplitudes approximately proportional to the dc current density I_0/A and saturates with increasing oscillation amplitude U_1 in a nearly parabolic manner at least up to amplitudes of about 12 V. In addition the shift of bias voltage U_0 behaves quadratically with amplitude, as suggested by (6.23) (again up to $U_1 \approx 12$ V) with, however, a negative back-bias factor B ("negative rectification"), i.e., the dc voltage increases with amplitude. Such behaviour has also been found experimentally and may arise from enhanced depletion of the avalanche region during the negative RF voltage half-swing, as described in [6.11 and 20]. But the sign of B is anyway of no significance for the conversion behaviour.

For a determination of the derivatives of device conductance and dynamic-bias voltage with respect to bias current and oscillation amplitude, as needed for the conversion-gain equations in Sect. 6.2, the curves in Fig. 6.4 are smoothed by matched low-order polynomials. In addition, a diode cross-section diameter of 20 μm and a maximum admissible current density of 1×10^4 A/cm^2 is assumed. With these parameters, the output power P_0 and matched conversion gain $g_{c,m}$ are calculated for series- or parallel-resonance matching as functions of oscillation amplitude or bias current, and are plotted in Fig. 6.5 as the $g_{c,m}P_0$-product versus output power P_0 for several different cases.

The dashed lines in Fig. 6.5 correspond to device operation at constant (maximum) current and detuning the RF circuit, whereas the solid lines show the behaviour for constant RF matching to maximum-power output and then reducing the bias current. The upper curves are found in each case if RF losses due to device contact-resistances and resonator losses are neglected, and in the lower curves, the RF losses expressed by a device series-resistance of $R_s = 1.0\ \Omega$ are taken into account. The corresponding effective-loss conductance G_1 is determined via the total device admittance including the imaginary part (not shown in Fig. 6.4).

It can be seen that losses reduce the output power and conversion gain. But even with losses appropriate for monolithically-integrated oscillators, a conversion-gain-power product of several dBm seems to be achievable with IMPATT devices, too. In these calculations, however, additional losses of the radiated RF power outside the resonator circuit are not regarded. They affect equally output power and received signal such that, e.g., 5 dB radiation loss will reduce the conversion-gain-power product by 10 dB.

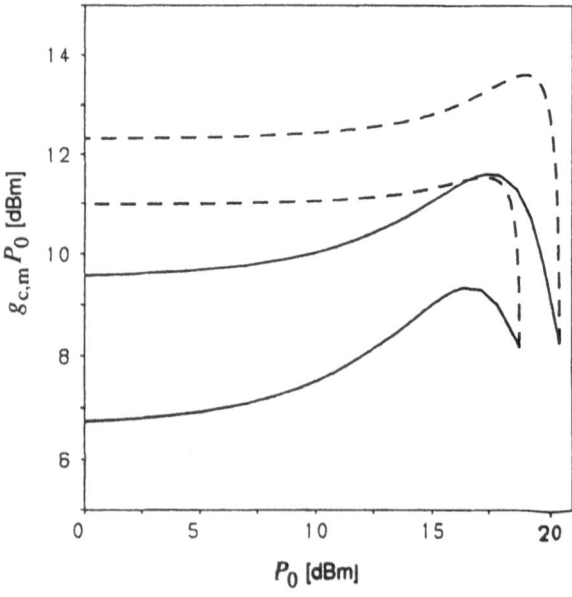

Fig. 6.5. Matched-conversion-gain-power product $g_{c,m}P_0$ versus output power P_0 for constant bias current and RF detuning (dashed lines) and for constant matching and current reduction (solid lines), each without losses (upper curves) and with series loss-resistance of 1 Ω (lower curves)

The numerical results in Fig. 6.5, which were derived from a rather complex large-signal computer simulation of the device, can be compared with the simplified analytic model of Sect. 6.2.3. by means of (6.43). Insertion of the dynamic-bias voltage shift of, e.g., $\Delta U_{0,m} \approx 1.2$ V from Fig. 6.4 in the power-matching point at $U_{1,max} \approx 12$ V (appropriate for a series loss resistance of 1 Ω) and regarding that $\hat{U}_{d,0}^2/P_s = 8g_{c,m}R_{dyn}$, with R_{dyn} calculated from (6.16) yielding $R_{dyn} \approx 140 \ \Omega$ at $P_0 = 0$ dBm, results in a conversion-gain-power product of $g_{c,m}P_0 \approx 7$ dBm. This is in reasonable agreement with the lower solid line in Fig. 6.5, as determined under equal conditions indicating that the dynamic-bias-voltage shift is indeed a good measure for the self-mixing properties of a device. The deviations from a constant value in the numerical calculations, shown in Fig. 6.5, are due to variations of R_{dyn} as well as deviations from the parabolic behaviour of $-R_D(U_1)$ and $U_0(U_1)$ but reach only a few dB. For the dashed lines, also the dependence of $\alpha(P_0)$, as shown in Fig. 6.3a has strong influence.

6.6 Measuring Techniques and Experimental Results

The characteristic quantities specifying the properties of a self-mixing oscillator are the conversion gain $g_{c,m}$, the mixer output-noise spectral density, the minimum-detectable signal and the matching impedance R_{dyn} for each operation condition of the oscillator characterized by its output power P_0. These will, in general, depend on the modulation frequency $f_d = \omega_d/2\pi$ resulting from the frequency set-off of the received signal and on the specific method of oscillator mismatch. To achieve reproducible results, it is most convenient to match the oscillator to maximum output power at the maximum admissable bias current and to control the power output only by bias-current changes with fixed RF circuit tuning. This procedure is applied in all experimental investigations reported below.

6.6.1 Measuring Set-up

All quantities characterizing the self-mixing properties of a device can be measured in a configuration that simulates Doppler-radar operation. Such a set-up is shown in Fig. 6.6: The active device is incorporated into an oscillator circuitry that transmits its output power into a waveguide system. Bias supply is connected to the device via an adjustable bias resistor R_B, at which the down-converted signal can be detected by a selective voltmeter.

The RF output power is measured and then passes an attenuator that replaces the damping of radiated power by spreading. Instead of being reflected at a moving object, the signal enters a pin modulator that is driven by a square-wave generator. Here, the signal is alternately transmitted to an absorber or is fully reflected. The reflected signal thus exhibits a 100% square-wave amplitude modulation with a modulation frequency given by the period of the driving generator. This adds a series of upper and lower sidebands to the reflected signal of which only the two

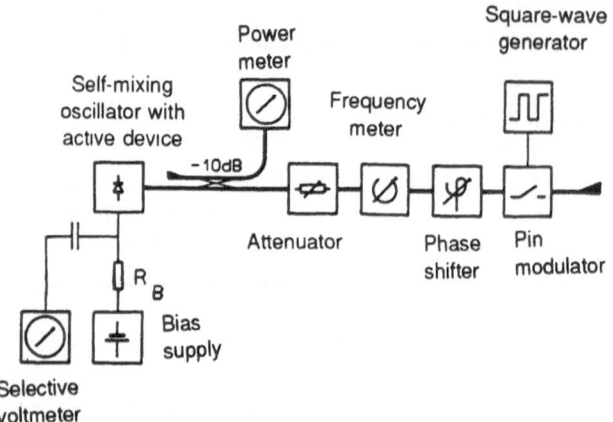

Fig. 6.6. Experimental set-up for measuring conversion gain, minimum-detectable signal, mixer output-noise, and matching impedance of a self-mixing oscillator in a Doppler-radar simulating configuration

lowest-order signals are further considered. The amplitude of these sidebands is $1/\pi$ times the amplitude of the incoming signal. Another possibility is to use the transmission-modulated signal behind the pin modulator by guiding it back to the oscillator through a circulator [6.4,5].

The reflected (or back-guided) signal passes the attenuator and is finally injected back into the oscillator. Here, the received signal superimposes onto the oscillation and is, by self-mixing, down-converted to the bias where it can be measured with the selective voltmeter.

The difference as compared to an actual Doppler-radar system is in each case, however, the simultaneous receipt of two signals with positive and negative frequency shift. The resulting amplitude distortion of the oscillator depends critically on the relative phase of the two sidebands. This can be adjusted by a phase shifter, e.g., between the pin modulator and the attenuator such that maximum amplitude modulation of the oscillator is achieved. This can be monitored via the output signal with the selective voltmeter. Then, both sidebands superimpose in such a way that their effective amplitude doubles, i.e., it simulates the effect of a single-sideband signal of four-times the power. The effectively received signal input power P_s of the set-up (Fig. 6.6) can, therefore, be determined from the oscillator output power P_0 by

$$P_s = \frac{4}{\pi^2} \frac{P_0}{\alpha_{att}^2}. \tag{6.74}$$

It can in its magnitude be adjusted by the variable damping α_{att} of the attenuator. The frequency shift f_d is given by the modulation frequency of the square-wave generator. Higher-order sidebands leading to signals at multiples of f_d may be ignored, as they are not measured with the selective voltmeter and do not affect the mixer output signal as long as the mixer operates in its low-signal linear range.

6.6.2   Experimental Results of a Si-IMPATT Device in the V-band

In the following, typical measurements of self-mixing properties of an IMPATT
V-band oscillator are presented. The device is a silicon epitaxial double-drift
IMPATT diode (type Mesa 5/72N6HU, fabricated at the Daimler Benz Forschungs-
institut Ulm) which is incorporated in a waveguide resonator and is operated up
to a bias current of 400 mA with $P_{0,\text{max}} = 100$ mW output power at 62.7 GHz.
As stated above, the RF-circuit tuning remained unchanged during the measurements and output-power variations were achieved by adjustment of the bias current
only.

The matched conversion gain $g_{c,m}$ was determined with a signal level P_s,
see (6.74), about 70 dB below the output power P_0 to avoid nonlinearities of the
mixing process but to remain well above the noise floor of the mixer. The resistor
R_B in the bias circuit acting as output load was adjusted to yield a maximum
down-converted signal power P_d for each operating condition which was achieved
with $R_B = 30\ \Omega$ at the higher modulation frequencies and with a higher value of
$R_B = 40\ \Omega$ at $f_d = 1$ kHz due to thermal effects.

The resulting conversion gain as the ratio of maximum mixer output to oscillator RF input $g_{c,m} = P_{d,m}/P_s$ is plotted in Fig. 6.7. versus oscillator power P_0
(square symbols, left scale). It is fairly independent of the modulation frequency
(± 1 dB, which is in the range of measuring accuracy). Therefore, only one value
is given at each operating point. As to be expected from Sect. 6.2, the conversion gain increases with decreasing oscillator power P_0 at low levels proportional
to $1/P_0$. The conversion-gain-power product $g_{c,m}P_0$ is at a low power several
dB below and at power maximum a few dB above 1 mW. This is not much less

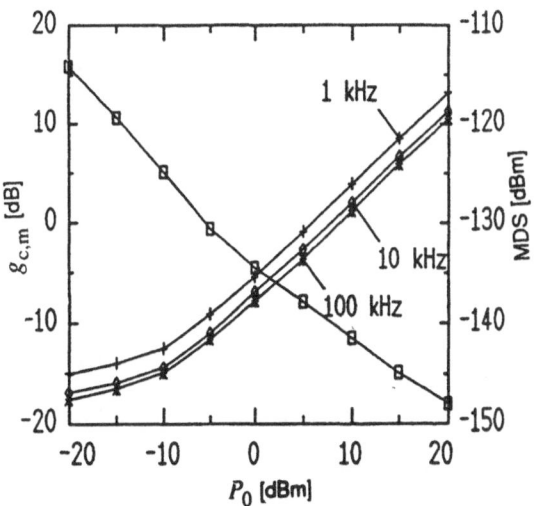

Fig. 6.7. Matched conversion
gain $g_{c,m}$ [dB] (square symbols, left scale) and Minimum-
Detectable Signal MDS [dBm]
(remaining curves, right scale)
versus oscillator output power
P_0 [dBm]; parameter modulation frequency

than found by numerical simulations in Sect. 6.5, though, however, with a different IMPATT device in a different circuit. The experimental device had a larger area but lower oscillation amplitude leading to an inferior shift of the dynamic current-voltage characteristic with respect to the nonoscillating device of only $\Delta U_{0,m} \approx 0.1$ V. Insertion into (6.43) and making use of $g_{c,m} = (\hat{U}_{d,0}^2/P_s)/8R_{dyn}$ results in a conversion-gain-power product of $g_{c,m}P_0 \approx \Delta U_{0,m}^2/2R_{dyn} = -8$ dBm (with $R_{dyn} = R_B = 30$ Ω) which is in reasonable agreement with the experimental values of about -5 dBm at low power levels.

The MDS is defined as that signal level P_s at the receiver input which generates a mixer output of equal power P_d as the noise in 1 Hz bandwidth. It can be measured in the set-up of Fig. 6.6 by increasing the damping α_{att} of the attenuator until the signal at the selective voltmeter is just 3 dB above the noise floor of the mixer, i.e. 3 dB above the output level that is measured without signal input. It can, however, also be derived by dividing the noise power at the bias resistor R_B by the conversion gain. In both cases, the result has finally to be divided by the bandwidth of the selective voltmeter.

The MDS is also depicted in Fig. 6.7. as a function of the oscillator power P_0 for modulation frequencies of 1 kHz, 10 kHz, and 100 kHz (remaining curves, right scale). It increases with P_0 at high power even more than inversely proportional to the conversion gain $g_{c,m}$ due to an enhancement of the low-frequency noise with bias current, but it saturates at low power slightly above MDS ≈ -150 dBm. This can be explained by down-conversion of the RF noise, as shown in Sect. 6.4.1, assuming an oscillator noise measure M of about 25 dB. In addition, the MDS exhibits at all power levels an increase towards lower modulation frequencies because of the beginning $1/f$-flicker-noise of the active device.

From a system point of view, the ratio of P_0/MDS is of importance as it determines the maximum admissable attenuation in the radiation field and thus the signal range, e.g., of a Doppler-radar sensor. This ratio reaches values of nearly 140 dB for the investigated self-mixing IMPATT oscillator that compare well or even exceed those of BARITT [6.4,5] and TUNNETT oscillators [6.6]. The same will be true for monolithically integrated versions of these oscillators or will even favour IMPATT devices, as the limiting influence of resonator losses, as discussed in Sect. 6.5, is less severe in IMPATT oscillators due to their higher impedance level.

References

Section 6.0

6.1   M. Kotani, S. Mitsui: Self-mixing effect of Gunn-oscillator. Electr. Commun. Jpn. B **55** 60–67, (1972)

6.2   Y. Takayama: Doppler signal detection with negative resistance diode oscillators. IEEE Trans. MTT-**21** 89–94 (1973)

6.3   J.R. East, H. Nguyen-Ba, G.I. Haddad: Design, fabrication, and evaluation of BARITT devices for Doppler system applications. IEEE Trans. MTT-**24** 943–948, (1976)

6.4   P.N. Förg, J. Freyer: Ka-band self-oscillating mixers with Schottky BARITT diodes. Electron. Lett. **16** 827–829, (1980)

6.5   U. Güttich: 60-GHz BARITT diodes as self-oscillating mixers. Electron. Lett. **22** 629–630, (1986)

6.6   M. Pöbl, J. Freyer, M. Claassen, M. Meinl: W-band TUNNETT diodes as self-oscillating mixers. Arch. Elektron. Übertrag. techn. **47** 57–60, (1993)

6.7   K.M. Strohm, J. Buechler, J.-F. Luy, F. Schäffler: A silicon technology for active high frequency circuits. ESDERC Conf. Bruxles (1992) published in Microelectr. Eng. **19** 717–720, (1992)

Section 6.1

6.8   K. Kurokawa,: Noise in synchronized oscillators. IEEE Trans. MTT-**16** 234–240, (1968)

6.9   K. Kurokawa: Injection locking of microwave solid-state oscillators. Proc. IEEE **61** 1386–1410, (1973)

6.10   H.J. Thaler, G. Ulrich, G. Weidmann: Noise in Impatt-diode amplifiers and oscillators. IEEE Trans. MTT-**19** 692–705, (1971)

Section 6.2

6.11   L. Gaul, M. Claassen: Stability and output power of GaAs PIN-avalanche diodes. Arch. Elektron. Übertr. techn. **45** 126–130, (1991)

6.12   M. Claassen, U. Güttich: Conversion matrix and gain of self-oscillating mixers. IEEE Trans. MTT-**39** 25–30, (1991)

6.13   W.T. Read: A proposed high-frequency negative resistance diode. Bell Syst. Tech. J. **37** 401–446, (1958)

Section 6.3

6.14   H.-G. Unger, W. Schultz: *Elektronische Bauelement und Netzwerke II* (Vieweg, Braunschweig 1969)

6.15   A. Vanoverschelde, G. Salmer, J. Ramaut, D. Meignant: The use of punch-through diodes in self-oscillating mixers. J. Phys. D **8** 1108–1114, (1975)

6.16   W. Harth: Conversion gain of self-oscillating BARITT-diode mixers. Arch. Elektron. Übertrag. techn. **34** 426–428, (1980)

Section 6.4

6.17   P. Blakey, R.A. Giblin, A.J. Seeds: Large-signal time-domain modelling of avalanche diodes. IEEE Trans. ED-**26** 1718–1728, (1979)

6.18   W.N. Grant: Electron and hole ionization rates in epitaxial silicon at high electric fields. Solid-State Electron. **16** 1189–1203, (1973)

6.19   C. Jacoboni, C. Canali, G. Ottaviani, A. Alberigi Quaranta: A review of some charge transport properties of silicon. Solid-State Electron. **20** 77–89, (1977)

6.20   M. Claassen: Conversion-gain model for a self-mixing Impatt oscillator. Arch. Elektron. Übertrag.-techn. **47** 156–159, (1993)

7 Silicon Millimeter-Wave Integrated Circuit Technology

K.M. STROHM

Daimler Benz Research Center, Wilhelm-Runge Str. 11, 89081 Ulm, Germany

A monolithic integrated circuit consists of a semiconductor single-crystal chip containing both active and passive elements and their interconnections. Since the invention of the integrated circuit in 1958 much progress has been made concerning packaging density, power consumption, speed and frequency performance. The concept of Microwave Integrated Circuits (MIC) was inaugurated in 1964. Prior to that nearly all microwave equipment utilized waveguide, coaxial or strip-line circuits. These systems have been costly, large and heavy. Especially for the millimeter-wave region (30–300 GHz) these systems became rather expensive due to the wavelength-determined small size and the necessary highly-precise machine tolerances.

The advantages of monolithic millimeter-wave integration are obvious:

- small size
- light weight
- low cost (due to the large quantities processed)
- high reliability
- large volume fabrication
- high yield
- improved reproducibility
- control and reduction of parasitic capacitance and inductance
- better thermal embedment of the active devices
- good control of circuit dimensions.
- overall improved performance

However, special requirements are still necessary for millimeter-wave integration. This concerns the substrate material, the fabrication process, and the integration technology. In this chapter silicon as a semiconductor substrate material for monolithic millimeter-wave integration is described and some basic technologies and fabrication processes for Silicon Millimeter-Wave Integrated Circuits (SIMMWICs) are discussed.

7.1 Technological Requirements for a Millimeter-Wave Substrate

The substrate for monolithic millimeter-wave integration has to be a low-loss dielectric used as a mechanical support for the circuit elements, as a waveguide

Springer Series in Electronics and Photonics, Vol. 32
Silicon-Based Millimeter-Wave Devices, Eds.: Luy et al.
© Springer-Verlag Berlin Heidelberg 1994

medium for interconnecting transmission lines, and as a technological medium for the fabrication of active devices [7.1]. Concerning the mechanical characteristics the substrate should be mechanically stable, shape stable, and long-term stable. It should have a high thermal conductivity and a thermal-expansion coefficient similar to the metallization. Concerning the wave-conducting characteristics it should have a large dielectric constant ε_r, which yields a high wavelength-reduction factor and favours miniaturisation. It should be homogeneous concerning ε_r, it should have a low dielectric loss tangent, a high resistivity, small thickness variations, and high electrical breakdown stability.

Concerning the technological characteristics it should be stable up to high temperatures for the various processing techniques, resistant against chemical treatments, flat, smooth, stable and defect free. Concerning the commercial aspects and manufacturing criteria it should be low cost, non-perishable, non-toxic and commercially available.

Table 7.1 lists some of the common substrates for millimeter-wave integration including the semiconductor materials gallium arsenide (GaAs) and silicon (Si).

As can be seen from Table 7.1 the semiconductor materials GaAs and Si fulfill most of the above-mentioned requirements for a monolithic millimeter-wave substrate material. GaAs is favoured by many groups due to its higher electron mobility and better semi-insulating properties. Theoretical and experimental investigations show that in the millimeter-wave region the attenuation losses of the substrate are not decisive. Here the conductor losses due to the skin effect and radiation dominate [7.2] (Chaps. 1 and 5). Therefore with respect to attenuation high-resistivity silicon is also a good choice as substrate material. This can be seen in Fig. 7.1 where the total attenuation of a 50-Ω microstrip line is given as a function of frequency for ceramics, semi-insulating GaAs and 10000 Ωcm Si.

Especially the flatness, homogeneity and mechanical stability of Si are excellent. In comparison to GaAs silicon offers the following advantages: Silicon is the most widely used material in semiconductor industry and has a mature and low-cost technology. Silicon substrates are cheaper and available with larger diameters (up to 200 mm diam). Silicon is mechanically more stable, it is about 2–3 times stronger, harder and less likely to break, it is over three times lighter when using a wafer of suitable thickness, has a three times higher thermal conductivity and a

Table 7.1. Properties of different substrate materials

Material	Relative permittivity ε_r	Loss tangent tan δ at 10 GHz	Thermal conductivity at 300 K [W/cm K]	Thermal expansion [10^{-6}/K]
Al$_2$O$_3$, pure	9.8	0.0001	0.37	6.3
Quartz	3.78	0.0001	0.017	0.55
Sapphire	9.4–11.6	0.0001	0.42	6.0
GaAs	12.91[a]	0.001[a]	0.46	6.86
Si	11.68[a]	0.002[a]	1.55	2.6

[a] at 100 GHz

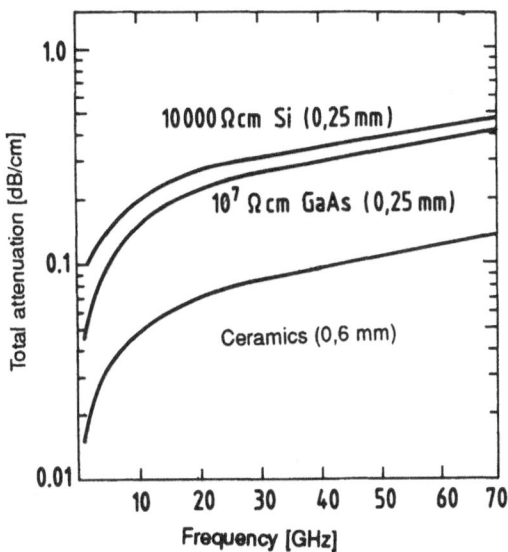

Fig. 7.1. Total attenuation of a 50 Ω microstripline on 10000 Ωcm Si compared to semi-insulating GaAs and ceramics as a function of frequency

three times lower thermal expansion [7.3]. Last but not least silicon is non-toxic, more abundant than gallium and arsenic and has a natural oxide.

7.1.1 Historical Background of SIMMWIC Technology

The concept of silicon monolithic Microwave Integrated Circuits (MICs) is not new. Its origin goes back to the US-funded MERA (Molecular Electronics for Radar Applications) program announced in 1964 [7.4]. *Hyltin* [7.2] demonstrated that 1500 Ωcm (boron-doped) silicon could provide around 0.5 dB/cm loss over a temperature range from +5°C to +110°C in the X-band. The most critical problem encountered during the MERA developments of silicon monolithic MICs was the conversion of high-resistivity p-type silicon to n-type lower-resistivity silicon during the high-temperature processing sequences [7.5]. This is depicted in Fig. 7.2: 800 Ωcm p-Si undergoes resistivity change to 1–10 Ωcm n-Si after approximately 6 hours at 1100°C. One speculation was that the inversion was due to a fast-diffusing interstitial donor such as sodium or copper [7.4].

Due to this inversion problem during high-temperature processing sequences very lossy substrates may result, which then will be unacceptable for microwave circuitry. This was stated by *Ertel*, who published the earliest example of a silicon monolithic MIC, an all-silicon X-band switch [7.6]. Therefore it was generally believed that the processing temperature should not exceed 800°C.

This was first fulfilled by a group at RCA in the beginning of the eighties (1981) using ion implantation and laser annealing. The group at RCA fabricated IMPATT diodes, and lateral and vertical PIN diodes by ion implantation and selective pulsed-laser annealing while keeping the substrate temperature below 800°C [7.7,8]. This

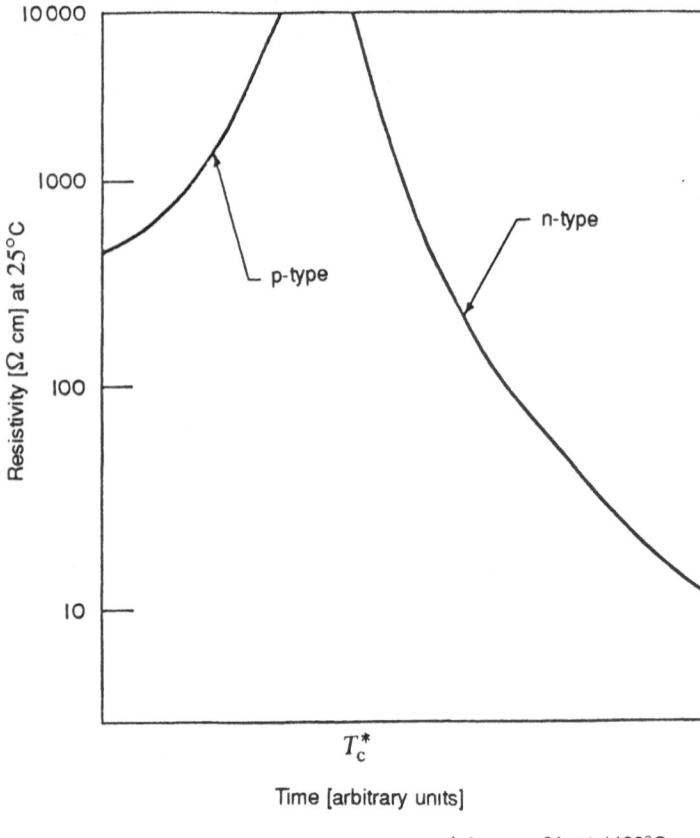

Fig. 7.2. Behaviour of high-resistivity silicon during high-temperature processing (after approximately 6 h. at 1100°C) as published in 1968 [7,5]

group reopened silicon to the world for monolithic millimeter-wave circuits [7.9] and introduced the term Silicon Millimeter-Wave Integrated Circuit (SIMMWIC) in the literature [7.10].

At the same time the group of E. Kasper at AEG had developed a low-temperature deposition technique, the silicon Molecular Beam Epitaxy (Si-MBE) for the growth of active, thin monocrystalline films with abrupt junctions at temperatures between 550°C and 750°C [7.11,12] and was testing X-ray lithography, a new lithographic method with submicrometer resolution [7.13]. These two VLSI techniques seemed to be very suited for the silicon monolithic millimeter-wave technology. Additionally, at that time the silicon wafer suppliers were able to deliver extremely pure silicon wafers with resistivity greater than 10000 Ωcm. Therefore, a group at AEG started extensive investigations on the characteristics and behaviour of these high-purity silicon wafers by applying

Si-MBE and X-ray lithography to check whether these techniques were suitable for a SIMMWIC technology [7.14–16].

7.1.2 Characterization of High-Resistive Silicon Substrates

The substrate losses of millimeter-wave transmission lines are connected with the electrical conductivity, because free carriers in the substrate may extract energy from the propagating electromagnetic field. Therefore it is extremely important to keep the substrate conductivity very low. The conductivity of a high-resistive substrate is determined by the dopants, the compensation degree and partly by deep impurity levels (often metals), because the Fermi level of 10000 Ωcm p$^-$-silicon is about 0.45 eV above the valence band.

The residual doping concentration may be evaluated from resistivity measurements. The resistivity is given by

$$\rho = \frac{1}{q\,(n\mu_n + p\mu_p)} \tag{7.1}$$

where ρ is the resistivity [Ωcm], q the electronic charge ($1.6 \cdot 10^{-19}$ As), n, p the free electron and hole concentration [cm^{-3}], and μ_n, μ_p the mobility of electrons (μ_n) and holes (μ_p) [cm^2/Vs].

The mobility is a function of carrier concentration and temperature, and has a value of 450 cm^2/Vs for low-doped p-silicon at room temperature. For a p-resistivity of greater than 10000 Ωcm the doping concentration should therefore be less than $1.4 \cdot 10^{12}$ cm^{-3}.

The measurement of the electrical characteristics of high-resistivity semiconductor materials is difficult due to effects which can be neglected in low-resistivity substrates. These are the contact resistance, the presence of space charge regions, which may disturb the internal electrical fields, and possible space-charge limiting currents.

A very convenient method for measuring the resistivity is the four-point probe technique [7.17,18]. However, for high-resistive silicon wafers ($\rho > 2000$ Ωcm) the measurement limit is normally out of range. Therefore, the slices have been investigated by Hall measurements [7.19] and the spreading resistance method [7.20].

Spreading resistance is a known method for dopant profiling. Because the spreading-resistance probe senses the resistivity in a microscopic sampling volume immediately under the probe tip, one can angle lap a silicon structure, obtaining a resistivity versus depth profile. Figure 7.3a shows the spreading resistance of a 100 mm diameter, $\langle 100 \rangle$ oriented, p$^-$-boron-doped silicon wafer with a specified resistivity $\rho > 4000$ Ωcm, Fig. 7.3b depicts that of an n$^-$-phosphorus-doped silicon wafer with the same specification. The surface was covered with a 150 nm thick pyrox layer for defining the measurement starting point and for reference reasons. From the spreading-resistance measurements doping profiles can be evaluated using calibration samples of known resistivity.

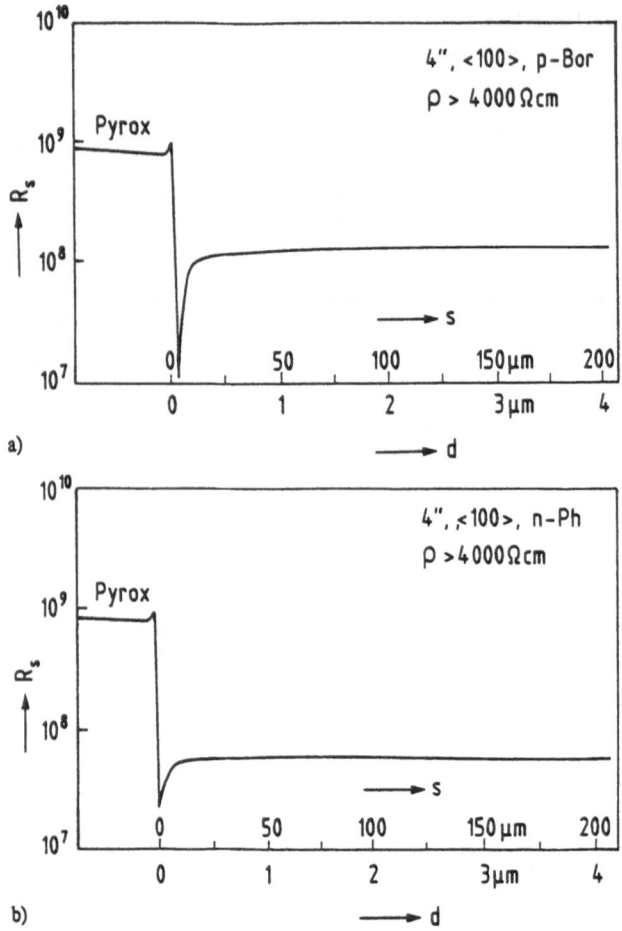

Fig. 7.3. Spreading resistance R_s of high-resistive p⁻ (a) and n⁻ (b) 4" silicon wafers. R_s is given as a function of the measurement distance s of the sample lapped with an angle of 1.20°. d gives the corresponding depth. The surface was covered with a 150 nm thick pyrox layer as a reference point

Hall measurements are used for investigating the resistivity, the carrier density and the Hall mobility. Figure 7.4a exhibits the temperature-dependent resistivity of a $\rho > 10000$ Ωcm specified silicon sample investigated by Hall measurements. The resistivity shows a maximum above room temperature (50°C) and decreases at higher temperatures due to the thermal activation of intrinsic charge carriers and decreases at lower temperatures due to the increase of mobility. Therefore, only within the temperature range between −170°C and 110°C the resistivity remains greater than 2000 Ωcm. This is in accordance with theory. Figure 7.4b displays the theoretical dependence of resistivity on temperature for different low-doped silicon samples. It depicts the importance of keeping substrate doping low.

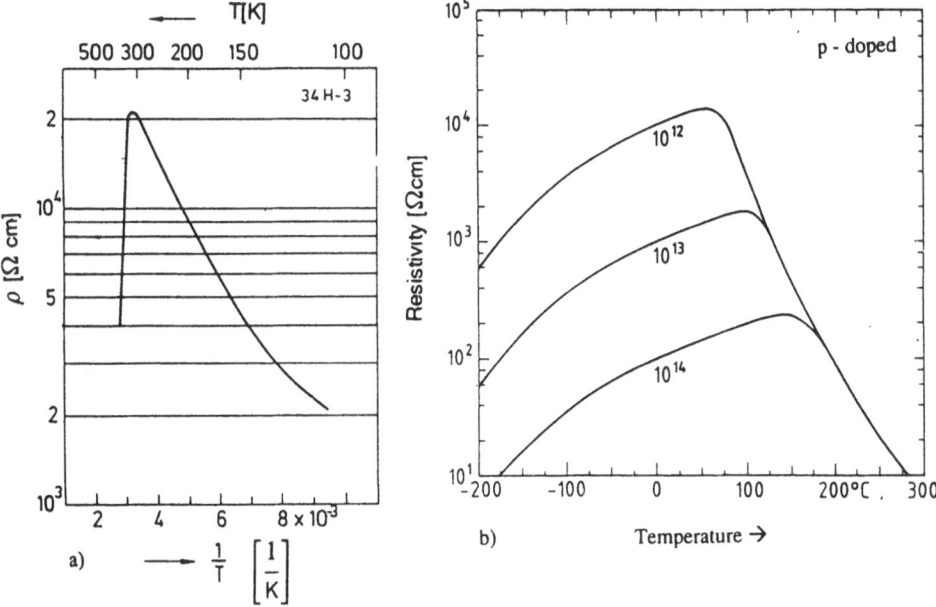

Fig. 7.4. (a) Measured temperature dependence of resistivity for a $\rho > 10000\ \Omega$ cm specified silicon sample investigated by Hall measurements. **(b)** Theoretical value of resistivity for different p$^-$-substrate dopings as a function of temperature

7.1.3 Behaviour of High-Resistive Silicon Substrates During Fabrication Processes

Because of the inversion problem found during the first investigations on silicon microwave integrated circuits [7.5] extensive experiments were carried out to check whether the resistivity is changed by common fabrication-process steps. The change in electrical resistance after wafer preparation, cleaning processes, molecular beam epitaxy, pyrolithic oxidation, thermal oxidation, wet chemical etching, plasma etching, and deposition and patterning of metal films was investigated with the spreading-resistance method and Hall measurements [7.15].

Figure 7.5a depicts the spreading resistance of a polished and cleaned 10000 Ωcm wafer. The wafer is bevelled by an angle of 6.84° and the spreading resistance is measured along the beveled surface for a distance of over 250 μm. This corresponds to a depth of more than 30 μm. The spreading resistance remains constant over the whole distance, with a value of about $10^8\ \Omega$.

The high-temperature thermal oxidation processes at 1100°C of *Battershall* and *Emmons* [7.5], in which a resistivity change from 800 Ωcm p-type to 1–10 Ωcm n-type was observed (Fig. 7.2), were repeated. Spreading resistance showed that even after 6 h of thermal oxidation at 1100°C the high resistivity of the bulk material is preserved (Fig. 7.5b). However, a lowering of the spreading resistance near the SiO$_2$/Si interface is observed. This lowering may be caused by charges in the SiO$_2$/Si interface which induce a carrier accumulation under the SiO$_2$/Si

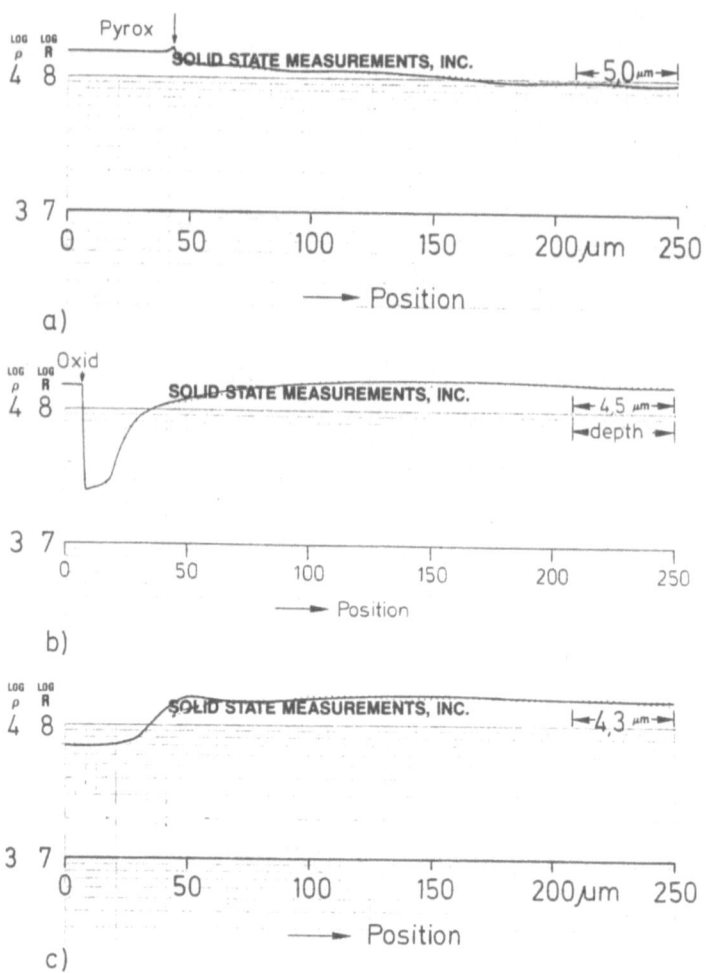

Fig. 7.5. Spreading resistance curve of a high-resistive silicon wafer: (**a**) after cleaning, (**b**) after thermal oxidation at 1100°C for 6 hours, and (**c**) after cleaning, metal evaporation, X-ray exposure, patterning and O_2 plasma etching steps

interface. Therefore nowadays high-temperature processes are also applicable for high-resistivity silicon, but still care has to be taken for proper processing and cleanliness.

Deposition of metal films, photolithographic patterning, plasma etching, and wet chemical etching show no effects on the high-resistivity silicon. This is depicted in Fig. 7.5c, where the spreading resistance is given for a sample on which microstrip resonators were fabricated. During this fabrication process, cleaning procedures, evaporation of metal films, exposure with X-rays, gold electroplating and O_2-plasma etching were performed. No drastic difference is found in comparison to the spreading resistance curve of the cleaned wafer shown in Fig. 7.5a.

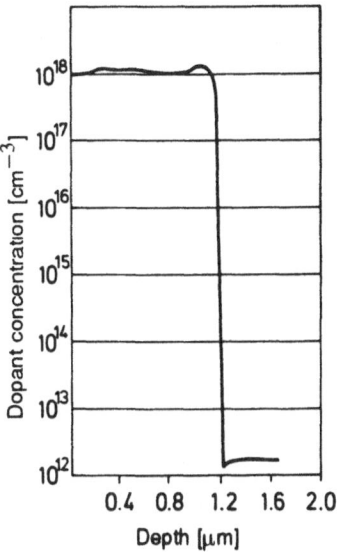

Fig. 7.6. Spreading-resistance profile of a Ga-doped MBE layer on a high-resitivity silicon wafer

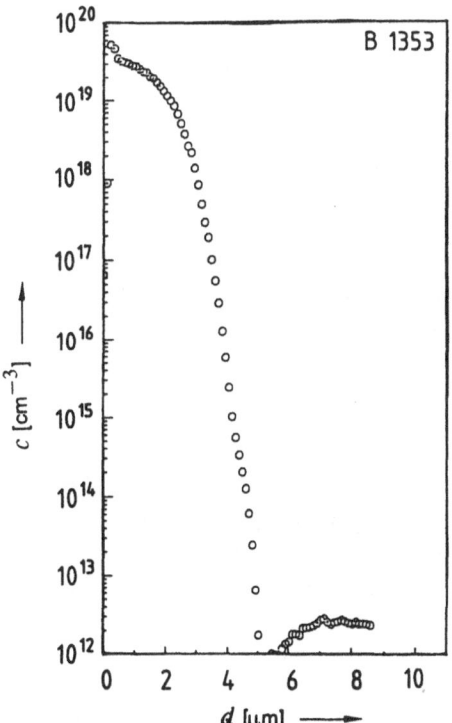

Fig. 7.7. Spreading resistance profile of a diffused n^+-layer (1186°C, 18 h) on high-resistivity silicon

The epitaxial growth of active layers with the MBE method [7.11] at temperatures between 550°C for about 120 min preceded by a thermal cleaning process at 900°C for 5 min produced no change in the electrical characteristics of the high-resistivity bulk material. To show this and the capability of MBE, a Ga dopant profile on high-resistivity silicon is exhibited in Fig. 7.6. The Ga dopant was grown at 550°C. A constant doping level of 1×10^{18} cm^{-3} is found in the MBE layer, and an abrupt junction to the high-resistivity silicon substrate is observed. The substrate doping is not influenced by the MBE process.

To achieve highly-conductive layers in silicon substrates, buried layer diffusion processes are often used (Sect. 7.2.1). Figure 7.7 shows an n$^+$-doping profile measured by the spreading resistance method. The n$^+$-diffusion was performed using As spin-on glass at 1184°C for 18 h with 3% O$_2$ followed by a second diffusion at 1050°C for 30 min with 50% O$_2$ to increase the surface concentration. The most important fact of Fig. 7.7 is that the high-resistivity characteristics of the substrate are maintained even after this long and high-temperature diffusion process (1184°C, 18 h), in contrast to the findings of *Battershall* and *Emmons* [7.5].

7.2 Basic Technologies

SIMMWIC technology is profiting greatly from standard integrated circuit technology and their progress. Excellent books and review articles have been written on semiconductor and IC technology [7.21–24] which may be referred to for detailed information on common technologies. In this chapter only some basic technologies for the fabrication of SIMMWICs are given. These are the formation of highly-doped buried layers, the epitaxial growth of the active-device layers, photolithographic and X-ray lithographic patterning techniques, metallization, air-bridge technology and via-hole technology.

7.2.1 Buried Layers

To obtain highly conductive layers on silicon substrates, buried-layer diffusion processes are often used. For the fabrication of monolithically integrated transmitter and receiver diodes, and of bipolar and heterobipolar transistors it is advantageous to form the highly conductive layer as a buried layer. In this way the devices can be manufactured in a quasi-planar configuration and only mesa etches have to be performed with the epitaxially grown films. Depending on device type either highly conductive p$^+$ or n$^+$ buried layers are needed. The requirements for the buried layers are:

– low sheet resistance to minimize gap resistance and spreading resistance,
– high surface concentration to achieve a low contact resistance,
– low defect density for the subsequent growth of defect-free epitaxial layers.

There are several methods for forming highly-doped diffusion layers:

- diffusion from a chemical source in vapour form at high temperatures,
- diffusion from a doped oxide source,
- diffusion and annealing from an ion implanted layer.

Annealing of ion-implanted layers is necessary for activating the implanted atoms and reducing the crystal damages resulting from ion implantation. When the annealing takes place at high temperatures, diffusion also occurs. Since ion implantation provides more precise control of total dopants from 10^{11} cm^{-2} to greater than 10^{16} cm^{-2}, it replaces the chemical or doped oxide source wherever possible.

Table 7.2 summarizes the results of different n$^+$-diffusion processes investigated for SIMMWIC technology and lists the value of sheet resistance R_s [$\Omega/\square$] the concentration N_0 [cm^{-3}] at the surface, and the specific Ti/Au contact resistance S_c [Ωcm^{-2}] measured with the TML method [7.25].

With an As diffusion from spin-on glass at 1187°C for 4 h and a subsequent diffusion at 1187°C for 0.5 h a sheet resistance of 7.5 [$\Omega/\square$] and a surface concentration of 2×10^{19} cm^{-3} is achieved.

With an As double implantation (1×10^{16} cm^{-2}, 100 keV) and a subsequent diffusion process at 1200°C for 5 h and 1 h, respectively, a sheet resistance of 4.7 [$\Omega/\square$] and a surface concentration of over 1×10^{20} cm^{-3} are achieved. This buried-layer process is performed in the following way:

First the wafers are thermally oxidized to a thickness of 1.3 μm. By a photolithographic step oxide windows are defined and etched wet chemically. After a cleaning step a 20 nm thick scatter oxide is grown. This scatter oxide prohibits channeling effects during ion implantation. Then, the first As implantation is performed with a dose of 1×10^{16} cm^{-2} and an energy of 100 keV. Annealing

Table 7.2. Characteristics of different n$^+$-buried layers

Process	Implantation	Diffusion	R_s[$\Omega/\square$]	N_0 [cm^{-3}]	S_C [Ωcm^2]
Spin-on As solution	–	a) 1187°C, 4 h	13.3		
		b) 1187°C, 0.5 h	15		
		a) + b)	7.5	2×10^{19}	1×10^{-6}
As-double implant.	a) As 1×10^{16}, 100 keV	a) 1200°C, 5 h	7.9		
	b) dito.	b) 1200°C, 1 h	9.2		
	a) + b)	a) + b)	4.7	1.3×10^{20}	3×10^{-7}
As-implantation + P diffusion	a) As: 1×10^{16}, 100 keV	a) 1200°C, 5 h	7.95		
		b) P-Diffusion: 1050°C 35'	4.65		
		+ 800°C, 130'			
		a) + b)	3.65	7×10^{20}	2.5×10^{-7}
As high-dose implantation	As: 2×10^{16}, 100 keV	1200°C, 5 h	4.3	9×10^9	

is performed at 1200°C for 5 h. This yields a sheet resistance of 7.9 [Ω/□]. After this the oxide grown during annealing is etched away and again a 20 nm thick scatter oxide is grown. The second As implantation is performed with the same parameters as the first one, but the annealing is performed at 1200°C for 1 hour only to achieve a higher surface concentration. The second implantation and annealing process alone yields a sheet resistance of 9.8 Ω/□ and a surface concentration of 1.3×10^{20} cm^{-3}. The combination of first and second implantation and annealing process yield a sheet resistance of approximately 4.7 Ω/□ and a surface concentration of 1.3×10^{20} cm^{-3}. The corresponding SIMS profile is shown in Fig. 7.8 (As-double implantation).

By replacing the second As-implantation process by a P-diffusion process an even lower sheet resistance and a higher surface concentration is achieved. The P diffusion is performed in two steps: first a pre-diffusion is performed at 1050°C for 35 min with PBr$_3$. then a post-diffusion is performed at 800°C for 20 min in pure N$_2$ atmosphere. This P diffusion alone gives a sheet resistance of 4.6 Ω/□ and a junction depth of 2.2 μm. The combination of the first As implantation $(1 \times 10^{16}$ cm^{-2}, 100 keV) and annealing at 1200°C for 5 h and the subsequent P diffusion yields a sheet resistance of 3.6 Ω/□ and a maximum surface concentration of 7×10^{20} cm^{-3}. The corresponding SIMS profile is also displayed in Fig 7.8 (As-implantation and P-diffusion). Unfortunately, as will be reported later

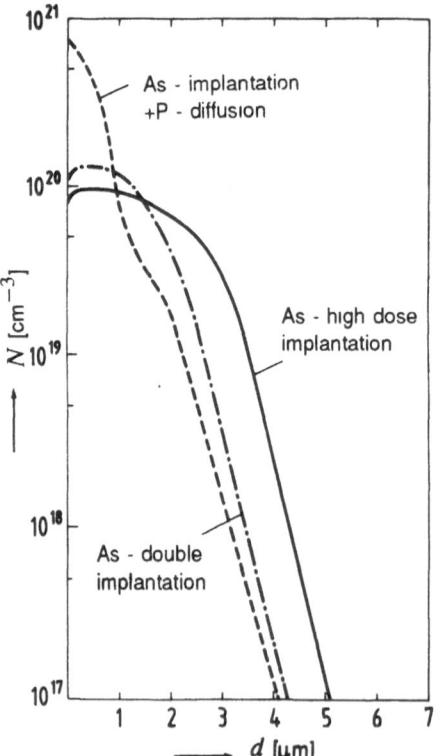

Fig. 7.8. SIMS profile of different processed n$^+$-buried layers

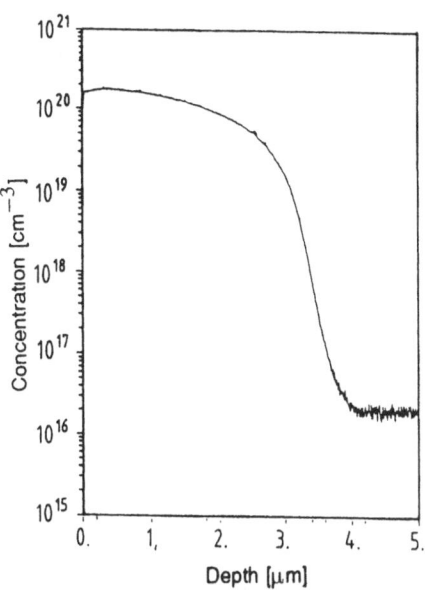

Fig. 7.9. SIMS profile of a p^+-buried layer formed by a BN diffusion process at 1150°C, 50 min

in Sect. 7.4.1, P outdiffusion is observed when growing epitaxial films on this highly conductive buried layer.

A technologically more simple process can be achieved by a high-dose As implantation. In this case 2×10^{16} As/cm^{-2} are implanted with 100 keV. When the annealing and diffusion process is performed at 1200°C for 5 h a sheet resistance of 4.3 $\Omega/\square$ and a surface concentration of 9×10^{19} cm^{-3} is obtained. Figure 7.8 shows the corresponding SIMS profile (As-high dose implantation). The homogeneity of the sheet resistance over a 4 inch wafer is better than 1%.

The defect densities of the buried layers were determined by etching with a Secco etchant. The defect density was found to be less than 2×10^4 cm^{-2} for all processes.

For the fabrication of monolithically-integrated IMPATT diodes it is advantageous to grow the active layers on an etch-stopping p^+ buried layer. For the formation of p^+ buried layers a diffusion process is used which was originally developed for the fabrication process of silicon X-ray mask membranes. The diffusion is performed at 1150°C for 50 min with ramping. The wafers are positioned between BN sources. Figure 7.9 exhibits the corresponding SIMS profile. A sheet resistance of 2.5 $\Omega/\square$ is obtained, the junction depth is 4.1 µm. The concentration near the surface lies over 10^{20} cm^{-3} within a depth of 2 µm.

7.2.2 Epitaxial Growth

Epitaxy, a transliteration of two Greek words epi, meaning "upon", and taxis, meaning "ordered", is a term applied to processes used to grow thin crystalline layers on a crystalline substrate. In the epitaxial process the substrate wafer acts as a seed crystal. Most epitaxial processes use Chemical-Vapor Deposition (CVD) techniques. A different approach is Molecular Beam Epitaxy (MBE) which uses an evaporation method. Since CVD is a high-temperature deposition technique (1050–1150°C) where the achievable doping profiles are limited by solid-state out-diffusion and autodoping effects [7.22], MBE is much more suitable for SIMMWIC technology and will be discussed in the following in more detail.

Si-MBE offers the following advantages:

- low growth temperature (450–750°C),
- precise control of thickness with submicrometer resolution,
- precise control of doping distribution,
- atomically abrupt interfaces and junctions,
- flexibility in the choice of material combination and layer structures,
- growth of heterostructures (Si/SiGe),
- growth of superlattices.

MBE ultilizes the clean environment of an Ultra High Vacuum (UHV) system for the growth of single-crystalline films on oriented substrates. The molecular beams of the matrix and dopant materials are either generated by effusion or Knudsen cells, by electron-beam evaporators or by gas sources. The beams are directed towards the substrate wafer which is heated to the desired growth temperature. Process conditions such as residual gas composition, temperatures, beam flux densities and surface reconstruction can be monitored in situ, and these informations are frequently used as inputs for process control systems. Figure 7.10 shows a schematic diagram of an industrial, single-slice, fully automatic, computer controlled Si-MBE apparatus [7.26].

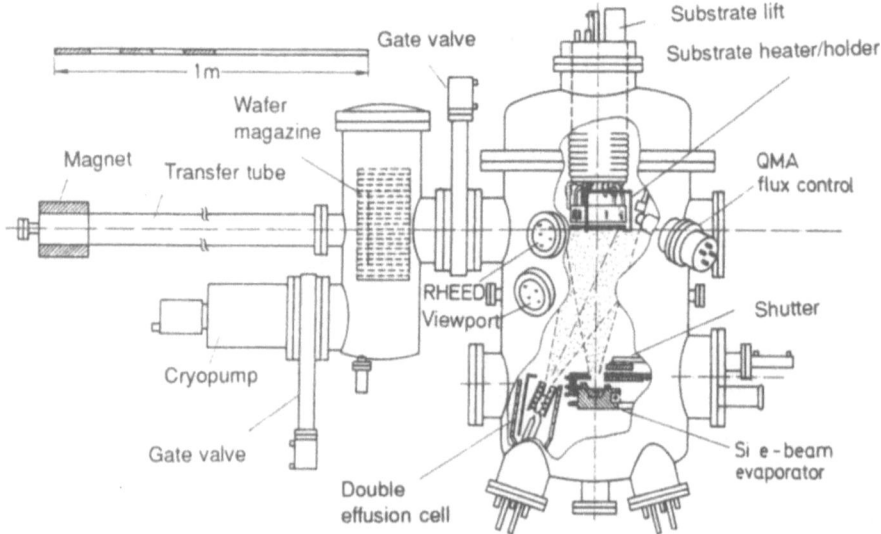

Fig. 7.10. Schematic view of a Si-MBE machine. Pumps and analyzing equipment of the growth chamber are not shown (after [7.26])

It consists of two independently pumped UHV chambers, a storage chamber for 25 Si wafers with 150 mm diameter and a growth chamber. The chambers are connected by a gate valve and separately pumped down to the 10^{-11} mbar range in the growth chamber and to the 10^{-9} mbar range in the storage chamber. A typical vacuum of 10^{-10} mbar during the growth process is achieved. The $\langle 100 \rangle$-oriented silicon substrates are transported by a horizontal wafer-transfer system from the storage chamber into the substrate heater. The matrix material silicon is evaporated from electron-beam evaporators with growth rates of typical 0.25 nm/s. For heteroepitaxial growth Ge is also evaporated from an electron-beam evaporator.

Flux control is accomplished by a quadrupole mass spectrometer and allows thickness control of better than 5%. As dopant materials antimony (Sb) is used as n-type dopant, and gallium (Ga) is used as p-type dopant. These are evaporated from conventional, thermally controlled effusion cells. The problem with these dopants is that they are found to exhibit low incorporation ratios, i.e. low ratios of dopant concentration in the grown film compared to the concentration in the corresponding surface adlayer. A properly designed beam aperture and a variable substrate voltage allow enhancement of Sb incorporation by a method called doping by secondary implantation [7.27]. Boron is favoured in p-type doping because of its high solubility and very low segregation. Special boron effusion cells are nowadays available [7.28]. Problems and solutions to doping in Si-MBE cannot be discussed here in more detail. Instead, the reader is referred to [7.11,29].

During growth the substrate is radiation heated by a large graphite meander. Temperatures of up to 900°C can be achieved. Typical growth temperatures are 550°C, for highly doped contact layers even 350°C are used.

Preparation of an atomically clean surface is essential in order to achieve a high-quality epitaxial growth [7.30]. There are two main steps required: a chemical precleaning and an in situ thermal cleaning. The standard chemical cleaning process (RCA clean) is based on oxidation and dissolution of residual organic impurities and certain metal contaminations in a mixture of H_2O-NH_4OH-H_2O_2 at 80°C, followed by dissolution and complexing of remaining trace metals and chemisorbed ions in H_2O-HCl-H_2O_2 at 80°C [7.31]. This cleaning forms a protective oxide, approximately 0.5 nm in thickness. The thinly oxidized substrate is then introduced into the growth chamber and heated to 900°C for 5 min prior to epitaxial film growth so that the thin oxide and residual carbon contaminations sublime, leaving a clean silicon surface. Another chemical pretreatment uses an HF dip immediately before insertion into the growth apparatus.

By using Si-MBE the active layers for basic SIMMWIC devices with rather sophisticated structures can be grown. Examples are: Schottky-barrier diodes [7.32,33], Planar Doped Barrier (PDB) diodes [7.34], Si/SiGe tunneling diodes [7.35], avalanche transit-time diodes [7.36], bipolar and heterobipolar transistors [7.37].

One of the most simple layer sequence is applied to Schottky-barrier diodes. Here 100–200 nm thick epitaxial layers with doping concentrations between 1×10^{16}–2×10^{17} cm^{-3} are grown on highly-doped buried layers. Both p- and n-type Schottky-barrier diodes can be grown. A strong outdiffusion of P is found on As/P buried layers, when the P concentration is higher than 5×10^{20} cm^{-3}.

Planar doped-barrier diodes need a n^+-i-p^+-n^+ layer sequence [7.34]. For symmetrical PDB diodes the p^+-doping spike is in the center. Figure 7.11a displays the SIMS profile of a symmetrical PDB diode. The p^+-doping spike was achieved with Ga doping by secondary implantation. The Ga doping spike consists of 1×10^{12} cm^{-2} doping atoms. About 9.75×10^{11} Ga/cm^{-2} are within 17.5 nm.

Examples for IMPATT diode layer structures are given in Chap. 2. Figure 7.11b shows the SIMS profile of a Double Low-High-Low (DLHL) structure grown on

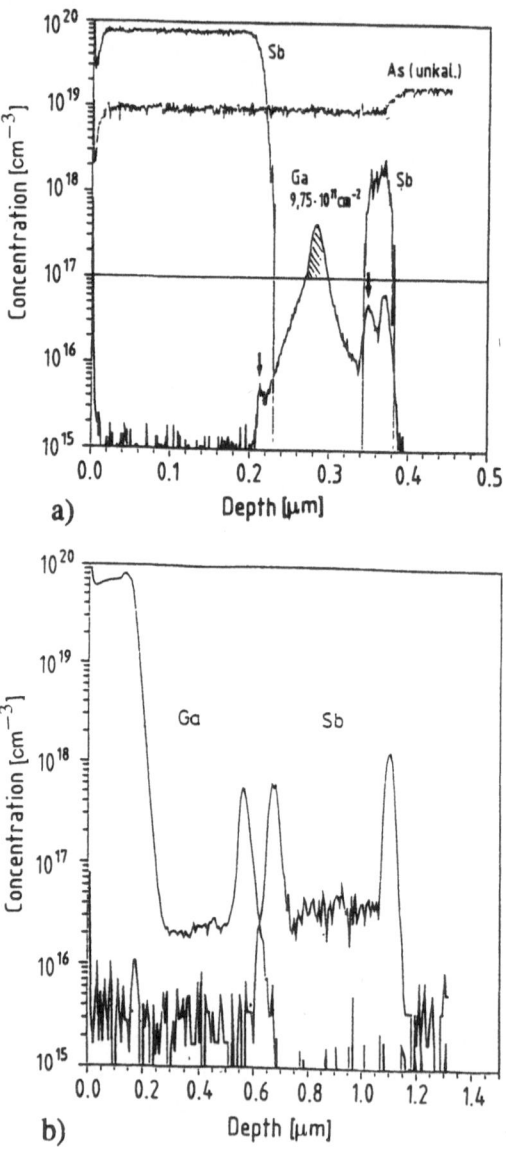

Fig. 7.11. SIMS profile of Si-MBE grown: (a) symmetrical Planar-Doped Barrier (PDB) diode strructures, (b) double low-high-low avalanche transit-time diode structures

an n^+ -buried layer for monolithic integration. The two doping spikes limit the avalanche region to increase the impedance compared to flat profile structures.

7.2.3 Lithography

Lithography is the process of transferring geometrical shapes on a mask to the surface of a silicon wafer. For this a photosensitive polymer film is applied to the silicon wafer, dried, and then exposed with the proper geometrical patterns through a mask to UltraViolet (UV) light or other radiation. After exposure, the wafer is soaked in a solution that develops the images in the photosensitive material. For a positive working polymer the exposed areas of film are removed in the developing process [7.38]. The wafer may then be placed in an ambient that etches surface areas not protected by polymer patterns. Because the polymeric materials resist the etching process, they are called resists. There are resists which are sensitive to UV light, electron beams, X rays or ion beams.

Optical lithography has been the dominant integrated circuit patterning technology for many years, and this dominance is expected to continue well into the submicrometer-feature-size regime. Optical lithography is normally used for the fabrication of SIMMWIC's. For very fine patterns electron-beam or X-ray lithography may be used. Especially X-ray lithography has been applied to SIMMWIC technology and will be described in more detail in the following subsection.

Principles of X-Ray Lithography

X-ray lithography is a promising technique for future high-volume and high-yield fabrication of submicrometer devices [7.13]. The main advantages of X-ray lithography compared with photolithography and electron-beam lithography are due to the short wavelength (0.2–2 nm), the low energy (0.6–6 keV), the reduced diffraction, the limited scattering, the absence of reflection, and the low absorption of the soft X-rays. Thus high lateral resolution down to 0.1 μm, high topographic resolution, unlimited depth of focus, uniform in-depth exposure, high aspect ratio, superior image fidelity and immunity to particulate defects are achieved and excellent resist profiles can be realized.

Some of these advantages of X-ray lithography can find useful applications in manufacturing monolithic millimeter-wave integrated circuits. The high resolution may be used for the fabrication of sensitive receiver diodes and high-frequency three-terminal devices. The uniformity of exposure depth and the high aspect ratio may be applied for the fabrication of low-capacitance air bridges. The high edge steepness and high aspect ratio may be used for the definition of low-loss microstrip lines and high-quality planar antennas.

Since there are no imaging optics for X rays, X-ray lithography is limited to shadow printing in a configuration similar to that of optical proximity printing. A schematic diagram of the exposure arrangement is shown in Fig. 7.12. Soft X rays,

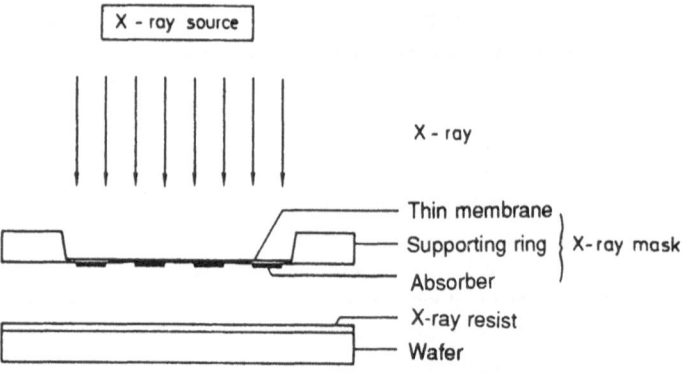

Fig. 7.12. Schematic diagram of X-ray exposure arrangement

generated by an X-ray source pass the thin absorber-free parts of the X-ray mask membrane and expose the X-ray resist on the wafer.

One of the most serious problems in X-ray lithography is the fabrication of low distortion and defect-free masks. In contrast to optical lithography, where stable glass or quartz substrates provide high transparency to visible or ultraviolet light, only thin membranes (2–4 μm) of material having a low atomic number are sufficiently transparent to X rays as well as to visible light for alignment. On the other hand, for blocking the X rays and achieving a high contrast the absorber has to be of highly dense materials and must be relatively thick. Therefore, the main problem with X-ray masks arises from the fact that they consist of thin membranes (2–4 μm) and of relatively thick absorbers (Au 0.5–1.0 μm). Stress-compensated B/Ge-doped Si membranes are a good choice for X-ray masks [7.39,40]. There are three different X-ray sources: X-ray tubes, synchrotron sources and plasma sources. Synchrotron sources are the most powerful ones [7.13].

X-Ray Exposure of High-Resistivity Silicon

In a study the effects of X-ray exposure on high-resistivity silicon substrates were investigated. This was done to check, whether the X-ray exposure increases the low conductivity of the substrate by possible activation of impurity atoms or surface charges and thus making the substrate unsuitable for millimeter-wave integration (increase of attenuation). Several 10000 Ωcm substrates were exposed to the BESSY (Berlin Electron Storage Ring for Synchrotron Radiation) spectrum [7.13] with different doses. The resistivity of the exposed samples was measured with the spreading-resistance probe and Hall measurements, and compared with unexposed samples. Even for an exposure of 100 J/cm^2, which is 100 times the exposure necessary for an insensitive X-ray resist (PMMA), no degradation of resistivity takes place. Hall measurements confirm these results.

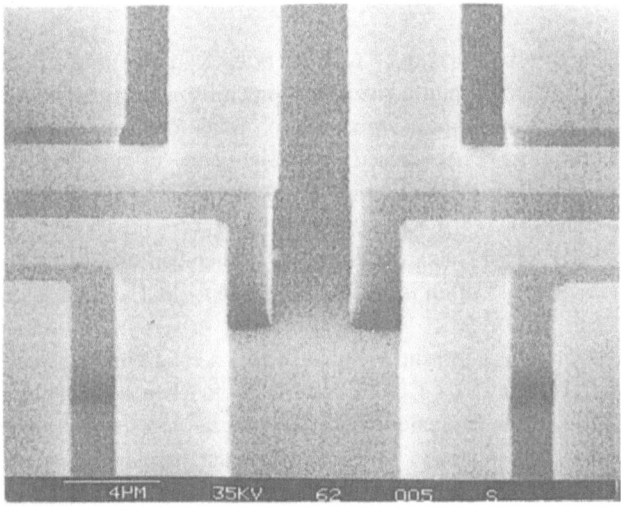

Fig. 7.13. Resist pattern of the experimental X-ray resist J 855 from Hoechst after development with vertical edges and a resist height of 8 μm

Application of X-Ray Lithography to SIMMWIC Technology

As an example for the application of X-ray lithography to SIMMWIC technology the fabrication of monolithically integrated Schottky-barrier diodes with X-ray lithography is described in Sect. 7.3.1.

Another useful application of X-ray lithography is in patterning thick resists suitable for the fabrication of air bridges, low-attenuation microstrip lines and high-quality antennas. Figure 7.13 shows a resist pattern with a height of 8 μm and vertical edges generated by X-ray lithography.

7.2.4 Pattern Transfer

The resist patterns defined by the lithographic process are not permanent elements of the final device but only replicas of circuit features. To produce circuit features, these resist patterns must be transferred once more into the underlying layers comprising the device. The pattern transfer may be accomplished by an etching process which selectively removes unmasked portions of a layer. In SIMMWIC technology passivation layers, epitaxially grown layers and metallization layers have to be etched. Both wet chemical etching and dry etching are used. A general description of etching techniques can be found in [7.21,22,41,42]. Some special techniques used in SIMMWIC technology will be explained in the following.

Etching of Epitaxial Layers

For the fabrication of Schottky-barrier diodes with air bridges thin epitaxial layers (100–200 nm) have to be patterned and etched very precisely. The simplest way to do this is by etching with a resist mask. Most wet chemical etchants attack the resist. The following solutions have been tested: the Wright etch (60 ml HF, 30 ml HNO_3, 30 ml CrO_3, 2 g $Cu(NO_3)_2 \cdot 3H_2O$/60 ml H_2O, 60 ml CH_3COOH 100%) and the permanganate etch (0.1–0.4 g $KMnO_4$, 25 ml HF, 1–5 ml CH_3COOH, > 20 ml H_2O) [7.41]. The Wright etch yields good results concerning surface quality, Si/resist etch ratio (Si:975 nm/min, resist: 90 nm/min), but a large lateral underetch is observed.

The permanganate etch yields good surface quality with a 1 : 4 diluted solution of 0.2 g $KMnO_4$, 25 ml HF, 1 ml CH_3COOH, 20 ml H_2O, when the etching is performed in an ultrasonic bath. Si etch rate is 800–1000 nm/min, resist etch rate is 100–500 nm/min. An increased etch rate is observed at resist edges which causes small trenches. The ratio of trench depth to etch depth is 0.1–0.2. The ratio of lateral underetch to etch depth is 3–4.

Fig. 7.14. Mesa-etched IMPATT diode structures. The top metallization is used as etching mask. A selective underetch is performed using the self-stopping etch mechanism of the aqueous KOH solution on highly-doped boron layers

Much better results are achieved by dry etching. Different machines are nowadays available using different physical and chemical mechanisms for etching (plasma etching, reactive ion etching, reactive ion beam etching, sputter etching, ion beam milling). Si may be etched in fluorine or chlorine plasmas. Isotropic and anisotropic etching is possible by proper adjusting of gas flow, pressure and RF power [7.42].

Mesa etch of avalanche transit-time diodes on p^+-buried layers is advantageously performed in aqueous KOH solutions. The following mixture is used:

300g	KOH
2g	$K_2Cr_2O_7$
1200 ml	H_2O
300 ml	Isopropanol

The etch rate is 1 μm/h at 25°C, 0.16 μm/min at 51°C and 1 μm/min at 80°C. This etchant is anisotropic (54.74°, $\langle 111 \rangle$ plane) and self stopping on the highly-doped boron diffusion layer ($> 7 \times 10^{19}$ cm^{-3}). This may be used for a selective underetch necessary for forming self-aligned contacts. Figure 7.14 depicts mesa etched IMPATT-diode structures, selectively underetched in aqueous KOH solution with the top metallization as etch mask.

Etching of Passivation Layers

Thermal oxide, low-temperature sputter oxide and plasma-enhanced CVD silicon nitride films are used as passivation layers in SIMMWICs. Thermal oxide layers and sputter oxide layers can be etched in buffered HydroFluoric (HF) acid using resist masks. CVD Si_3N_4 layers have a much lower etch rate in HF solution. Therefore only thin layers (up to 300 nm) can be wet chemically etched without large undercut. Thick layers have to be dry etched using fluorine gas mixtures.

Etching of Via Holes

In MMICs based on microstripline design via holes are used for grounding. Also for microstrip oscillators via holes may be used. This means, holes have to be etched from the backside to the frontside of the wafer and covered with metallization. The problems encountered with via-hole technology are:

- the whole process has to be performed on wafers with finished devices and thinned to a thickness of 100 or 150 μm.
- A back-side to front-side alignment has to be performed.
- 100–150 μm thin substrates have to be handled.
- The finished devices on the front side have to be protected during the etch process.

- A low-temperature passivation layer has to be used withstanding the etch process.
- The via holes have to be properly metallized.

These problems can be solved by the following processes:

- Use of an anisotropic etchant containing ethanolamine, gallic acid, water, pyrazine, peroxide, and a surfactant [7.43].
- Use of low-temperature CVD silicon nitride (> 600 nm) or sputter oxide (> 1000 nm) for passivation of front and backside of the wafer.
- Use of a photolithographic exposure machine with an additional infrared alignment equipment.
- Evaporating of Ti/Au into the etched via holes in a rotating evaporation system, followed by an Au electroplating process.

The fabrication steps for via holes are depicted in Fig. 7.15. Figure 7.16a shows a metallized via hole formed by the processes described above in a 100 μm thick silicon substrate. This technology may be used for the fabrication of monolithic microstrip oscillators with integrated heat sink (Fig. 7.16b).

The anisotropic etching of silicon is also applied for the fabrication of integrated horn antennas for millimeter-wave applications [7.44]. The etching process forms pyramidal holes, bounded by the crystal planes and produces a horn flare angle of 70.6° [7.45].

A via-hole formation with selctive plasma etching of Si using a mixture of CF_4 and O_2 is described in [7.46]. 0.1 μm thick Ni was used as etch mask.

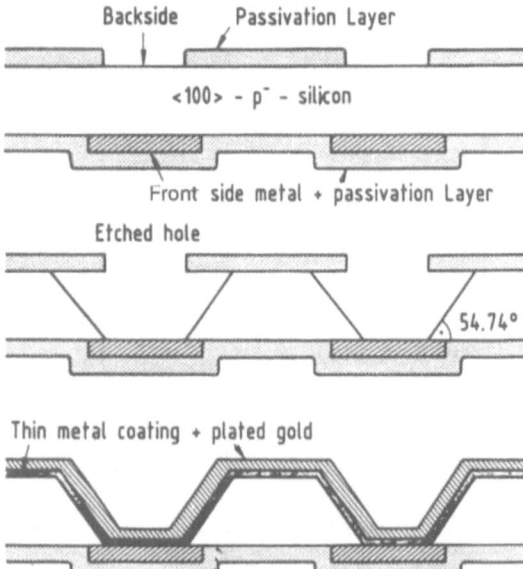

Backside   Passivation Layer

<100> - p⁻ - silicon

Front side metal + passivation Layer

Etched hole

54.74°

Thin metal coating + plated gold

Fig. 7.15. Fabrication steps for via holes

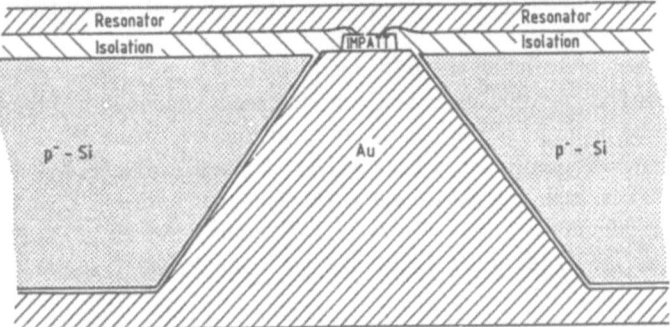

Fig. 7.16. (a) Metallized via hole in a 100 μm thick silicon substrate formed by anisotropic etching.
(b) Cross section of a monolithic microstrip oscillator with integrated heat sink

7.2.5 Metallization and Air Bridge Technology

Since the dominant transmission-line losses in the millimeter-wave region are due
to conductor losses a proper metallization is essential for MMIC's. Losses by skin
effect can be minimized by a high-conductive material and a conductor thickness
of 5 to 10 times the skin depth (0.25 μm for gold at 94 GHz). Radiation losses
can be minimized with vertical and straight transmission lines. For MMICs gold
metallization is most widely used because of its high conductance, its stability and
inertness and its suitability for electroplating processes.

The adhesion of gold on silicon and silicon dioxide is poor. Therefore an
adhesion layer is needed. Thin Ti or Cr layers (10–100 nm) are well suited. To
prevent interdiffusion between Si, Ti and Au an interdiffusion barrier metal is
recommended. Pt, Pd, TiW or TiN are proposed for this [7.47].

Air Bridge Technology

Air bridges are indispensable to monolithic millimeter-wave integrated-circuit design in coplanar waveguide technique [7.48]. They have to ensure the biasing of active areas on the chip and suppress multimode propagation along the RF signal paths. But they may also be used for the fabrication of Metall-Insulator-Metall (MIM) capacitors, for spiral inductors and active devices. The air bridges using air as a dielectric medium offer low parasitic interconnect capacitance and low inductance. Furthermore, immunity to edge profile problems and the ability to carry substantial current are additional advantages of this technique [7.49].

Figure 7.17 depicts a schematic diagramm of the air-bridge fabrication technique. First a thick layer of photoresist is spun on and patterned. After a careful bake of the resist for removing all solvents a thin Ti/Au metallization layer is evaporated under swaying motion for covering the edges of the resist. This thin metallization layer serves as a plating base. Then a second layer of photoresist is spun on and patterned. This upper resist is used as mask for gold electroplating to a thickness of several microns. Finally the upper resist is removed, the thin Ti/Au plating base is etched away and finally the lower resist is removed.

Figure 7.18 shows a scanning electron micrograph of a test array with air bridge connections. The gold thickness is 2.7 μm, the headroom width is 5.0 μm. Figure 7.19 gives the depth profile of the air bridge of Fig. 7.17 measured with a Dektak profiler. The measurements have been made with a needle weight of 10 mg. The air bridges are very stable. Even for the 144 μm long air bridge of Fig. 7.18 only a maximum bending of 0.7 μm is observed. The widths of the air bridges are 8, 18, 28 and 48 μm, the length of the short bridge is 75 μm, the length of the long bridge is 144 μm.

Figure 7.20 exhibits a test pattern of a spiral inductor with an air bridge. The width of the conductor is 13 μm, the length is 90 μm. Examples for air bridges in active devices are given in Sects. 7.3.4.

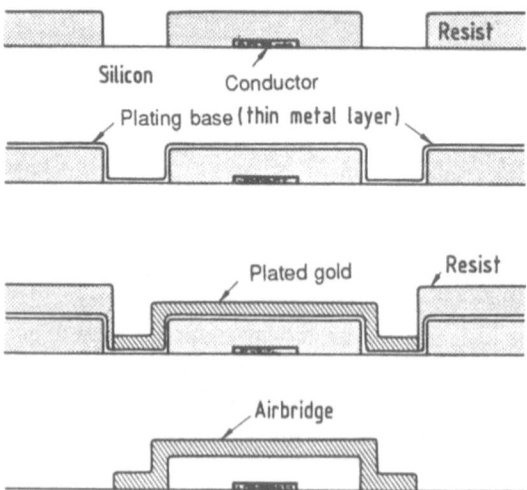

Fig. 7.17. Fabrication steps for air bridge

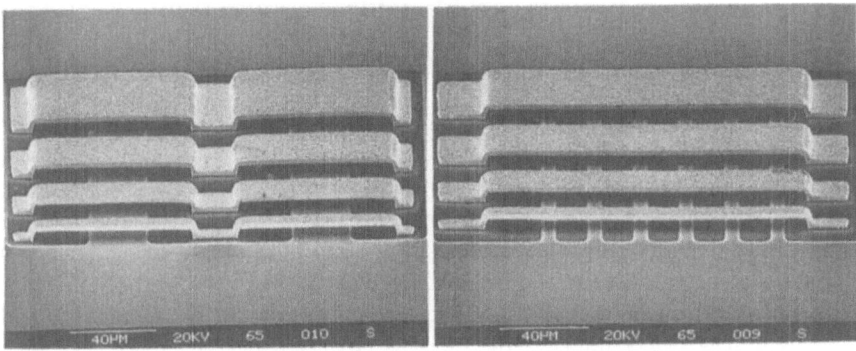

Fig. 7.18. Scanning Electron Micrographs (SEM) of a test array with air bridge connections. The widths of the air bridges are 8, 18, 28 and 48 μm respectively, the lengths are 75 μm and 144 μm

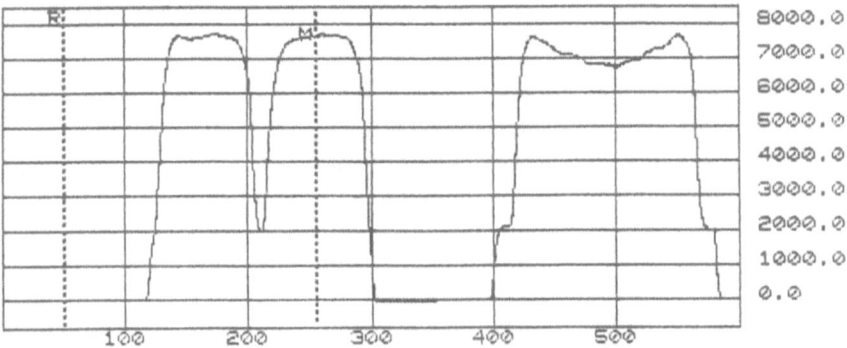

Fig. 7.19. Depth profile of the air bridges of Fig. 7.18 measured with a Dektak profiler. The needle weight is 10 mg

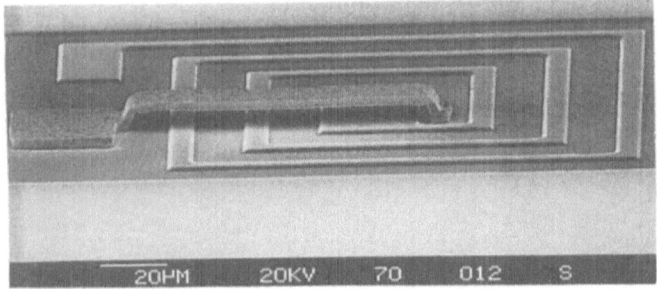

Fig. 7.20. SEM photograph of a spiral inductance test pattern. The tilt angle of observation is 70°. The width of the air bridge is 13 μm, the length 90 μm

7.3 Fabrication Process and Monolithic Integration of Two-Terminal Devices

Up to now most SIMMWICs have been realized with two-terminal devices (Chaps. 1 and 5). Elements with a nonlinear current-voltage relationship are used as rectifying devices for detector, receiver and mixer circuits. Schottky-barrier diodes are very suitable. The physics of Schottky-barrier diodes are described in Chap. 3. The fabrication process for monolithic integration of Schottky-barrier diodes or related diodes will be explained in Sect. 7.3.1.

Silicon avalanche transit-time diodes are up to now the most efficient solid-state millimeter-wave sources. The physics, characteristics and power capability of silicon transit-time diodes are treated in Chap. 2. The monolithic integration of these diodes in transmitter circuits is described in Sect. 7.3.2.

7.3.1 Fabrication Process of Coplanar Schottky-Barrier Diodes

Schottky-barrier diodes are efficient detector elements for millimeter waves. In order to use them in a receiver or mixer chip they have to be realized in coplanar form (Fig. 7.21). For proper working of these diodes in the millimeter-wave region, the series resistance and capacitance should be as small as possible. The quality factor of diodes for mixer applications is the cut-off frequency given by

$$f_{co} = 1/2\pi R_s C_j \qquad (7.2)$$

where f_{co} denotes the cut-off frequency, R_s the series resistance, and C_j the junction capacitance at zero bias.

To achieve best mixer characteristics the cut-off frequency should be 10 times the detecting frequency. Junction capacitance is related to the diode area, doping

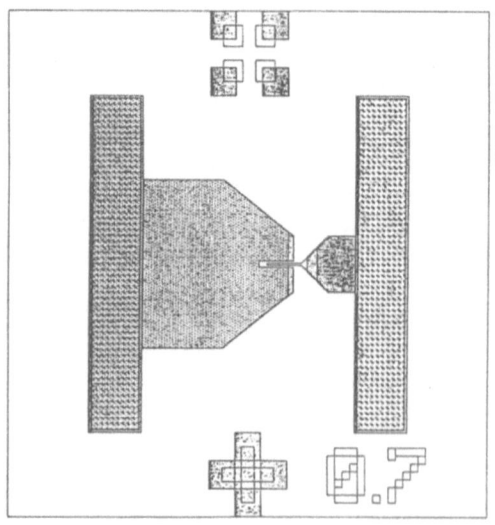

Fig. 7.21. Layout of a coplanar Schottky-barrier diode

of epitaxial layer and diffusion voltage (barrier height). Series resistance is also related to the diode area and doping of epitaxial layer but in opposite sense. Therefore, an optimization of geometrical parameters (small area and minimal parasitic components) and doping parameters has to be performed to achieve a high cut-off frequency [7.50].

Fabrication Process with Lift-Off Technology

Figure 7.22 depicts schematically the fabrication process of coplanar n-type Schottky-barrier diodes on high-resistivity silicon substrates. In detail the following process steps have to be performed:

- thermal oxidation to grow 1.3 μm SiO_2.
- Lithographic process for defining oxide windows (mask level 1).
- SiO_2 etching in buffered HF.
- Growth of 20 nm scatter oxide.
- As implantation with a dose of 2×10^{16} cm^{-2} at 100 keV.
- Annealing and diffusion process at 1200°C for 5 h (n^+-buried layer formation).
- SiO_2 etching in buffered HF to a final thickness of 600 nm.
- RCA clean.
- Growth of epitaxial layer by Si-MBE.
- Lithographic process for defining Schottky anode finger (mask level 2).
- Evaporation of Schottky-anode-finger metallization.
- Lift-off process for defining Schottky-anode-finger metallization.

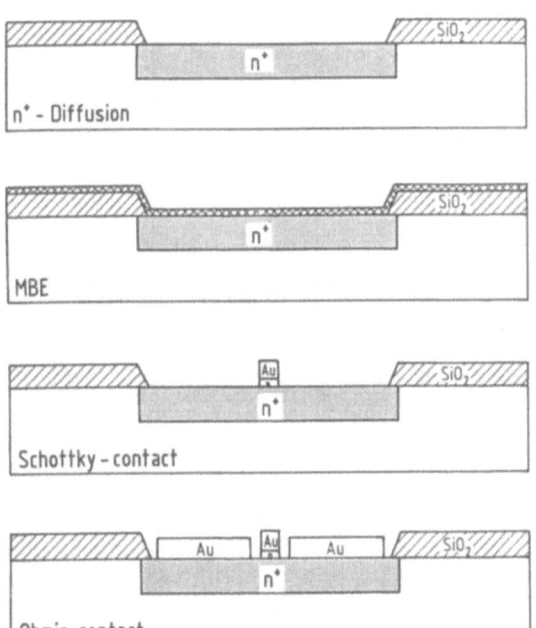

Fig. 7.22. Fabrication process of Schottky-barrier diodes

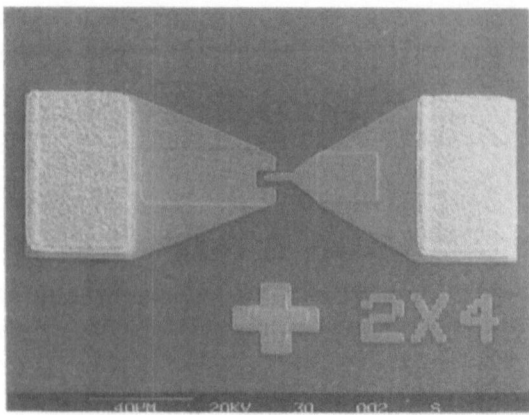

Fig. 7.23. Passivated coplanar Schottky-barrier diode with gold plated bond pads

- Etching of epitaxial layer using Schottky-anode-finger metallization as mask.
- Lithographic process for defining ohmic and pad contacts (mask level 3).
- Evaporation of ohmic and pad contacts.
- Lift-off process for defining ohmic and pad contacts.
- Passivation.
- Lithographic process for defining pad contact openings (mask level 4).
- Etching of passivation layer in pad openings and transmission lines.
- Evaporation of plating base.
- Lithographic process for pad and transmission lines (mask level 5).
- Electroplating of pad contacts and transmission lines.
- Removal of resist and plating base.
- Thinning of substrate

Figure 7.23 illustrates a processed Schottky barrier diode with a passivated Schottky anode finger and gold-plated pad contacts. These diodes are integrated in receiver circuits (Chaps. 1 and 5), but may also be used as discrete diodes for detectors or as discrete diodes in hybrid circuits mounted upside down.

Fabrication Process of Schottky-Barrier Diodes with X-Ray Lithography

For the definition of the critical Schottky-anode finger X-ray lithography was successfully applied in a mix-and-match mode combined with optical lithography. The employed X-ray mask is based on a stress-compensated Si membrane, which was fabricated by means of a simultaneously B- and Ge-doped epitaxial layer [7.39]. The patterning of the X-ray mask absorber was performed by e-beam writing (Philips EBPG3), three-level resist technique and Au electroplating [7.51]. From the X-ray master mask working masks were fabricated by X-ray copying using the *Bessy* radiation, Hunt resist HPR 204 and B/Ge wafers with Ti/Au plating base. The minimal mask dimension of the critical Schottky anode finger is 0.6 μm (Fig. 7.24). The exposures were performed at BESSY with a Süss X-ray stepper and HPR 204

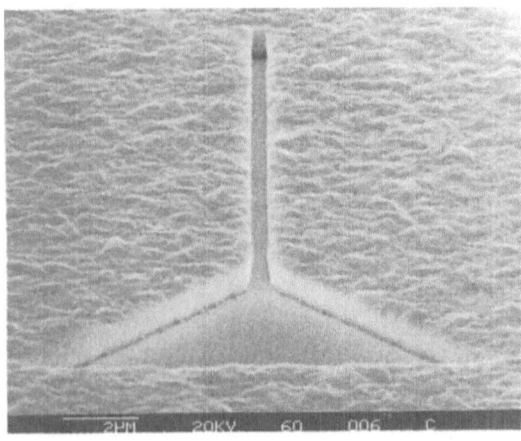

Fig. 7.24. SEM of an X-ray mask absorber pattern for the Schottky anode finger. Minimum dimension is 0.6 μm

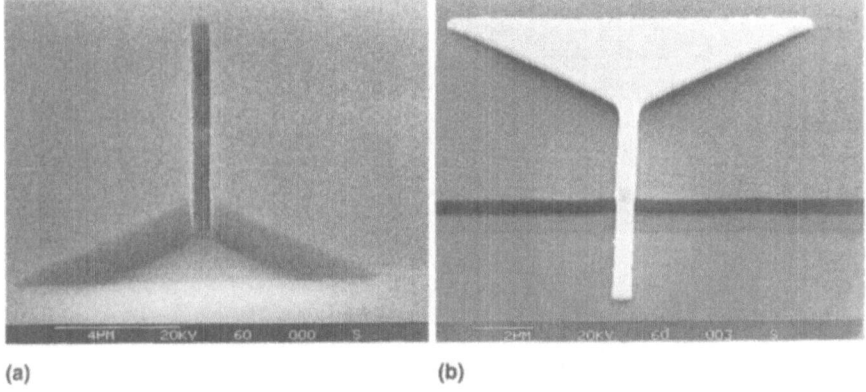

(a)   (b)

Fig. 7.25. SEM of the Schottky anode finger. (a) X-ray resist pattern, (b) Schottky anode metallization (100 nm Ti, 50 nm Pt, 200 nm Au) after a lift-off process with the resist pattern of (a)

as process resist. The transferred X-ray resist patterns of the Schottky anode finger show vertical side walls (Fig. 7.25a) and can directly be used for the definition of the Schottky-finger metallization by a lift-off process (Fig. 7.25b). Minimum Schottky-finger length of 0.6 μm have been achieved.

Fabrication Process with Air-Bridge Technology

Air-bridge technology was also applied for the fabrication of Schottky-barrier diodes. The following advantages are expected:

– lower parasitic capacitance due to the low dielectric constant of air ($\varepsilon_r = 1$),
– better isolation,
– lower leakage currents,
– lower interconnection resistance due to larger conductor cross-section.

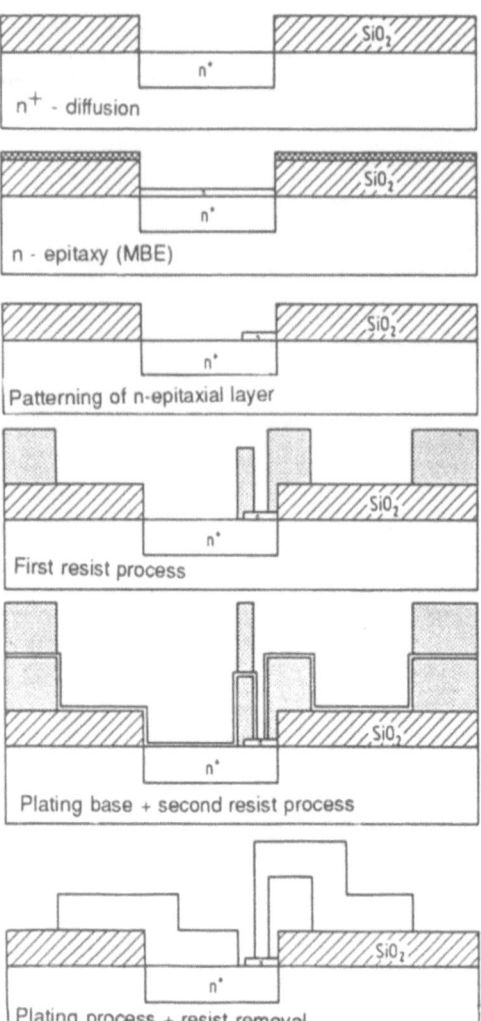

Fig. 7.26. Fabrication steps for Schottky-barrier diodes with air bridge

The fabrication process of Schottky-barrier diodes with air bridges is schematically depicted in Fig. 7.26. For the formation of the n^+-buried layer and the growth of the epitaxial layer the same processes are used as in the fabrication process with lift-off technology. But then the epitaxial layer is patterned by a photolithographic and a dry-etching process. Then air-bridge technology is applied, as described in Sect. 7.2.5.: A first thick photoresist is spun on and patterned for defining Schottky and ohmic contacts. After evaporation of the plating base (50 nm Ti, 200 nm Au) a second lithographic process defines the air bridge and contact areas. With this second resist pattern the gold plating process is performed. Gold is plated to a thickness of 3–5 μm. After removal of the resists and the plating base diodes with a Schottky-anode contact connected by an air bridge, as shown in Fig. 7.27, are

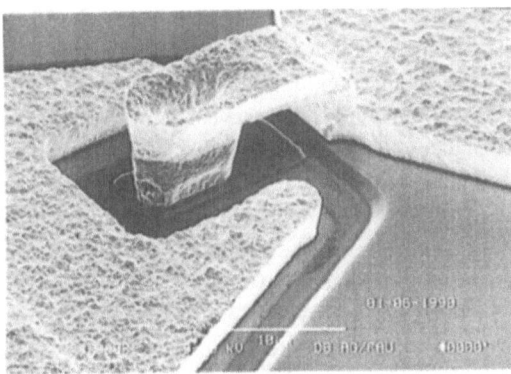

Fig. 7.27. Schottky-barrier diode
with air bridge

achieved. Figure 7.27 depicts the oxide window with the n^+-buried layer, the ohmic
contact on the left, the Schottky-anode contact on the plasma-etched epitaxial layer
and the air-bridge connection.

DC and RF Characteristics

DC characteristics of the diodes are evaluated from I-V and C-V measurements.
The most interesting parameters are series resistance R_s, junction capacitance C_j,
cut-off frequency f_{co}, barrier height Φ and ideality factor n. R_s, Φ, and n are determined by fitting the forward I-V curve to the ideal thermionic-emission equation
[7.52]. These parameters depend on n^+-buried layer, n-epitaxial layer thickness and
doping, junction area, ohmic contact resistance, Schottky metallization and diode
geometry.

Some results of different wafers with different n^+-buried layers and n-epitaxial
layers are summarized in Table 7.3. The Schottky contact area is quite large and
amounts to more than 30 μm^2. For all wafers Ti with a theoretical barrier height
of 0.5 V was used as the Schottky-contact metallization.

Table 7.3. Schottky-barrier diode parameters of differently processed wafers

MBE-Nr.	n^+-diffusion	n-epitaxy	Area [μm^2]	$R_s[\Omega]$	$\Phi[V]$	n	C_j [fF]	f_{co}[GHz]
C0695	As	150 nm 2×10^{16} cm^{-3}	30	10.4	.5	1.04	32	480
C0551 Air bridge	As	100 nm 2×10^{17} cm^{-3}	32	6.1	.5	1.09	47	550
C0767 LTD, 300°C	As/P	100 nm 2×10^{17} cm^{-3}	33	2.6	.48	1.09	59	1030
C0769 550°C, P-outdiffusion	As/P	100 nm $> 10^{18}$ cm^{-3}	33	1.4	.38	1.29	204	520
C0904	As/As	50 nm 1×10^{17} cm^{-3}	33	2.3	.49	1.09	85	800

The diodes of wafer C0695 have a low-doped n-epitaxial layer of 150 nm thickness. This yields high series resistances and a low junction capacitance. The diodes of wafer C0551 were fabricated with air-bridge technology. Wafer C0767 has an As/P buried layer and the n-epitaxial layer is grown by low-temperature doping at 300°C. Low series resistances are measured and cut-off frequencies higher than 1 THz are evaluated. The diodes of wafer C0769 are fabricated on As/P buried layers with nominally undoped n-epitaxial layers grown at 550°C. Due to the outdiffusion of P the n-epitaxial layer has a doping level larger than 10^{18} cm^{-3}. This causes a high junction capacitance, lowers the barrier height and degrades the ideality factor. Wafer C0904 has a very thin epitaxial layer (50 nm). This yields a low series resistance but a high junction capacitance.

With the above-described technology the following rectifying silicon diodes have been fabricated on high resistivity silicon:

- N-type Schottky barrier diodes [7.32]. These diodes are fabricated with n$^+$- buried and n-epitaxial layers. With low-doped ($< 1 \cdot 10^{16}$ cm^{-3}) and thin (70 nm) epitaxial layers these diodes behave like Mott diodes, and show a conversion loss of 6.5 dB in single-ended mixers at 94 GHz in zero-bias operation with an LO power of 10 dBm and Ti as Schottky-barrier metal ($\Phi = 0.5$ V).
- P-type Schottky-barrier diodes [7.33]. These diodes are fabricated with p$^+$- buried and p-epitaxial layers. Cut-off frequencies up to 380 GHz have been obtained. By introducing a p$^+$-spike on the epitaxial p-layer a barrier-height reduction down to 0.3 V has been achieved. These diodes have been successfully applied in zero-bias detectors.
- Camel diodes [7.53]. For these diodes n$^+$p$^+$n-epitaxial layers are used. The barrier height can be adjusted by the epitaxial layer sequence. Barrier heights between 0.2 and 0.56 V were obtained.
- Planar-doped barrier diodes [7.34]. Symmetrical PDB diodes are fabricated with n$^+$-i-p$^+$-i layers on n$^+$-buried layers. Barrier heights between 0.3 and 0.57 V were obtained. These diodes are tested in subharmonic pumped mixers. A minimal conversion loss of 10.8 db is measured at 126 GHz.
- Si/SiGe resonant tunneling diodes [7.35,54]. Double-barrier hole-resonant tunneling structures are grown on p$^+$-buried layers. Full wave rectifying is possible with these diodes.

7.3.2 Fabrication Process of Monolithically Integrated Transit-Time Diodes

Different approaches are made for monolithic integration of transit-time diodes. The most successful fabrication process uses a self-stopping etchant, self-aligned contacts, silicon-nitride passivation and air-bridge technology [7.36,55]. Figure 7.28 shows schematically the main fabrication steps.

First the high-resistive ($\rho > 4000$ Ωcm) silicon substrates are thermally oxidized and highly doped p$^+$-layers are formed by B diffusion. A sheet resistance smaller than 2.5 Ω/$\square$ and a surface concentration greater than 10^{20} cm^{-3} has been achieved (Fig. 7.9). Next the active layers of sophisticated IMPATT profiles are

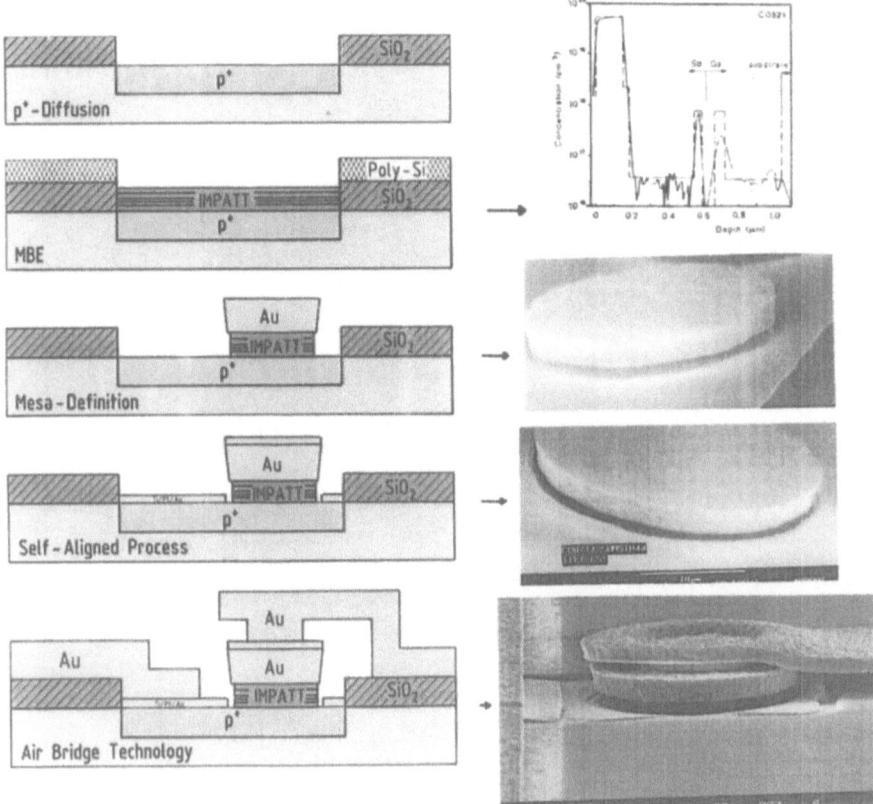

Fig. 7.28. Fabrication process of monolithically integrated IMPATT diodes

grown by Si-MBE (Fig. 7.11b). The growth temperature is 550°C. Then the top contact of the IMPATT diode is defined, and the diode is mesa etched in a self-stopping etchant (aqueous KOH solution). A slight undercut is performed. This undercut enables self-aligned technology for defining the lower contact and reduces series resistance. To reduce surface-leakage currents the mesa edges are passivated by a 150 nm thick, low-temperature (300°C) plasma-enhanced deposited Si_3N_4 film. Finally, air-bridge technology is applied for forming the top contact of the diode.

With this technology monolithically integrated IMPATT diodes with different diameters (5–40 μm), different geometrical-shape (circular, annular, linear) and evenly distributed diodes have been manufactured. Figure 7.29 shows circular, annular, linear and distributed IMPATT diodes monolithically integrated in a slot line resonator (Chap. 5). The scanning electron micrograph depicts the p^+-buried layer, the self aligned lower contact and the upper contact with air-bridge connection.

The dc characteristics of the diodes are good. Figure 7.30 exhibits the IV curve of a double-low-high-low transit-time diode. A sharp breakdown is observed at

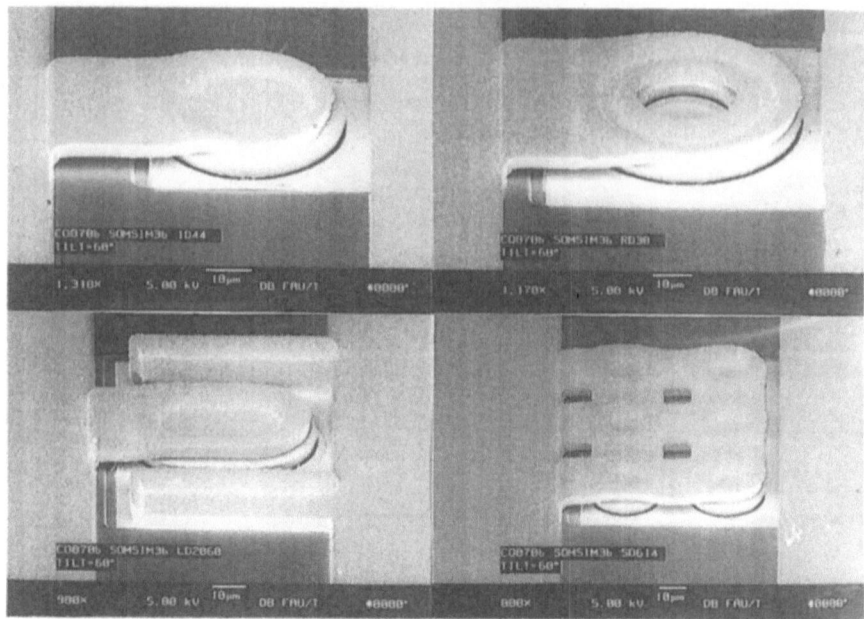

Fig. 7.29. Monolithically-integrated circular, annular, linear and distributed IMPATT diodes with an air bridge

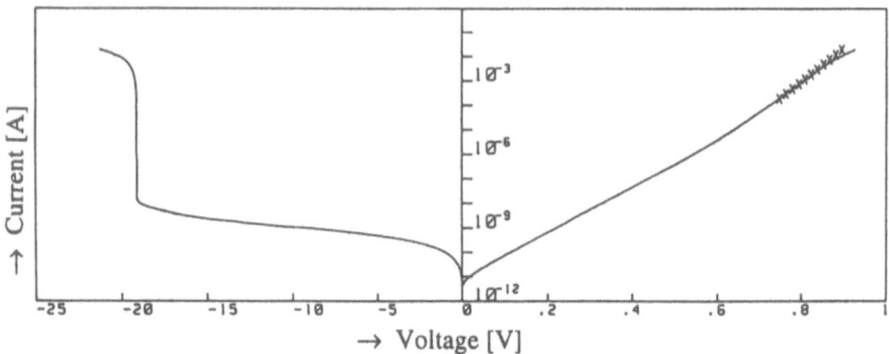

Fig. 7.30. I-V curve of an integrated diode with 24 μm diameter. The deviation from the crosses (ideal curve) is due to the series resistance.

19.2 V. Due to passivation low leakage current is achieved. Due to the self-aligned contact the series resistance is low (1.7 Ω).

Two other fabrication processes have been studied which use either oxide isolation or polysilicon isolation for the monolithic IMPATT diodes (Fig. 7.31). In the oxide isolation process (Fig. 7.31a) thick, low-temperature deposited oxide is used for passivation. In the polysilicon isolation process a so-called differential epitaxy

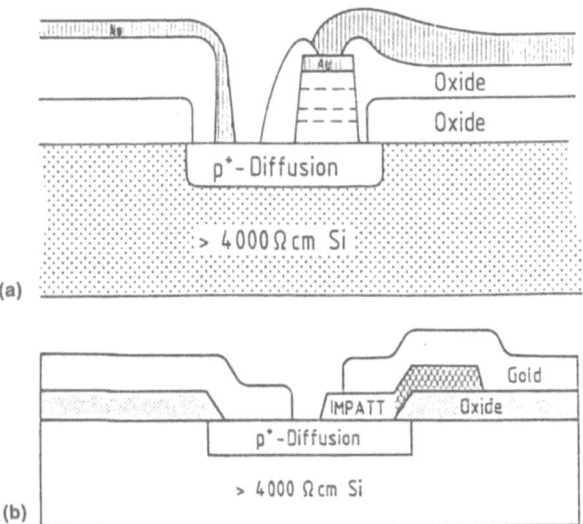

Fig. 7.31. Schematic drawing of monolithic integrated IMPATT diode with oxide isolation (**a**) and with polysilicon isolation (**b**)

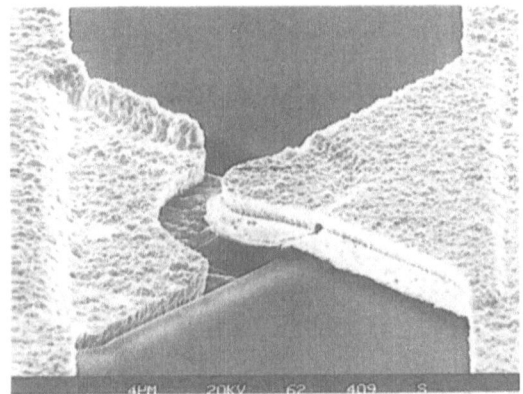

Fig. 7.32. Scanning electron micrograph of an integrated IMPATT diode with polysilicon isolation

process is employed, where the diode is partly embedded in polycrystalline silicon. Here, the fact is used that silicon grows monocrystalline on a buried layer but polycrystalline on oxide. Low-doped polysilicon itself is isolating [7.56]. Therefore the mesa etch is made in a way that the remaining polysilicon serves as isolation for the top contact. The advantage of this technology is that only three masks are necessary for the fabrication process.

Figure 7.32 illustrates a scanning electron micrograph of an integrated IMPATT diode with polysilicon isolation. The drawback of this technology is that it is not applicable to high-power diode structures containing doping spikes like Quasi Read Double Drift (QRDD) or Double-Low-High-Low (DLHL) diode structures. The highly doped spikes prevent isolation. RF-results with these diodes are reported in Chap. 5.

7.3.3 Fabrication Process of Lateral PIN Diodes

PIN diodes are commonly used in microwave/millimeter-wave switches, limiters and attenuators [7.8,57,58,]. The switch operation is based upon the difference in RF impedances between the forward bias and the reverse bias states. The silicon PIN diodes have been the dominant control devices due to their high performance. They provide low forward resistance due to the large carrier lifetime of silicon.

For the fabrication of monolithically-integrated lateral PIN diodes the high resistivity, undoped or "intrinsic" silicon substrate itself is used for the intrinsic (I) zone. The p^+ (P) and n^+ (N) region are formed by standard buried-layer processes. Selective ion implantation and simultaneous annealing is used. Therefore the fabrication process of lateral PIN diodes is very simple. No epitaxy process is needed. The p^+ and n^+ contacts are also formed simultaneously with Ti-Au metallization.

Figure 7.33 shows a schematic cross section through the PIN diode, an interference microscopic photograph of the lapped p^+, i, n^+ regions, a SEM of the diode with electroplated gold metallization, a photo of an integrated diode and the IV-curve of a diode with a very small i-zone (1 μm). These diodes have been tested

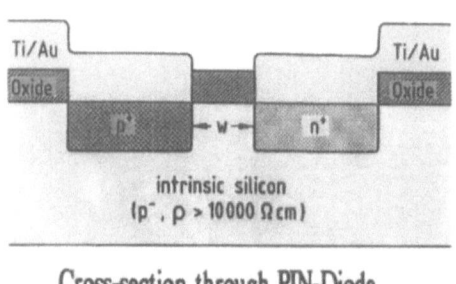

Cross-section through PIN-Diode

Doping of actual PIN-Diode

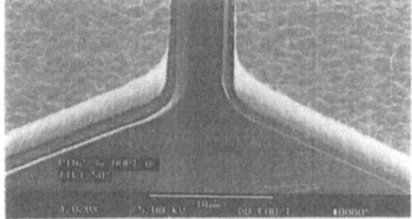

SEM of PIN-Diode

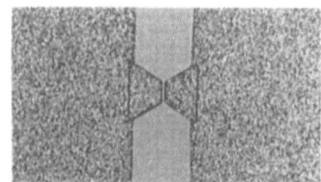

Photo of integrated Diode

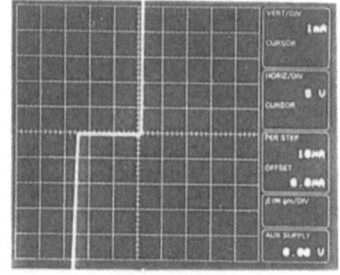

IV-Curve

Insertion loss: 1.5 dB
Isolation: 30 dB   (26 – 40 GHz)

Fig. 7.33. Monolithically integrated lateral Si-PIN diode in high-resistivity silicon: cross-section, active doping zones, SEM, photo and IV curve

in Ka-band (26.5–40 GHz). Insertion loss of 1.5 dB and isolation of 30 dB were measured. These diodes are also integrated in a microstrip line designed PIN switch.

7.4 Fabrication Process of Three-Terminal Devices

In recent years significant enhancement in silicon transistor performance and speed has been demonstrated. For Si-MOSFETs, scaling the channel length into deep-submicrometer regime leads to a higher cutoff frequency f_T due to a reduced carrier transit-time across shorter device active regions [7.59]. With a 0.15 μm channel length MOSFET a unity-current-gain cutoff frequency (f_T) of 89 GHz at room temperature has been achieved [7.60].

For bipolar transistors the speed improvement has been accomplished by scaling the vertical profile, and reducing the parasitic resistances and capacitances associated with the extrinsic portion of the device. f_T as high as 30 GHz has been achieved by implantation [7.61], a value of 44 GHz was reported for bases formed by diffusion from doped glass [7.62] and even 64 GHz was measured using in-situ phosphorus-doped polysilicon emitter technology [7.63].

The most significant progress has however, been achieved by the advent of the silicon-germanium (SiGe) heterojunction bipolar transistor. Steady progress especially in adjusting the thin and highly doped, low resistance base has pushed the unity current gain cutoff frequency f_T from 75 GHz [7.64] to 91 GHz [7.65] and f_{max} to 65 GHz [7.65]. Because of this three-terminal devices are now available for silicon that work at millimeter-wave frequencies too.

7.4.1 Bipolar Transistors

The last ten years have witnessed rapid progress in bipolar technology, leading to a great number of new transistor configurations with good results [7.66]. The driving forces are distinctly coming from CMOS technology, mainly thin-film deposition techniques and dry etching of the deposited films. These two techniques in combination enable self-aligned process sub-sequences.

The critical dimension of the vertical npn transistor is not given by a lateral length as in the case of the MOS transistor, but by the vertical doping profile. Therefore, the lithography is not so important for bipolar devices as for MOS devices. The most critical dimensions of the vertical npn device for transistor performance are the width of the active base region and the emitter depth, both resulting from well-controlled crystal-growth techniques like deposition and diffusion. Dimensions below 100 nm can be achieved.

The conventional bipolar device, usually denoted as the "Standard Buried Collector" (SBC) suffers from a large parasitic capacitance, a large extrinsic base resistance and limited vertical scaling (Fig. 7.34a). By introducing the so-called self-aligned polysilicon technology a further vertical scalability is achieved (Fig. 7.34b). This technology uses a thin polysilcon film as the emitter contact. The

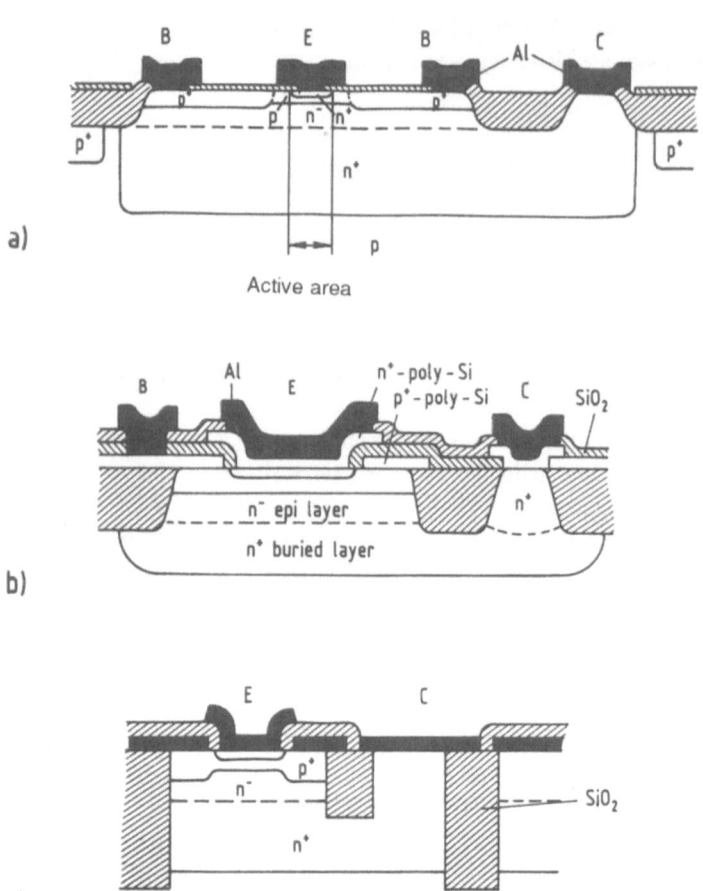

Fig. 7.34. (a) Configuration of the conventional bipolar transistor (Standard Buried Collector). (b) Configuration of the self-aligned polysilicon-emitter bipolar transistor. (c) Schematic figure of the deep trench isolation

polysilicon film also serves as the diffusion source for forming shallow and steep n^+ emitter junctions smaller than 100 nm. A self-aligned emitter-base configuration reduces the contacting area, external base resistance and capacitance of the base collector junction. Further reduction of the external base resistance, which is one of the most important parameters for high-speed operation, can be achieved by introducing polysilicon and silicide conducting layers.

A further reduction of parasitic capacitance is achieved by an advanced isolation scheme, a deep trench or U-groove isolation shown in Fig. 7.34c. This yields a strong reduction of device area. With this technology a cut-off frequency f_T of 64 GHz has been reported [7.63]. The basic process technology was as follows:

- self-aligned double polysilicon bipolar transistor SICOS technology with U-groove isolation,

- low-temperature in-situ phosphorus doped polysilicon deposition technology to obtain an ultra shallow emitter junction below 30 nm,
- thin emitter polysilicon thickness of 80 nm due to reduce transit time,
- low-energy BF_2^+ implantation to achieve a sub 100 nm base width,
- precisely controlled sub-micrometer epitaxial growth,
- pedestal collector fabrication technology.

7.4.2 Hetero Bipolar Transistors for SIMMWICs

The advent of the silicon germanium heterojunction bipolar transistor provides a three-terminal device which is well behaved and capable of working at millimeter-wave frequencies. Theoretical modelling yield f_T and f_{max} well beyond 100 GHz [7.67]. SiGe HBTs are described in detail in Chap. 4. In this subsection only a brief description of the fabrication process is given.

Compared with the bipolar junction transistor the SiGe HBT offers the possibility of a high base doping for a low base resistance, an accelerating field by grading the Ge content, a short base transit time by reducing the base thickness and the elimination of the emitter delay. The growth of the complete transistor structure by MBE meets the demand of precise control of thickness, doping and Ge content, and avoids high-temperature processing [7.68].

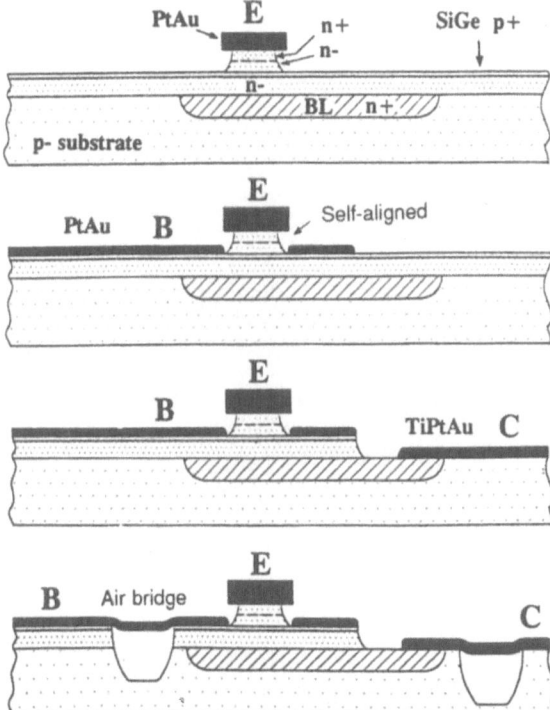

Fig. 7.35. Fabrication technology for Si/SiGe HBTs (by courtesy of A. Gruhle [7.68])

The complete HBT layers are grown on 4-inch p^- substrates with As implanted and subsequently diffused buried layers with $5-10$ $\Omega/\square$. The wafers received an RCA cleaning and an in-situ 900°C flash-off. The 200 nm thick collector was doped 3×10^{17} cm^{-3} by Sb doping using secondary-ion implantation. The SiGe base is grown at 530°C by coevaporation of Si and Ge and starts with a 10 nm undoped spacer layer followed by a 20 nm, 2×10^{19} cm^{-3} boron-doped layer. The boron flux is switched off shortly before the germanium in order to introduce a $1-2$ nm thick emitter spacer layer. The 2×10^{18} cm^{-3} Sb-doped emitter layer is 70 nm thick and grown by spontaneous dopant incorporation at relatively low temperatures between 425°C and 450°C followed by the 230 nm thick n^+ emitter cap grown at 320°C to achieve $1-2 \times 10^{20}$ cm^{-3} electrically active Sb concentration.

The device fabrication sequence is shown in Fig. 7.35. The PtAu emitter metal is defined by lift off and acts as a mask for the following selective wet-chemical etch which stops at the SiGe layer. The base metallization is self-aligned with respect to the emitter because of the controlled overetch. A dry etch gives access to the buried layer. After collector metallization a second dry etch forms deep trenches that separate the contact pads from the active area leaving air bridges.

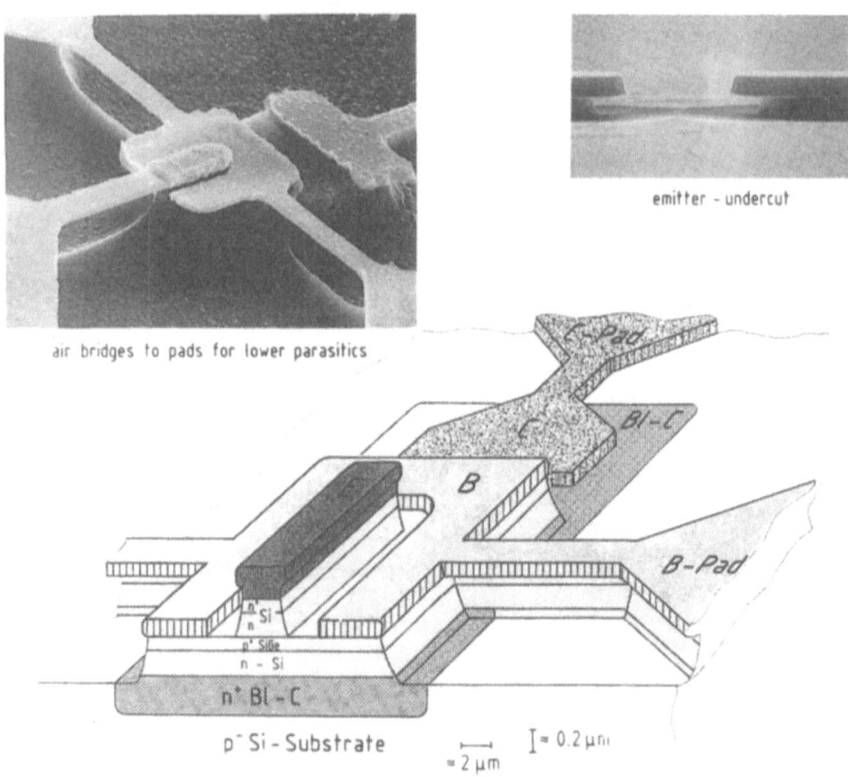

emitter - undercut

air bridges to pads for lower parasitics

Fig. 7.36. Layout and SEM of a SiGe HBT with air bridge and emitter undercut (by courtesy of U. König [7.69])

Figure 7.36 depicts the layout and scanning electron micrographs of air bridges and emitter undercut. With these transistors f_T of 91 GHz and f_{max} of 65 GHz was measured [7.65].

The excellent FM-noise behaviour makes them attractive for future monolithically integrated millimeter-wave oscillators. A first step toward this was a hybrid microstrip DRO for X-band frequencies with the Si-SiGe HBT as active oscillator device. At 9.6 GHz an oscillator output power of 10 mW with a conversion loss of 17.5% is measured. Phase noise N/C_{FM} of -85 dBc (1 Hz) is determined at 100 kHz off carrier [7.70].

7.5 Summary and Prospects

Silicon technology is mature for monolithic millimeter-wave integration. Due to advanced technologies like Si-MBE and improved lithographic and patterning techniques not only two-terminal devices but also three-terminal devices are now available for the millimeter-wave range.

Schottky-barrier and related nonlinear diodes are available for monolithic integration as rectifying elements in receiver circuits, transit-time diodes as power-generating devices in transmitter circuits, and lateral PIN diodes in switching circuits. Due to low silicon wafer costs the whole millimeter-wave circuits including antenna patterns are cost effectively integrated on a single silicon slice. This improves system performance and lowers overall costs. Owing to progress in Si/SiGe HBTs first microwave oscillators with excellent noise-to-carrier ratio have been realized on high-resistivity silicon [7.70]. Further Si/SiGe HBT microwave and millimeter-wave circuits will follow.

The next step will be the co-integration of SIMMWICs with preamplifier and signal-processing circuits. Thus complete millimeter-wave sensor systems may be realized. Integration concepts for this, called heterointegration have already been discussed [7.69]: SIMMWIC devices can be arranged monolithically in reserved areas of complex Si-ICs thus merging millimeter-wave circuits with signal-processing circuits.

Acknowledgement. The financial support of the BMFT grants NT 2686 B/4, NT 2709 C9 and NT 2769 C0 are gratefully acknowledged. The author would like to thank many inhouse coworkers listed alphabetically, J. Büchler, A. Gruhle, J. Hersener, E. Kasper, M. Kuisl, J.F. Luy, H. Kibbel, E. Sasse, F. Schäffler.

References

Section 7.1

7.1   R.K. Hoffmann: *Handbook of Microwave Integrated Circuits* (Artech House, Norwood 1987)
7.2   T.M. Hyltin: Microstrip transmission on semiconductor dielectrics. IEEE MTT-*13*, 777–781 (1965)

7.3 P.H. Saul: Comparison of GaAs and silicon ICs in high speed digital applications. Mil. Micr. Conf. (1986), pp. 460–466

7.4 D.N. Mcquiddy, J.W. Wassel, J.B. Lagrange, W.R. Wisseman: Monolithic microwave integrated circuits: An historical perspective. IEEE Trans. MTT-32, 997–1008 (1984)

7.5 B.W. Battershall, S.P. Emmons: Optimization of diode structures for monolithic integrated microwave circuit. IEEE J. SC-3, 107–112 (1968)

7.6 A. Ertel: Monolithic IC techniques produce first all-silicon X-band switch. Electronics, 76–81 (January 1967)

7.7 A. Rosen, M. Caulton, P. Stabile, A.M. Gombar, W. J- Janton, C.P. Wu, J.F. Corboy, C.W. Magee: Millimeter-wave device technology. IEEE Trans. MTT-30, 47–55 (1982)

7.8 P. Stabile, A. Rosen, W.M. Janton, A. Gombar, M. Kolan: Millimeter wave silicon device and integrated circuit technology. Proc. IEEE MTT-Symp. (1984) pp. Digest, pp. 448–450

7.9 A. Rosen, M. Caulton, P. Stabile, A.M. Gombar, W.J. Janton, C.P. Wu, J.F. Corboy, C.W. Magee: Silicon as a millimeter-wave monolithically integrated substrate - A new look. RCA Rev. 42, 633–656 (1981)

7.10 P. Stabile, A. Rosen: A silicon technology for millimeter-wave monolithic circuits. RCA Rev. 45, 587–605 (1984)

7.11 E. Kasper, J.C. Bean (eds.): Silicon Molecular Beam Epitaxy. (CRC, Boca Raton 1988)

7.12 M.A. Herman, H. Sitter: Molecular Beam Epitaxy, Springer Ser. Mater. Sci., Vol. 7, (Springer, Berlin, Heidelberg 1989)

7.13 A. Heuberger: X-ray lithography. Microelectr. Eng. 3, 535–556 (1985)

7.14 K.M. Strohm, J.F. Luy, E. Kasper, J. Buechler, P. Russer: Silicon technology for monolithic millimeter wave integrated circuits. Mikrowellen & HF Magazin, 14, No. 8, 750–760 (1988)

7.15 K.M. Strohm, J. Buechler, P. Russer, E. Kasper: Silicon high resistivity substrate millimeter-wave technology. Monolithic Circuits Symp. IEEE 1986 Microwave and Millimeter-Wave (Baltimore, MD), (1986) pp. 93–97

7.16 J. Buechler, E. Kasper, P. Russer, K.M. Strohm: Silicon high resistivity substrate millimeter wave technology. IEEE Trans. MTT-34, 1516–1521 (1986)

7.17 W.R. Runyan: Semiconductor Measurements and Instrumentation. (McGraw-Hill, New York 1975)

7.18 D.K. Schroder: Semiconductor Material and Device Characterization. (Wiley, New York 1990)

7.19 L.J. van der Pauw: A method for measuring specific resistivity and hall effect of discs of arbitrary shape. Phil. Res. Rep. 13, 1–9 (1958)

7.20 R. Brennan: Determination of diffusion characteristics using two and four point probe measurements. Solid State Technol., 127–132 (December 1984)

Section 7.2

7.21 S.M. Sze: Semiconductor Devices-Physics and Technology. (Wiley, New York 1985)

7.22 S.M. Sze (ed.): VLSI Technology. (McGraw-Hill, New York 1983)

7.23 W.C. O'Mara, R.B. Herring, L.P. Hunt (eds.): Handbook of Semiconductor Silicon Technology. (Noyes, New Jersey 1990)

7.24 D.J. Elliot: Integrated Circuit Fabrication Technology (McGraw-Hill, New York 1982)

7.25 G.K. Reeves, H.B. Harrison: Obtaining the specific contact resistance from transmission line model measurements. IEEE EDL-3, 111–113 (1982)

7.26 E. Kasper, H. Kibbel F. Schäffler: An industrial single-slice Si-MBE apparatus. J. Electrochem. Soc., 136, 1154–1158, (1989)

7.27 H. Jorke, H. Kibbel: Doping by secondary implantation. J. Electrochem. Soc., 133, 774–778, (1986)

7.28 H. Kibbel, E. Kasper, P. Narozny, H.U. Schreiber: Boron doping of SiGe base of heterobipolar transistor. Thin Sol. Films, 184, 163–169 (1990)

7.29   E. Kasper, F. Schäffler: Low temperature molecular beam epitaxy of silicon (Si-MBE). Physica Scripta, T29, 147–151 (1989)

7.30   P.H. Singer: Trends in wafer cleaning. Semiconductor Int. 15, No. 13, 36–39 (1992)

7.31   W. Kern, A.D. Puotinen: Cleaning solutions based on hydrogen peroxide for use in semiconductor technology. RCA Rev. 31, 187–206 (1970)

7.32   K.M. Strohm, J.F. Luy, J. Buechler, F. Schäffler, A. Schaub: Planar 100 GHz silicon detector circuits. Microelectronic Eng., 15, 285–288 (1991)

7.33   K.M. Strohm, J. Buechler, J.F. Luy: 90 GHz SIMMWIC rectennas. 22nd Europ. Microwave Conf. (1992), pp. 608–613

7.34   U. Güttich, K.M. Strohm, F. Schäffler: D-band subharmonic mixer with silicon planar doped barrier diodes. IEEE Trans. MTT-39, 366–368 (1991)

7.35   J.F. Luy, K.M. Strohm, J. Buechler: A 91 GHz Si/SiGe resonant tunneling detector. Archiv Elektr. Übertrg., 46, 370–373 (1992)

7.36   K.M. Strohm, J. Buechler, J.F. Luy, F. Schäffler: A silicon technology for active high frequency circuits. Microelectronic Eng. 19, 717–720 (1992)

7.37   A. Gruhle, H. Kibbel, U. König, U. Erben, E. Kasper: MBE grown Si-SiGe HBT's with high β, f_T and f_{max}. IEEE EDL-13, 206–208 (1992)

7.38   L.K. White: Positive-resist processing for step- and -repeat optical lithography. RCA Rev., 47, 345–379 (1986)

7.39   K.M. Strohm, J. Hersener, H.J. Herzog: Stress compensated Si-membrane masks for X-ray lithography with synchrotron radiation. Eurocon 86 Proc., Paris, Paper AI.4 (1986)

7.40   K.M. Strohm, J. Hersener, E. Piper: X-ray lithography for monolithic millimeter wave integration. Microcircuit Eng. 9, 131–134 (1989)

7.41   W. Kern: Chemical etching of silicon, germanium, gallium arsenide and gallium phosphide. RCA Rev. 39, 278–309 (1978)

7.42   D.M. Manos, D.L. Flamm, (eds.): Plasma Etching- An Introduction. (Academic, Boston 1989)

7.43   H. Linde, L. Austin: Wet silicon etching with aqueous amine gallates. J. Electrochem. Soc. 139, 1170–1174 (1992)

7.44   G.M. Rebeiz, L.P.B. Katehi, W.Y. Ali-Ahmad, G.V. Eleftheriades, C.C. Ling: Integrated horn antennas for millimeter-wave applications. Radioscientist 3, 68–77 (1992)

7.45   K.E. Peterson: Silicon as a mechanical material. Proc. IEEE 70, 420–457 (1982)

7.46   G.W. Turner, C.L. Chen, M.K. Connors, L.J. Mahoney, W.L. McGilvary: Selective Plasma Etching of Si from GaAs-on Si wafers for microwave via-hole formation. Electron. Lett. 26, 854–855 (1990)

7.47   J.M. Poate: Diffusion and reactions in gold films. Sol. State Tech., 227–234 (April 1982)

7.48   N.H.L. Koster, S. Koßlowski, R. Bertenburg, S. Heinen and I. Wolff: Investigations on air bridges used for MMICs in CPW technique. 19th Europ. Microwave Conf. (1989) pp. 666–671

7.49   J.K. Singh, O.P. Daga, H.S. Kothari, B.R. Singh, W.S. Khokle: Air bridge and via hole technology for GaAs based microwave devices. Microelectr. J. 19, 23–27 (1988)

Section 7.3

7.50   J.M. Dieudonné, B. Adelseck, K.-E. Schmegner, R. Rittmeyer, A. Colquhoun: Technology related design of monolithic millimeter wave Schottky diode mixers. IEEE Trans. MTT-40, 1466–1474 (1992)

7.51   J. Hersener, E. Piper, A. Wilhelm, G. Birkenstock: Application of X-ray lithography for manufacturing a metal-oxide semiconductor field effect transistor tetrode, J. Vac. Sci. Technol. B5, 253–256 (1987)

7.52   S.M. Sze: Physics of Semiconductor Devices. (Wiley, New York 1981)

7.53   K.M. Strohm, J.F. Luy, J. Buechler, J. Hersener, H. Kibbel, F. Schäffler, A. Schaub, A. Wilhelm: Submikron-Technologie für Millimeterwellensensoren auf hochohmigem Silizium. BMFT-Abschlußbericht NT 2769C0 (1991)

7.54   J.F. Luy, K.M. Strohm, J. Büchler: Monolithic Si/SiGe millimeter-wave detector circuits. Int. Semicon. Dev. Res. Symp. (December 1991), pp. 155–158

7.55   B. Bayraktaroglu: Monolithic IMPATT technology. Microwave J. 73–86 (April 1989)

7.56   M. Kuisl, U. König, F. Schäffler, R. Lossos: Characterization of MBE-grown polysilicon, in *Polycrystalline Semiconductors*, ed. by H.J. Möller, H.P. Strunk, J.H. Werner, Springer Proc. Phys *35*, 192–197, (Springer, Berlin, Heidelberg 1989)

7.57   A.K. Sharma: Solid-state control devices: State of the art. Microwave J., 95–112 (1989)

7.58   P.J. Stabile, A. Rosen, P.R. Herczfeld: Optically controlled lateral PIN diodes and microwave control circuit. RCA Rev. *47*, 443–456 (1986)

Section 7.4

7.59   S.M. Sze: *High Speed Semiconductor Devices*. (Wiley, New York 1990)

7.60   R.-H. Yan, K.F. Lee, D.Y. Jeon, Y.O. Kim, B.G. Park, M.R. Pinto, C.S. Rafferty, D.M. Tennant, E.H. Westerwick, G.M. Chin, M.D. Morris, K. Early, P. Mulgrew, W.M. Mansfield, R.K. Watts, A.M. Voshschenkov, J. Bokor, R.G. Swartz, A. Ourmazd: 89-GHz f_T room-temperature silicon MOSFETs. IEEE EDL-*13*, 256–258 (1992)

7.61   T. Gomi et al.: A sub-30 psec Si bipolar LSI technology. IEDM Techn. Dig. (1988) pp. 744–747

7.62   M. Sugiyama et al.: A 40 GHz f_T Si bipolar LSI technology. IEDM Techn. Dig. (1989) pp. 221–224

7.63   M. Namba T. Kobayashi, T. Uchino, T. Nakamura, M. Kondo, Y. Tamaki, S. Iijima, T. Kure, M. Tanabe: A 64 GHz Si bipolar transistor using in-situ phosphorus doped polysilicon emitter technology. IEDM Techn. Dig. (1991) pp. 443–446

7.64   G.L. Patton, J.H. Comfort, B.S. Meyerson, e.F. Crabbé, G.J. Scilla, E. de Fresart, J.M.C. Stork, J.Y.-C. Sun, D.L. Harame, J.N. Burghartz: 75 GHz f_T SiGe-base heterojunction bipolar transistors. IEEE EDL-*11*, 171–173 (1990)

7.65   A. Gruhle, H. Kibbel, U. Erben, E. Kasper: 91 GHz SiGe-HBT's grown by MBE. Electronics Lett. *29*, 415–417 (1993)

7.66   L. Treitinger, M. Miura-Mattauch (eds.): *Ultra-Fast Silicon Bipolar Technology*. Springer Ser. Electron. Photon. Vol. 27, (Springer, Berlin, Heidelberg 1988)

7.67   S.A. Campbell, A. Gopinath: Modeling of Ge-Si heterojunction bipolar transistors for use in silicon monolithic millimeter-wave integrated circuits. IEEE Trans. MTT-*37*, 2046–2050 (1989)

7.68   A. Gruhle, H. Kibbel, E. Kasper: The influence of MBE-layer design on the high frequency performance of Si/SiGe HBTs. Microelectronic Eng. *19*, 435–438 (1992)

7.69   U. König: Electronic Si/SiGe devices: basics, technology, performance. *Festkörperproblem/ Advances in Solid State Physics*, *32*, 199–220 (Vieweg, Braunschweig 1992)

7.70   U. Güttich, J.F. Luy, A. Gruhle: A Si-SiGe HBT dielectric resonator stabilized microstrip oscillator at X-band frequencies. IEEE Microwave and Guided Lett. *2*, 281–283 (1993)

8 Future Devices

M. WILLANDER, Y. FU, and Q. CHEN

Department of Physics and Measurement Technology, Linköping University,
581 83 Linköping, Sweden

In the last few years, several excellent demonstrations of Silicon Monolithic Millimeter-Wave Integrated Circuits (SIMMWICS) as oscillators and receivers have been demonstrated [8.1], using Schottky-barrier diodes or IMPATT diodes selectively grown by silicon Molecular Beam Epitaxy (Si-MBE). Operation frequencies above 90 GHz have been obtained. Compared with III-V devices, silicon and silicon-germanium devices have: no drop of the high-field saturation velocity, a smaller drop of the differential ionization rate with respect to the electric field, a smaller carrier injection by interband tunneling, a longer carrier lifetime (smaller on-resistance for PIN diodes), a higher thermal conductivity and a lower thermal expansion coefficient. These properties are important, not only for the IMPATT and Schottky-barrier diodes, but also for the operation of Heterojunction Bipolar Transistors (HBTs), double- and single-barrier tunneling devices and Field-Effect Transistors (FETs).

In this chapter, we will discuss some aspects of tunneling devices and FETs based on Si/SiGe heterojunctions and δ-doping induced potentials. Since most of these devices are based on hole transport, we will start with a description of the valence-band structure of strained or relaxed $Si_{1-x}Ge_x$ to show its restriction on the devices. However, recently Negative Differential Resistance (NDR) in electron Resonant Tunneling Diodes (RTDs) has been demonstrated at room temperature (in contrast to only low temperature operation of hole resonant structure) [8.2]. Therefore, we will also analyze oscillators based on electron resonant tunneling. The most promising application of Si/SiGe double barriers, the millimeter-wave detector, will briefly be discussed. Finally, the single-barrier Si/SiGe varactor diode will be analyzed as a multiplier.

Over the last three years, excellent dc properties have been obtained for Si/SiGe MODFETs and MOSFETs. However, results from high-frequency characterization on these devices are still missing. This motivates us to make a prediction of the high-frequency properties for these devices as well as the δ-doped FET and to forecast the application potential of Si/SiGe MODFETs and MOSFETs in the high-frequency and high-speed fields.

8.1 Physics and Applications of Si/SiGe, Double-Barrier Structures

Our objective is to illustrate the calculation of the general features of the tunneling process in Si and SiGe related tunneling structures. In this way we will indicate

Springer Series in Electronics and Photonics, Vol. 32
Silicon-Based Millimeter-Wave Devices, Eds.: Luy et al.
© Springer-Verlag Berlin Heidelberg 1994

how the physical properties will restrict device performance and which opportunities the physical properties will give for a better device performance. Since many of the tunneling phenomena features in these structures are essentially of a one-dimensional nature, we will describe some basic concepts about tunneling by studying one-dimensional problems, i.e., the resonant tunneling diodes.

The traditional resonant tunneling diode consists of two contacts, a quantum-well sandwiched between two barriers. Usually, the barriers are made of AlGaAs, while the well and contacts are both GaAs. In the past few years, successful efforts in preparing double-barrier resonant tunneling structures using the p-type Si/SiGe system have given new dimensions to device physics. This is due to this system's large valence-band offset which always leads to a situation that holes are confined in the Ge rich layers. Quite different from the electron resonant tunneling structures in single Γ-valley semiconductors like GaAs, the degeneracy of the valence bands and the strain induced band splittings in SiGe or in Si make the hole tunneling more complicated. Efforts have been made, both theoretically and experimentally, to realize good device performances [8.3–9].

8.1.1 Band Structure of Si/SiGe

One should bear in mind the relationship between the band structure and the carrier effective mass, and that the smaller the effective mass of the tunneling carrier is, the higher is the maximum oscillation frequency of a Double Barrier Resonant Tunneling Diode (DBRT). Also, the temperature performance of a DBRT is enhanced with a smaller effective mass. However, the valence-band structure of SiGe is very complicated because of the intermixing of the heavy hole (hh), light hole (lh) and spin split-off (sso) bands (Appendix 8.A).

Details of the valence band structures close to the band edges have already been studied by, for example, the $k \cdot p$ method in the 1950's and 1960's [8.10–12], and the work was revived recently due to the new development of novel quantum devices using the p-type Si/SiGe system [8.13–15]. For most quantum applications layer structures which are used, e.g., the single- or double-barrier structures, the carrier transport can be approximated as one dimensional along the growth direction, and the motion of the free carriers in the plane perpendicular to the growth direction are in the form of plane waves due to the translational symmetry there. Assigning the z axis to the growth direction, the motion of the holes in this direction can by described by the single-particle effective mass Schrödinger equation

$$\left(-\frac{\hbar^2}{2m_j} \frac{d^2}{dz^2} + E_j + V \right) \psi_{ij}(z) = E_{ij} \psi_{ij}(z), \tag{8.1}$$

where j is the band index ($j = $ hh, lh, sso), i is the subband index, m_j is the z-direction effective mass, E_j is the band edge, and E_{ij} and $\psi_{ij}(z)$ are the ith eigenenergy and eigenfunction (usually called envelope function) of band j. A similar equation can be written for the electron motion in the z-direction and will be discussion in later sections.

The envelop function ψ and $m_j^{-1} d\psi/dz$ are continuous to ensure the uniqueness of differentiation in the Hamiltonian (8.1). (The effective mass m_j becomes z-dependent when dealing with the Si/SiGe layered structure.) The valence band structure is calculated self-consistently with Poisson's equation

$$\nabla \cdot (\varepsilon \Delta \Phi) = -e \left(-N_A + \sum_{i,j} f_{ij} |\psi_{ij}|^2 \right), \tag{8.2a}$$

$$f_{ij} = \frac{\mu_j kT}{\pi \hbar^2} \ln \left[\exp(\frac{E_{ij} - E_f}{kT}) + 1 \right], \tag{8.2b}$$

where e is the unit charge, N_A is the doping concentration (when p-type doped, the charge of the ionized impurity is negative), μ_j is the two-dimensional density of states effective mass in the xy-plane, and E_f is the Fermi level. (The concept of the Fermi level is valid in most quantum devices where tunneling current is very small compared with other kinds of currents, e.g., the drift diffusion current. When the current in the system is small, deviation from the equilibrium state can be approached as perturbation so that the concept of Fermi level remains intact) Details about carrier concentration will be presented later. It should be remembered that the valence-band structure is presented here in such a way that the hole state above the Fermi level is filled according to (8.2b). The carrier density in the well will directly influence the value of the negative differential resistance and hence also the speed and output power of an oscillator. One should also remember that the velocity of the carriers in the depletion region after the second barrier as well as the width of the depletion region will influence the maximum oscillation frequency and the output power of the oscillator.

Poisson's equation (8.2a) is solved as a boundary-condition problem: $e\Phi = -E_{fe}$ in the emitter far away from the double-barrier region, and $e\Phi = -E_{fc}$ in the collector far away from the double-barrier region. The two local Fermi levels, E_{fe} and E_{fc}, are obtained here by assuming that the two electrodes are independent bulk materials at thermal equilibrium.

The two-dimensional density of states effective mass (μ) in the xy-plane can be approximated, using the *Luttinger* Hamiltonian [8.16,17], by the effective masses in the z direction. This allows:

$$\frac{1}{m_{hh}} = \alpha - 2\beta, \quad \frac{1}{m_{lh}} = \alpha + 2\beta, \quad \frac{1}{\mu_{hh}} = \alpha + \beta \quad \text{and} \quad \frac{1}{\mu_{lh}} = \alpha - \beta. \tag{8.3}$$

In addition, $\mu_{SSO} = m_{SSO}$ is usually assumed. The m_{hh}, m_{lh} and m_{SSO} are effective masses in the z direction, which are calculated from the energy-band curvatures. They are energy dependent and are different from those determined experimentally or calculated by other theoretical works, where both the z direction and xy plane band structures are involved. (E.g., the density of states effective mass which is used when calculating the density of states, or for the carrier concentration effective mass, which is employed when calculating the free carrier concentration [8.18]).

Parameters used in the calculation are listed in Table 8.1 for Si and Ge (some of the parameters listed in the table are defined in the Appendix 8.A). For $Si_{l-x}Ge_x$,

Table 8.1. Parameters used in calculations

	Units	Si	Ge
L	eV Å^2	-25.51	-143.32
M	eV Å^2	-15.17	-22.90
N	eV Å^2	-38.10	-161.22
Δ	eV	0.044	0.282
a	eV	2.1	2.0
b	eV	-1.5	-2.2
d	eV	-3.4	-4.4
a_0	Å	5.4309	5.6561
c_{11}	10^{11} dyn cm^{-2}	16.56	12.853
c_{12}	10^{11} dyn cm^{-2}	6.39	4.825
m_t	m_0 (free electron)	0.98	1.64
m_l	m_0	0.19	0.082
ε	ε_0	11.8	16

linear interpolation is utilized. The deformation potential parameters (l, m and n) are defined by the deformation parameters a, b and d where $a = (l + 2m)/3$, $b = (l - m)/3$, $d = n/\sqrt{3}$. The electron effective masses in Si and Ge, which will be employed later, are obtained from [8.19]; others were tabulated by *Hinckley* and *Singh* [8.13].

At steady state, when the sample is subjected to an external voltage V_{ex}, the emitter and collector far away from the active layer are, at local thermal equilibrium, described by two quasi-Fermi levels E_{fe} and E_{fc}. The carriers between $z = 0$ and $z = L$ are also at steady states. Based on these facts, *Sofo* and *Balseiro* [8.20] assumed that the carrier wave functions are of three different kinds for $z = (0, L)$. Below the band edge of the emitter, $E = (-\infty, 0)$, the motion is unbounded at both ends ($z < 0$ and $z > L$). The envelop functions for (8.1), far from the active layer, correspond to running waves, and there are two states for each E: one incident from the left-hand, $\Psi^{(+)}$, and the other from right-hand, $\Psi^{(-)}$. For the energy between the band edges of the collector and emitter, $E = (0, eV_{ex})$, the motion is bounded at one end ($z < 0$) and there is one solution $\Psi^{(0)}$ for each E. The carrier density in (8.2) is calculated by summing over all states weighted with occupation probabilities:

$$
p = \frac{\mu kT}{\pi \hbar^2} \left\{ \sum_0^{eV_{ex}} |\Psi^{(0)}|^2 \ln \left(e^{[E - E_{fc}]/kT} + 1 \right) \right.
$$

$$
+ \sum_{-\infty}^0 \left[|\Psi^{(+)}|^2 \ln \left(e^{[E - E_{fe}]/kT} + 1 \right) + \right.
$$

$$
\left. \left. |\Psi^{(-)}|^2 \ln \left(e^{[E - E_{fc}]/kT} + 1 \right) \right] \right\},
$$
(8.4)

where $E_{fc} - E_{fe} = eV_{ex}$.

The carrier density in the double barrier and well region calculated using (8.4) is much lower than the one in the electrodes since the well width is usually quite thin. Very recently, *Yoshimura* et al. [8.21] reported on the photoluminescence (PL) and PL excitation spectra in an n-type GaAs/AlGaAs double-barrier tunneling diode showing clearly the large carrier confinement. It was also indicated by the capacitance-voltage (CV) measurements showing a peak at resonance [8.22]. It was interpreted [8.23] by introducing a quasi-bound subband in the well (coordinated $a < z < b$) with the carrier density N_s

$$N_s(z) = \frac{\mu kT}{2\pi\hbar^2} \ln\left[1 + \exp\left(\frac{E_{fe} - E_r}{kT}\right)\right] \sum \frac{|\Psi_r|^2}{\int\limits_a^b dz |\Psi_r|^2}. \tag{8.5}$$

Here, E_r is defined as the resonant state inside the quantum well, and Ψ_r is corresponding envelope wave function. The charge build-up processes are neglected on the resonant states in the well, and the charge build-up is assumed to be instant. The filling of this quasi-bound state is by various scatterings, so that the holes (p-type) or electrons (n-type) are relaxed while transporting through the well.

It was shown [8.23] that for a double barrier structure and when the charge confinement in the well, N_s in (8.5), is taken into account, the external voltage drops largely on the collector barrier. Since the charge density in the well is quite large compared with the doping levels in the electrodes (in [8.23], the electrodes of the n-type GaAs/AlGaAs sample under investigation are doped 2×10^{17} cm^{-3}, while the mobile carrier density N_s in the well is about 1.5×10^{18} cm^{-3}), its Coulomb potential cancels partly the electric field in the emitter barrier while enhancing the electric field in the collector barrier. When N_s is neglected in the self-consistent calculation of (8.1,2), the emitter barrier and the collector barrier are almost equally biased.

While it is important to introduce N_s to explain the experiments of [8.21–22], N_s only shifts the resonant peak to higher external voltage for the general IV spectral feature [8.23], it seems reasonable to drop N_s when only calculating the IV characteristics to simplify the self-consistent calculation.

8.1.2 Tunneling Current Calculation

The Si/SiGe structures commonly studied are (following the growth sequence) Si substrate (collector), buffer layer, SiGe spacer, Si barrier, SiGe well, Si barrier, SiGe spacer and contact (emitter). The buffer layer and contact can either be SiGe or Si. The emitter, buffer layer and collector are doped p-type (10^{19} cm^{-3}). Two coordinates are defined: $z = 0$ where carriers are emitted into the double barrier and well region, and $z = L$ where carriers are collected. The definitions of $z = 0$ and $z = L$ are not precise. Since they are the turning points when tunneling processes are started or are terminated, they are usually set at the boundaries between heavily doped electrodes and the lightly doped or undoped spacer layer. The error so introduced is negligible, since the band edge in these regions is almost flat. The region between $z = 0$ and $z = L$ is simply referred to as the active layer.

The typical band structures at thermal equilibrium and zero-bias condition are shown in Fig. 8.1 for two strain situations where E_f is referred to as the zero-energy level. It is found that the band structure profiles of the active layer, the emitter, and the collector are almost the same in different sample geometrical arrangements. For example, the band structures of the active layer are just like the one in Fig. 8.1b when the sample has either a Si contact or a SiGe emitter. Therefore, Fig. 8.1 also gives information about the band structures of other samples through proper combination of the band-profile segments in that figure.

The large band bending at interfaces is due to the fact that the valence bands of SiGe are lower than the Si valence bands so that there are large amounts of charges in the spacer and well region.

With a Si buffer layer, the compressive strain in GeSi spacers pushes the lh band down so that only the hh band is filled with holes. On the other hand, the hh and lh bands are degenerate and available to be filled with holes, when the sample has a lattice relaxed GeSi buffer layer and the GeSi spacers are unstrained.

Self-consistent calculations of (8.1,2) have shown that when an external voltage is applied to induce a current flow, the valence band behaviour of Si/GeSi can be approximated as at thermal equilibrium because of the small tunneling current, while the external voltage, V_{ex}, distributes linearly across that part of the active layer where the carrier concentration is low. These approximations are applicable since the external voltage is largely distributed in the high-resistance region where the carrier density is low. The electrodes are heavily doped, and for p-type Si/SiGe system there are large amounts of holes confined at interfaces between the active layer and the electrodes.

Knowing the band-edge profile, it is easy to calculate the tunneling probability T. For a parabolic band, or the approximated bands where the motion of the z direction is separated from the one in the xy plane (Appendix 8.A), it is expected that $T = T(E_z)$. The tunneling current density is calculated by

$$I = \frac{\mu e k T}{2\hbar^3 \pi^2} \int_{E_v(0)}^{-\infty} dE_z T(E_z) \ln \left(\frac{e^{[E_f(0)-E_z]/kT} + 1}{e^{[E_f(L)-E_z]/kT} + 1} \right), \tag{8.6}$$

where E_f is the local Fermi level, and it is easy to see that $E_f(0) = E_{fe}$ and $E_f(L) = E_{fc}$. For a p-type Si/GeSi system, the total current density is the summation over hh, lh and sso contributions.

One direct conclusion from Fig. 8.1 is the dominant hh contribution to the tunneling current in the sample with a Si collector layer, which causes bad temperature performance as well as poor oscillator and detector properties. Here the lh band is lower than E_f by 30 meV, and the hh band is higher than E_f by about 10 meV, when $x = 0.25$. However, the hh and lh bands are both populated in the spacer regions, when a lattice relaxed SiGe is introduced. In this case, the tunneling current consists of both hh and lh. The ratio between lh and hh currents depends on the strain in the Si barrier and the voltage applied in the active layer. Therefore the opportunity is opened for light-hole tunneling and hence one can obtain strongly improved oscillator and detector properties.

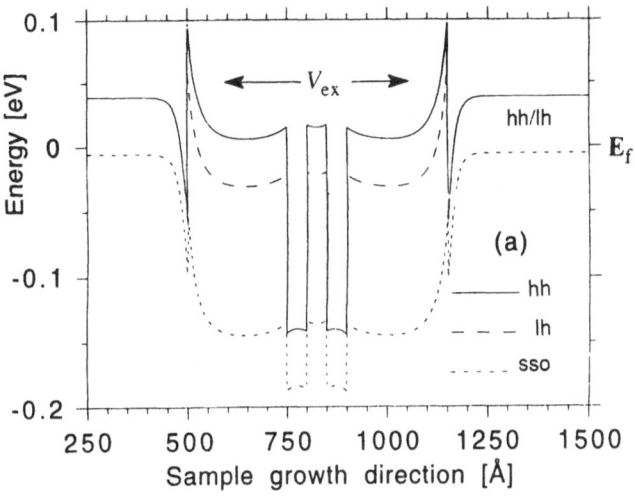

(a)

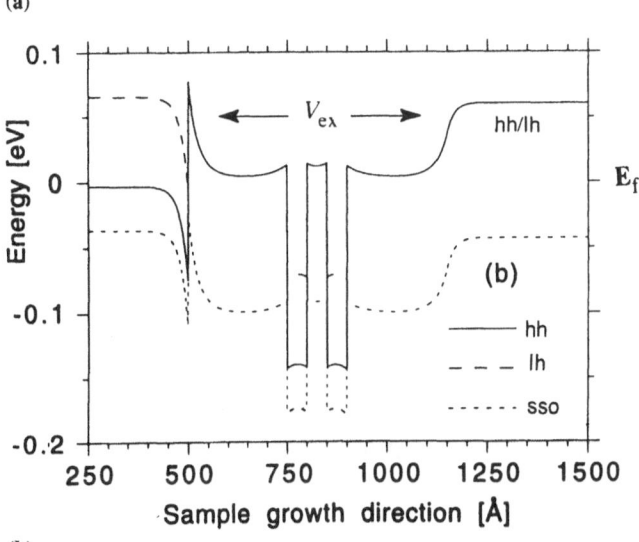

(b)

Fig. 8.1. (a) Valence-band diagrams of a Si/Si$_{0.75}$Ge$_{0.25}$ resonant tunneling structure with a Si buffer layer and a Si contact layer at thermal equilibrium. (b) Same as (a) but with a Si$_{0.75}$Ge$_{0.25}$ buffer layer and a Si contact layer [8.14]

The tunneling characteristics of the samples in Fig. 8.1 are shown in Fig. 8.2 (where the small sso current is negligible). It is mainly the hh tunneling current in Fig. 8.2a, while in Fig. 8.2b both the hh and lh make the tunneling current. The tunneling current at low bias possesses mainly lh characteristic.

The calculations here show a very small mixing of hh, lh and sso bands (Appendix 8.A), which differs from previous works [8.3,6]. The z-direction

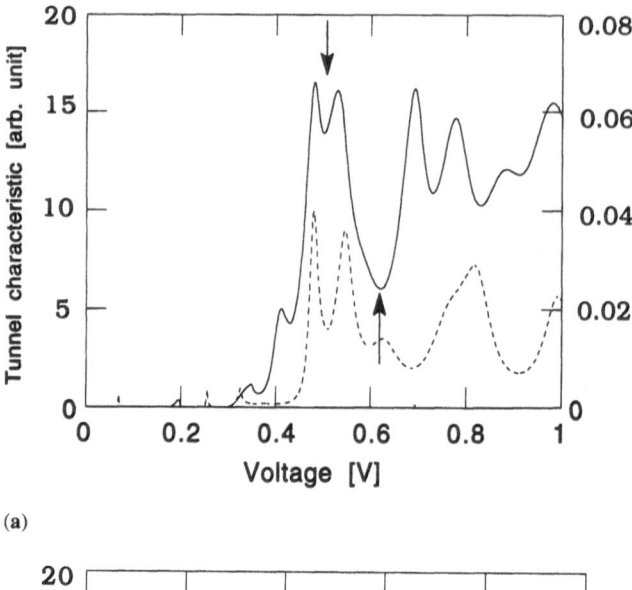

(a)

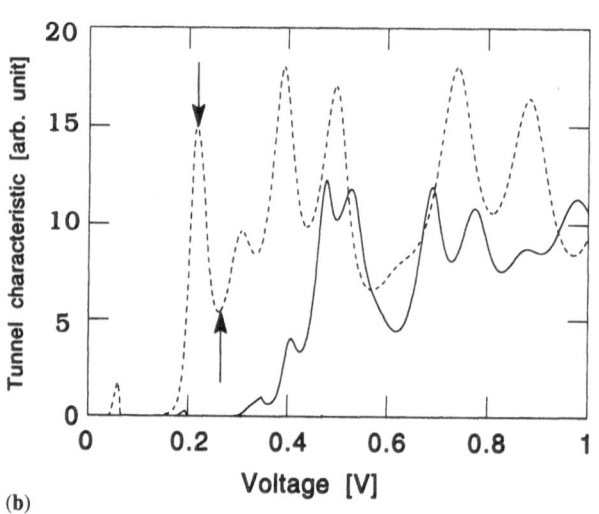

(b)

Fig. 8.2. (a) The tunneling characteristic (arb. unit) for the sample in Fig. 8.1a. **(b)** The tunneling characteristic (arb. unit) for the sample in Fig. 8.1b. The calculation temperature is 4.2 K [8.12]

resonant tunneling process is assumed in this work and scattering is neglected so that the moments k_x, k_y and the kinetic energy K_{xy} are conserved. Since the total kinetic energy of incident electron from $z = 0$ can be greater than E_{fe}, (which is less than 20 meV in the usual experimental devices), the K_{xy} is less than 20 meV, while the K_z can gain energy from the z-direction electric field. This is usually larger than 0.1 eV, when the tunneling current is significant. It is then expected (see details in Appendix 8.A) that the motion in the xy-plane is decoupled from the one in the z-direction. Various scattering processes should be included to account on the energy transfer from the z-direction electric field to the carrier's motion in the xy-plane.

An important factor that influences the IV characteristics was discussed as the lateral nonuniformity. Due to the surface kinetic processes at the interface between two binary materials such as AlAs/GaAs, a certain degree of both vertical and lateral intermixings of the two materials is inevitable. Thus, the barrier and the well can be thicker or thinner than their designed thickness. The resonant peak position is shifted to higher bias and enhanced if the well is narrowed, or is shifted to lower bias and becomes weaker if the well is widened [8.24]. The statistical average will show that the resonant peak is broadened and weaken while the background current will increase due to this lateral nonuniformity. This was demonstrated by *Gennser* [8.25] who increased the substrate temperature during the growth of the Si/SiGe sample to above 600 degree and the resonant peak disappeared from the IV characteristics.

8.1.3 The Quantum-Mechanical Concept of Electromagnetic Oscillations from Resonant-Tunneling Double Barriers

Since the maximum useful frequency f_{max} for the negative differential resistance in the IV curve is very important for many applications of the resonant tunneling devices, various time scales related to f_{max} have been studied. The equivalent circuit of a tunneling diode consists of the series inductance, the series resistance (R_s), the diode capacitance, and the negative resistance ($-R$) [8.26]. The real part of the input impedance R of the equivalent circuit is given by

$$R_{in} = R_s + \frac{-R}{1 + (\omega RC)^2}. \tag{8.7}$$

From (8.7), R_{in} will be zero at a certain frequency denoted as the resistive cutoff frequency f_r at which the diode no longer exhibits negative resistance:

$$f_r \equiv \frac{1}{2\pi RC} \sqrt{\frac{R}{R_s} - 1}. \tag{8.8}$$

In addition to the limitation due to circuit components, f_{max} is also limited quantum mechanically by the transit time τ_t through the active layer and the lifetime τ_w of a carrier in the quantum well. The frequency limit due to these factors can be approximated by $f \approx (2\pi\tau)^{-1}$. While τ_t has been discussed very comprehensively in the literature [8.27], we still do not know much about τ_w [8.28]. There are also other time scales studied which are related to the tunneling [8.27–32].

Generally speaking, the microscopic potential inside the device does not respond to the external bias change instantaneously. There are: (a) the time delay τ_q (where q denotes "quantum") that the time-dependent wave functions (either single particle wave functions or correlated ones) need to evolve from the form corresponding to the initial potential to the form of the final potential, (b) the time for the re-distributions of carriers in those wave functions via various scattering processes (this largely determines the life time τ_w), (c) the transit time τ_t, and

(d) the external circuit time delay. This is a self-consistent and dynamic process. The speed of this self-consistent process can be defined as f_{max}. The high frequency operation fails when the self-consistent process can not follow the external bias.

The response of the tunneling structure to small ac perturbation has been calculated in [8.32], where the local equilibrium is assumed which means that the charge relaxation is instantaneous. The basic idea is to investigate τ_q, the speed of the wave function evolution. For a GaAs/AlGaAs tunneling structure, the calculated resonant current density reaches its strongest response to the external ac perturbation when the ac frequency is $\omega_{max} = 4.6 \times 10^{12}$ (rad/s). When the ac frequency is further increased and is higher than $\omega' = 1.5 \times 10^{13}$ (rad/s), the current density approaches its original dc value. It is interpreted that the speed of the wave function evolution is of the order of $\tau_q = 1/\omega' = 6.7 \times 10^{-14}$ s.

In the active layer, where the doping level is usually quite low, the scattering rate is very low compared to ω_{max}. If other relaxations in the active layer and leads are almost instantaneous, the ω_{max} is then the dominant factor in determining the f_{max}. The calculated transit time ($\tau_q = 6.7 \times 10^{-14}$ s) matches well with the experimental interpretation of *Sollner* et al. [8.33] where $\tau_q = 6.5 \times 10^{-14}$ s.

It should be understood that the f_{max} is a very complicated parameter. This is because the measurement is the macroscopic average of all the microscopic processes. Thus, we cannot test a theory which deals about only one specific micro-process. Since it is usually very difficult to make the first-principle calculation, caution should be observed when interpreting and calculating this f_{max}.

8.1.4 Calculation of f_{max} for n-type Si/SiGe Tunneling Diodes

Due to the heavy masses for holes compared with the light electron mass, electron tunneling should be more promising if one can obtain large enough band offsets. Therefore, the electron resonant tunneling diode to be studied has the following structure [8.2]: relaxed $Si_{0.7}Ge_{0.3}$, 150 Å Si, 75 Å relaxed $Si_{0.7}Ge_{0.3}$, 50 Å Si, 75 Å relaxed $Si_{0.7}Ge_{0.3}$, 150 Å Si, and relaxed $Si_{0.7}Ge_{0.3}$. The band structure is schematically shown in Fig. 8.3. The structure is grown in the [001] direction. The growth direction [001] is set as the z axis for the following theoretical analysis.

First, let us discuss the conduction band calculation for the n-type Si/GeSi. Since the sample growth direction is along the (001) which is one of the principal axis of the ellipsoid of constant energy in Si, the description of the electron states by the effective mass approximation is greatly simplified [8.34]

$$\psi_{ij}(z, x, y) = \xi_{ij}(z) \exp(jk_x x + jk_y y), \tag{8.9}$$

$$\left(\frac{-\hbar^2}{2m_j} \frac{d^2}{dz^2} + E_c \right) \xi_{ij}(z) = E_{zij} \xi_{ij}(z), \tag{8.10}$$

$$E_{ij} = \frac{\hbar^2 k_x^2}{2\mu_{xj}} + \frac{\hbar^2 k_y^2}{2\mu_{yj}} + E_{zij}. \tag{8.11}$$

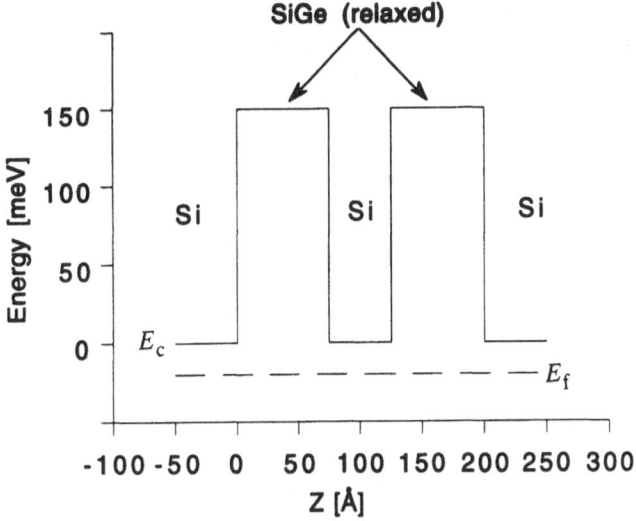

Fig. 8.3. Conduction-band diagram of an n-Si/SiGe resonant tunneling diode

The Si conduction band consists of six {100} ellipsoids of revolution. The principal effective masses in the ellipsoids are m_t, m_t and m_l. The six ellipsoids are divided into two groups, $j = 1, 2$. $j = 1$ marks the two ellipsoids (therefore its degeneracy $n_1=2$) along the z axis whose effective masses are $m_1 = m_l$, $\mu_{x1} = m_t$, $\mu_{y1} = m_t$. $j = 2$ marks the four ellipsoids ($n_2 = 4$) along the x and y axes whose effective masses are $m_2 = m_t$, $\mu_{x2} = m_t$, $\mu_{y2} = m_l$.

Other quantities in (8.9–11) are: the conduction band bottom E_c, the total electron energy E_{ij}, its z-component, E_{ijz}, $\psi_{ij}(z, x, y)$ (the total envelop wave function) and $\xi_i(z)$ (the one in the z direction). Here i is the subband index.

It should be noticed that the effective masses along the x and y axis are different for the bands of group 2. It is the density-of-states effective mass that should be used when calculating the carrier density. Therefore, the electron density is given by

$$n_{ij} = \frac{n_j kT \sqrt{\mu_{xj}\mu_{yj}}}{\pi\hbar^2} \ln\left[1 + \exp\left(\frac{E_f - E_{zij}}{kT}\right)\right], \tag{8.12}$$

where n_j in the above equation is the band degeneracy (as defined earlier).

Knowing the distribution of electrons along the z direction, the electric potential is obtained by Poisson's equation

$$\nabla(\varepsilon\nabla\Phi) = -e\left(N_D - \sum_{ij,e} n_{ij}|\xi_{ij}(z)|^2\right). \tag{8.13}$$

Here, N_D is the ionized impurity density and ε the dielectric constant. $\Phi_{(emitter)} = 0$ and $\Phi_{(collector)} = V_{ex}$ (the externally applied voltage). Knowing the potential, the conduction band bottom is easily obtained.

Since the current density of the first resonance is considerable small compared with the drift-diffusion current in bulk materials, the local equilibrium status is assumed. The Fermi levels in (8.13) are then the local Fermi levels.

It is the self-consistent calculation of (8.9–13) that determines the electron distribution and the potential.

The static dc tunneling current is calculated in the common way

$$J = \sum_j J_j,$$

$$J_j = \frac{n_j \sqrt{\mu_{xj}\mu_{yj}} ekT}{2\hbar^2 \pi^2} \int\limits_{E_{ce}}^{\infty} dE_{zj} T(E_{zj}) \ln \left(\frac{e^{(E_{fe}-E_{zj})/kT}+1}{e^{(E_{fc}-E_{zj})/kT}+1} \right), \tag{8.14}$$

where $T(E_{zj})$ is the tunneling probability which can be calculated by standard transfer matrix method or other numerical method when the potential is known.

The calculated IV characteristics are exhibited in Fig. 8.4. Contributions from the two groups of electrons are included. For the electronic state whose effective mass along the z direction is 0.98 (here, and later on, the effective mass is in the unit of bare electron mass, m_0), the tunneling probability and thus the tunneling current contribution is negligible relative to when the effective mass is 0.19. Therefore, the dc IV curve in Fig. 8.4 is virtually the tunneling current for $m = 0.19$. As indicated in [8.2], the onset voltage of the first resonant peak is 75 meV (see the curve in logarithmic scale), while the measured onset voltage is 0.45 V. *Ismail*, et al. [8.2] attributed this to the series parasitic resistance due to the ohmic contacts. We have neglected this problem in our calculations.

In order to study the response of the device to the external condition variation, a small ac perturbation signal is applied, in addition to the dc voltage. The

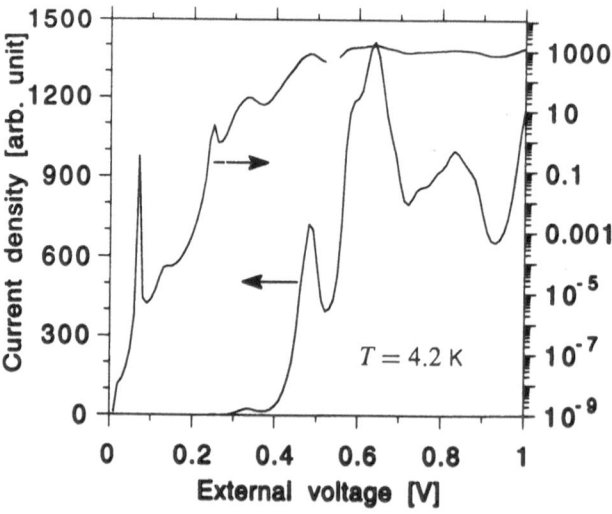

Fig. 8.4. Calculated IV characteristics

time-dependent effective mass Schrödinger equation becomes

$$j\hbar \frac{\partial \xi(z,t)}{\partial t} = \left(\frac{-\hbar^2}{2m} \frac{\partial^2}{\partial z^2} + E_c + V_{ac} e^{-j\omega t} + V_{ac}^* e^{j\omega t} \right) \xi(z,t), \qquad (8.15)$$

where V_{ac} is the amplitude of the ac perturbation and ω is its frequency. Mathematically, (8.15) is equivalent to the phonon-assisted tunneling problem discussed in [8.35–37]. Therefore, the ac perturbation couples the electronic state E with other states $E \pm n\hbar\omega$. It should be remembered that in the case of phonon-assisted tunneling, the coupling interaction depends on the phonon population. Thus at low temperature the rate of phonon-absorption is very low, and only the phonon-emission-assisted tunneling is observable. However for the ac perturbation in (8.15), the state, E, interacts equally with the states $E + n\hbar\omega$ and $E - n\hbar\omega$. When the electronic state allies with the resonant level, E_r, in the well, resonant tunneling occurs. Tunneling at $E_r \pm n\hbar\omega$ is called ac-assisted tunneling sidebands.

In (8.15) we have neglected various scattering processes so that only the collective response of the electronic states coupled with each other by the ac perturbation signal are investigated. Basically, the time response can be divided into two steps: the evolution of the electronic states following the external condition variation and the occupation of these states by electrons via various relaxation processes. The two steps form a self-consistent loop. Neglecting various relaxation processes in (8.15) assumes that the dissipation is instantaneous. Thus, the so studied response time determines, in principle, the maximum frequency under which the device can operate.

Since the effect of the ac perturbation on the current density is very small (see calculation results below), the self-consistently calculated potential in (8.13) is approximated as unaffected. Perturbation theory is used here to calculate the tunneling probability T, and the current density is obtained from (8.14) when T is known.

Plotted in Fig. 8.5 are the responses of the four lowest resonant peaks ($V_{ex} = 0.07, 0.25, 0.34$ and 0.48 V) in the IV curve to the ac perturbation signal. For the perturbation theory to be valid, $V_{ac}/\hbar\omega$ should be smaller than 1. In Fig. 8.5, the current density values at $\hbar\omega = 0$ are the ones in the dc IV curve. Current densities for $\hbar\omega \geq 1.0$ [meV] and $\hbar\omega \geq 1.5$ [meV] are calculated for the situations when $V_{ac} = 0.5$ [meV] and $V_{ac} = 1.0$ [meV], respectively. The current densities are re-scaled for comparison.

As discussed in [8.32] about the response of the resonant tunneling in GaAs/AlGaAs system, it is also observed here in Fig. 8.5 that the larger the amplitude V_{ac} of the ac perturbation, the more the resonant current densities are affected. However, the frequencies, $\hbar\omega_i$ (here, i is the index of the resonance peak, i.e., $i = 1$ stands for $V_{ex} = 0.07$ V and $i = 2$ for $V_{ex} = 0.25$ V and so on), of the ac perturbation with which the resonant currents are mostly affected are quite independent of V_{ac}.

Since the tunneling probability decays exponentially when the electronic energy shifts away from the resonant state, the resonance is affected by the ac perturbation

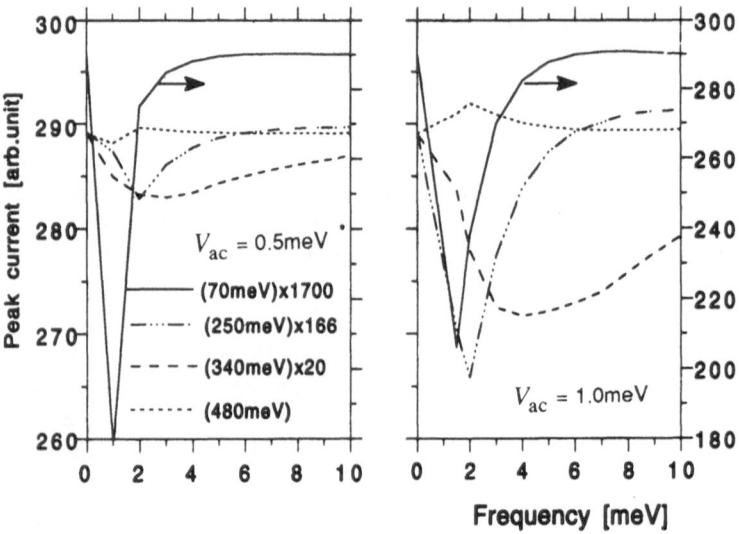

Fig. 8.5. Response of the resonant peak currents to a small ac signal. The horizontal arrows indicate that the two solid lines use the scale in the right side of the figure, while others use the left scale

only when $\hbar\omega$ is small so that the ac-assisted sidebands must be very close to the resonance for these sidebands in order to have significant tunneling probabilities. Since the integrated total tunneling probability is independent of the ac perturbation (this is affirmed in our calculations as well as reported by *Wingreen* et al. when discussing the electron-phonon interaction in resonant tunneling [8.35]), the tunneling probability at resonance is largely affected when the ac-assisted sidebands are significant. It is well known that the quantum confinement in the well between two barriers diminishes when the electronic energy increases. The resonance width which is closely connected with the quantum confinement then increases with increasing resonance energy. When the resonance width is large, the decaying of the tunneling probability is slow when the electronic energy shifts away from resonance. In this way, the ac-assisted sidebands at strong resonance can still be significant enough for the resonance being affected when $\hbar\omega$ is increased. This explains the observation of Fig. 8.5. that $\hbar\omega_i \neq \hbar\omega_j$ when $i \neq j$ and $\hbar\omega_i > \hbar\omega_j$ for $i > j$.

It is noticed that in GaAs/AlGaAs system, $\hbar\omega_1 = 3$ meV [8.32], whereas here $\hbar\omega_1 = 1$ meV. This, of course, can be explained by the difference in the electronic effective masses, m^*, in the z direction. In Si/SiGe, $m^* = 0.19$ while in GaAs/AlGaAs it is 0.067. We have also calculated $\hbar\omega_1$ for the electrons in Si/SiGe whose effective mass is 0.98 (the six ellipsoids in Si are divided into a 4-fold degenerate group whose effective mass in the z direction is 0.19, and a 2-fold degenerate group whose effective mass is 0.98) and the result is $\hbar\omega_1 \approx 0.2$ meV. Keeping the barrier height (ΔE_c) and the widths (d) of the well and barriers constant, $\hbar\omega_1$ is calculated for different effective masses. The relation between

$\hbar\omega_1$ and the effective mass can be well approached as $\hbar\omega_1 \approx 0.2/m^*$ [meV], when $\Delta E_c = 0.2$ eV and $d = 50$Å (for both the well and barrier thicknesses). It is also found that the above relation between $\hbar\omega_1$ and m^* is valid within the variational ranges of ΔE_c and d for the commonly used experimental Si/SiGe and GaAs/AlGaAs systems.

One should remember that the above results are obtained from the ac perturbation calculation along with the exclusion of various relaxation processes existing in reality. Therefore, the so obtained results are, in principle, the maximum time response that the device can operate. For the n-type Si/SiGe sample studied here, this quantum frequency limit is $f_{max} \approx \omega_1/2\pi = 150$ GHz.

8.1.5   Equivalent-Circuit Analysis of Oscillation Frequency and Output Power from an n-type Si/SiGe Double-Barrier Diode

In Fig. 8.6, and n-type Si/SiGe RTD is shown with its corresponding equivalent circuit. The maximum oscillation frequency can be approximated to [8.38] (Appendix 8.B)

$$f_{max} = \frac{W + l_{qw}}{4\pi A\varepsilon(R_s + R_{spch}/2)} \tag{8.16}$$

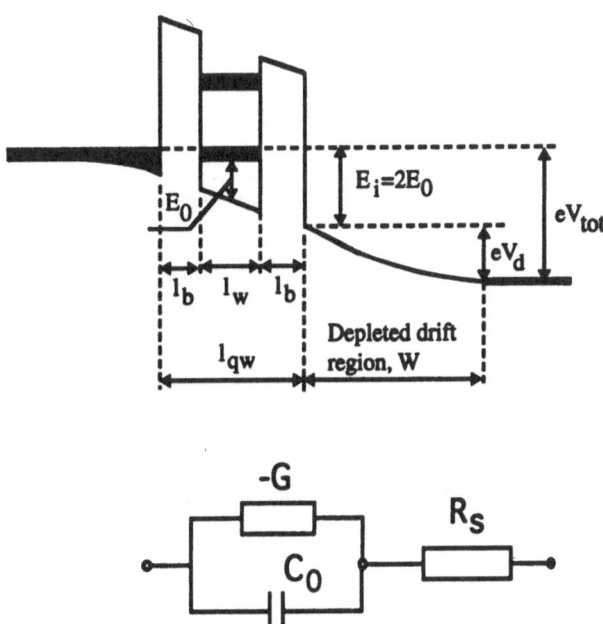

Fig. 8.6. The schematic conduction band diagram of a Si/SiGe QW diode. (a) The electron potential. (b) The equivalent circuit

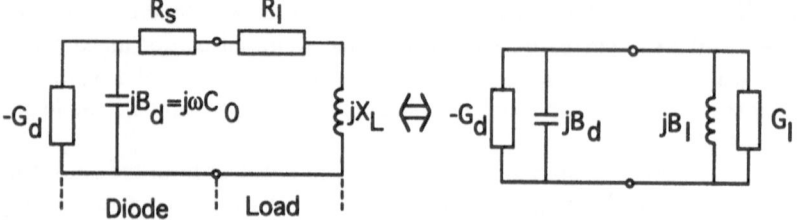

Fig. 8.7. The equivalent circuit of a QW diode oscillator and the equivalent parallel circuit

with

$$R_{spch} = \frac{W^2}{2A\varepsilon V_s},$$ (8.17)

where A is the diode area and V_s is the saturation velocity. The other parameters are defined in Fig. 8.6. For $R_s = 0.015$ Ω, $V_s = 1 \times 10^5$ m/s, $W = 300$ Å, $l_{qw} = 200$ Å, and $A = 1.25 \times 10^4$ μm², we obtain $f_{max} \approx 150$ GHz, which is of the same order as that due to the quantum-mechanical calculation. By changing doping concentration and substrate thickness, R_s could be decreased or increased and f_{max} can be both lower or higher than 150 GHz.

The maximum power delivered to the load (Fig. 8.7) can be written [8.38] (Appendix 8.B)

$$P_{del.\ max} = \frac{\Delta I^2}{16\pi} \frac{W + l_{qw}}{A\varepsilon} \left(\frac{1}{f} - \frac{1}{f_{max}} \right),$$ (8.18)

for $0 \ll f \ll f_{max}$. We see that we need a large ΔI^2 for the Si/SiGe RTD to obtain high power delivered to the load.

Finally it should be mentioned that with an optimum design higher f_{max} can be obtained. However, the values of f_{max} and $P_{del.\ max}$ depend also strongly on the material quality.

8.1.6 Millimeter-Wave Detection by Si/SiGe Double Barriers

A successful demonstration of a 91 GHz Si/SiGe resonant tunneling detector has been made [8.39]. In Fig. 8.8 a schematic drawing on the structure as well as the receiver design is shown.

The detector worked at 77 K and had a maximum sensitivity of 85 mV/mW/cm², when biased with 460 mV and working in the second resonant state. The device was monolithically integrated on a silicon substrate including the antenna.

Compared with a monolithically integrated Schottky detector, the performance of the Si/SiGe RTD detector is better.

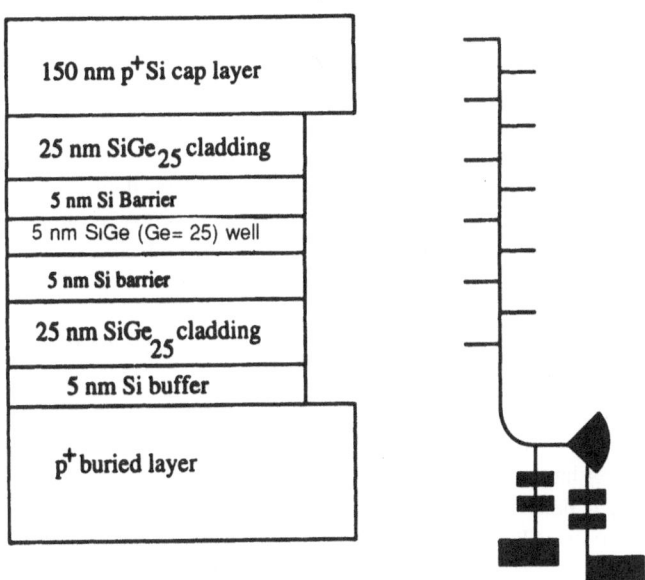

150 nm p⁺ Si cap layer
25 nm SiGe₂₅ cladding
5 nm Si Barrier
5 nm SiGe (Ge= 25) well
5 nm Si barrier
25 nm SiGe₂₅ cladding
5 nm Si buffer
p⁺ buried layer

Fig. 8.8. Layer sequence and receiver design [8.39]

8.2   The Si/SiGe Quantum Barrier Varactor Diode

As proposed by *Kollberg* and *Rydberg* [8.40], the single barrier varactor can be used as a multiplier for millimeter waves. The main advantage is that a tripler without an idler circuit can be used, which improves the tunability.

When a voltage is applied, the single barrier electrons are prevented from passing through the barrier (i.e., a capacitance is created behind the barrier). Since the capacitance-voltage characteristic is symmetric, only odd harmonics are obtained. The value of the capacitance is determined by the width of the depletion region and the barrier width. The barrier can also be created by the δ-doping technique. These structures are alternatives to the Schottky-barrier varactors. Compared with RTDs the single barrier varactor gives higher output power for the harmonics.

8.3   Field-Effect Devices: Si/SiGe MODFET and MOST, δ-Doped Si FET

In last two decades, the performance of microwave semiconductor devices and circuits has been improved greatly, due to the development of new device-physics concepts, the advance on material-growth methods and processing techniques, and the minimization of critical sizes and parasitic effects of devices [8.41]. However, those improved microwave devices are almost based on III–V compounds. It is desirable to introduce the concepts of bandgap engineering and heterostructure which have been making great contribution to the advance of III–V microwave devices [8.41] into the well developed Si technology. These are needed to develop new types of Si microwave devices and to improve microwave performance of existing Si devices; therefore, to enhance the application of Si-based devices in microwave field [8.41]. Naturally, such desires will be met only in Si-based heterojunction material systems. Towards this goal, the Si/SiGe system has emerged as a very promising material system, owing to the progress in Si/SiGe MBE [8.42] and other growth techniques such as LRP-CVD (Limited-Reaction Processing Chemical Vapor Deposition) [8.43], UHV-CVD (Ultra-High Vacuum CVD) [8.44] etc. Additionally, advances in understanding the effects of strain on the band structure [8.45] and transportation process of the carriers [8.13] in the strained layer and the band line-up at such pseudomorphic heterointerface [8.45] has been helpful. A lot of Si/SiGe heterostructure devices (many of which are the analogues of devices realized in III–V compound systems) have been experimentally demonstrated. Among them, Si/SiGe modulation-doped field-effect transistors (MODFET, both n- and p-type) [8.46–48], Si/SiGe quantum-well channel p-MOST [8.49–51], Si δ(planar)-doped MESFET (δFET) [8.52], Si/SiGe HBT [8.53–55], and the Double-Barrier Resonant Tunneling Diode (DBRTD) [8.5, 56] are the most interesting because of their potential high speed and multifunctionality over the conventional Si counterparts. It is helpful to explore, theoretically, their performance for exploiting the application of those Si/SiGe electric devices in the high-frequency and high-speed fields.

In the first part of this section, using an unified analytical dc and ac modeling method [8.57] together with a quantum-mechanical study of the gate charge-control model [8.58], we perform dc and first-order small-signal high-frequency (HF) modeling on Si/SiGe p-MOST, Si/SiGe n-MODFET and Si n-δFET. Comparisons between Si/SiGe n-MODFET and Si n-MOST, Si/SiGe and Si p-MOSTs are made. Also the processing steps for those devices including growth and post-growth processing are briefly described.

8.3.1 dc and HF Modeling

In an analytical and HF modeling for FETs, the first step is to develop an analytical model for the Gate-Charge Control (GCC) process (i.e., how the density n_s (p_s) of the channel carriers in a FET changes with the applied gate voltage V_{gs}). In some cases [8.41], (e.g., for a conventional MESFET/JFET and for the strong and weak inversion operation regions in a MOST), it is easy to develop the GCC model based on the approximate solution of one-dimensional Poisson-Boltzman equation. It is, however, not a simple route for the devices we consider here because of the existence of a quantum channel. Such quantum-channel results from either heterostructure in Si/SiGe p-MOST and n-MODFET or strongly localized doping profile (planar-doping) in Si n-δFETs. This results in the quantization of the channel carriers in the direction perpendicular to the heterointerface or the doping plane to form so-called two-dimensional electron (hole) gas (2DEG or 2DHG). Therefore, it is necessary to develop an accurate model from the viewpoint of quantum mechanics and analyze the charge-control process in order to understand the operation and properties of FETs.

Figure 8.9 exhibits the schematic structures and band profiles of a Si/SiGe n-MODFET, a Si/SiGe p-MOST and a Si n-δFET. All of these structures are grown on (100) substrate. The unique quantum-channel feature in those device structures directs us to use a self-consistent solution of the Schrödinger and Poisson's equations for studying the GCC process.

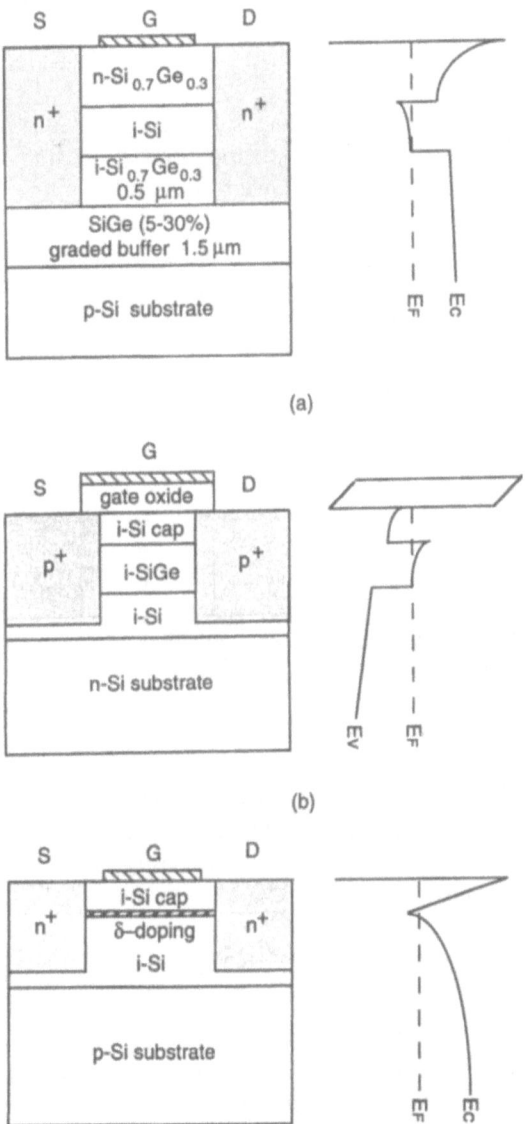

(a)

(b)

(c)

Fig. 8.9. Schematic structures and band profiles (not scaled) of (a) Si/SiGe n-MODFET, (b) Si/SiGe p-MOST, and (c) Si n-δFET

If the x-axis is chosen as the direction perpendicular to the heterointerface/doping layer and the origin of x-axis is set at the gate contact, the electronic structure of the system is described by the effective-mass Schrödinger equation [8.58]

$$\left[\frac{-\hbar^2}{2m_{x,k}} \frac{d^2}{dx^2} - e\Phi(x) + \Delta E_{\text{off},k}(x) + E_{\text{xc}}(x) \right] \zeta_{i,k}(x) = E_{i,k}\zeta_{i,k}(x), \quad (8.19)$$

where $k = 1, 2$ is the index of the set of subbands, $\zeta_{i,k}(x)$ is the envelope wave function of an electron (or hole) in the ith subband of eigen-energy $E_{i,k}$, $m_{x,k}$ is the effective mass of the electron (or hole) in the x-direction, $-e\Phi(x)$ is the electrostatic potential energy, $\Delta E_{\text{off},k}(x)$ is the band off-set, and $E_{\text{xc}}(x)$ is the local electron exchange-correlation energy.

Here, it is worth noting the strain-induced band splitting. In Si/SiGe n-MODFET, the tensile strain in the Si layer causes a strong downward shift of the two-fold conduction-band valleys normal to the interface [8.45]. For the Si channel layer, therefore, we may neglect the four-fold conduction-band minima in the growth plane because of their small electron occupancy, i.e. set $k = 1$. However, the strain relaxed SiGe alloy layer (for a Ge fraction less than 0.85) has Si-like electron band structure (i.e., six-fold degenerate (001) conduction-band minima) [8.60]. Electrons in the SiGe barrier and buffer layers have two different effective masses along the x-axis, in accordance with two-fold band minima normal to and four-fold minima within the plane, respectively. Thus $k = 2$. This is also true for a Si n-δFET. For Si/SiGe p-MOSTs, the compressive stress in the SiGe channel layer reduces the valance band edge degeneracy between Heavy Hole (HH) and Light Hole (LH) bands. The upper valance band will be HH in character. Such splitting between the HH and LH is dependent on the Ge content and not always high enough to neglect the LH band. Thus the inclusion of both HH and LH bands ($k = 2$) may be necessary.

The electrostatic potential $\Phi(x)$ is given by Poisson's equation

$$\frac{d^2\Phi(x)}{dx^2} = \frac{e}{\varepsilon} \left[\sum_{i,k} N_{i,k}|\zeta_{i,k}(x)|^2 + N_{\text{A}}(x) - N_{\text{D}}(x) - \text{p}(x) \right] \quad (8.20)$$

for n-type FETs. Here, ε is the dielectric permittivity of Si, $N_{\text{D}}(x)$ is the density of ionized donors, $N_{\text{A}}(x)$ is the concentration of ionized acceptors, $\text{p}(x)$ is the free (three-dimensional) hole concentration, and $N_{i,k}$ is the sheet density of electrons in the ith subband. $N_{i,k}$ is expressed as

$$N_{i,k} = \frac{g_k k_{\text{B}} T (m_{y,k} m_{z,k})^{1/2}}{\pi\hbar^2} \ln\left[1 + \exp\left(\frac{E_{\text{f}} - E_{i,k}}{k_{\text{B}}T} \right) \right]. \quad (8.21)$$

where, g_k is the corresponding band edge degeneracy, $m_{y,k}$ and $m_{z,k}$ are the electron (or hole) effective masses for any of the two sets of subbands in the y and z direction, respectively, and E_{f} is the Fermi energy (which is constant throughout the system in equilibrium or neglecting the gate current and is set to zero as an energy reference point). $\text{p}(x)$ is given by the usual form

$$p(x) = N_v \exp \left(\frac{E_v(x) - E_f}{k_B T} \right), \tag{8.22}$$

where N_v is the effective density of states for the valence band in Si, and $E_v(x)$ is the valence-band maximum at position x. For p-MOSTs (8.20–22) are still kept valid by just exchanging the electron with the hole.

For (8.19), the boundary conditions are defined so that the envelope function is zero at both the gate and the interface between the buffer layer and substrate. The boundary conditions for (8.20) are defined such that the potential is the sum of the barrier height, the applied gate voltage at the gate contact and the thermal equilibrium Fermi potential for the position deep into the substrate. Once the boundary conditions for the whole system are specified, (8.19–22) are self-consistently solved. The solution yields all the quantities of interest for the system in which the charge control process is studied.

Figure 8.10 provides some typical numerical results for the GCC processes in Si/SiGe n-MODFETs, Si/SiGe p-MOSTs and Si n-δFETs. In calculations the band line-up was taken from [8.45 and 14] for MODFETs and MOSTs, respectively. The electron effective masses are assumed to be equal to those of Si. This is accurate within a few percent. The longitudinal and transverse hole (HH and LH) effective masses was obtained from [8.14].

One important feature in the GCC for SiGe MODFETs and MOSTs is the formation of an undesirable parasitic channel with lower mobility. This reduces the usable gate voltage swing range in which the carriers are of higher mobility and the device has better dc and ac performances. Hereafter, we will limit the gate voltage in such a region. In MODFETs, such parasitic channels are due to the dopants neutralization and the higher density of states (six-fold degeneracy in the SiGe barrier layer instead of two in the Si channel). The parasitic channel in p-MOSTs arises from the Si cap layer sandwiched between the gate oxide and SiGe layer. This layer serves to reduce the density of interface states existing at the oxide/SiGe interface. In Si δFETs such nonlinearity might be purely a quantum effect.

As shown in Fig. 8.10, the numerical results of the GCC for the desirable channel carriers are all fitted well by a quadratic form

$$q n_s = C_G \left[(V_{gs} - V_{th}) + B(V_{gs} - V_{th})^2 \right], \tag{8.23}$$

where C_G and B are two constants dependent on the device material structure, V_{gs} is the gate-source voltage and V_{th} the threshold voltage. C_G may be called an effective gate capacitance per area and can be approximated as

$$C_G = \frac{\varepsilon_g \varepsilon_0}{d_{eff} + c}, \tag{8.24}$$

where c is a correction constant and around ten Å, ε_g the gate dielectric constant (ε_{SiGe}, ε_{ox} and ε_{Si} for MODFETs, MOSTs and δFETs, respectively) and d_{eff} the effective gate barrier thickness (d_b, $d_{ox} + d_{cap}\varepsilon_{ox}/\varepsilon_{Si}$ and d_{caps} respectively).

The linear term in (8.23) is commonly used in modeling Si MOSTs and III–V MODFETs. The quantum correction constant c in (8.24) has been pointed out in

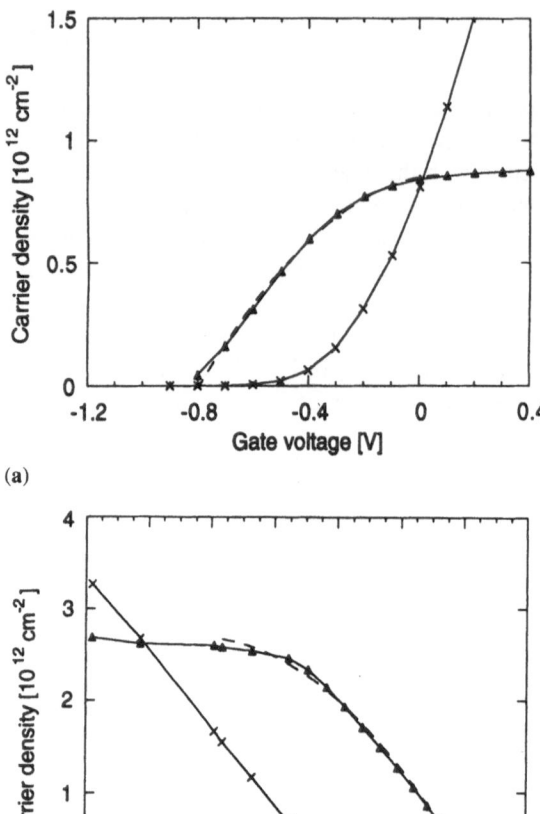

(a)

(b)

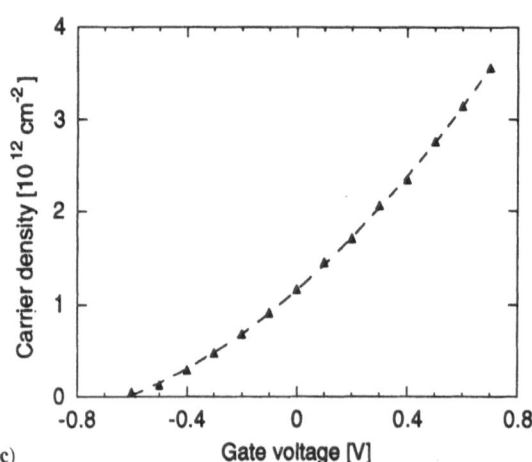

(c)

Fig. 8.10. Gate charge control process in (**a**) the Si/SiGe n-MODFET (d_{well} = 200 Å, d_b = 250 Å, N_b = 5 × 10^{18} cm^{-3}), (**b**) the Si/SiGe p-MOST (d_{well} = 100 Å, d_{cap} = 50 Å, d_{ox} = 250 Å, N_{sub} = 5 × 10^{16} cm^{-3}), both with Ge content of 0.3, and (**c**) the Si n-δFET (d_{cap} = 200 Å, $N_δ$ = 4 × 10^{12} cm^{-2}). Solid lines and data marks are numerical results: triangles are for carriers in the quantum well channel layer, crosses for carriers in the parasitic channel [SiGe barrier layer in (**a**) and Si cap layer in (**b**), respectively]. Dashed lines are the quadratic fits to carriers in the quantum well channel

[8.61 and 62] for MODFETs and MOSTs, respectively. However, the quadratic term ($B > 0$) has only been considered recently [8.57] for Si δFETs. For Si/SiGe MODFETs and MOSTs, as shown in Fig. 10a and b, this term cannot be neglected and is responsible for the saturation of the channel carrier density ($B < 0$) under high gate voltage.

After the establishment of the GCC model, we use the following velocity-field dependence of a nonlinear two-piece form [8.63]

$$v = \frac{\mu E_y}{1 + \dfrac{E_y}{2E_s}}, \quad \text{for } E_y \leq 2E_s, \tag{8.25a}$$

$$= v_{\text{sat}}, \quad \text{for } E_y > 2E_s \tag{8.25b}$$

in dc and ac modeling. $E_y = -\partial V/\partial y$ is the lateral electric field along the channel, and v_{sat} is the channel carrier saturation velocity. For MODFETs and δFETs, μ is the low-field drift mobility, $\mu = \mu_{\text{eff}}[1 + \theta(V_{\text{gs}} - V_{\text{th}})]^{-1}$ and μ_{eff} is the effective mobility for MOSTs. The GCA (gradual channel approximation) is used in the portion of the channel where the carrier drift velocity is smaller than the carrier saturation velocity v_{sat}.

In the model, the operation of the device is divided into two regions. In the linear region ($V_{\text{ds}} \leq V_{\text{ds, sat}}$), no channel carriers have been accelerated up to the saturation velocity, GCA ia valid everywhere in the channel. In the saturation region ($V_{\text{ds}} > V_{\text{ds, sat}}$), the point where $E_x = 2E_s$ and the carrier velocity reaches the saturation value, v_{sat} moves closer to the source. The portion of the channel in which no carrier has been speeded up to v_{sat} will be reduced. The channel can be divided into two sections: one is in which $E_y \leq 2E_s$ and GCA is applied, and the other where $v = v_{\text{sat}}$ and GCA is not valid due to the rapid variation of the electric field in both x- and y-directions and the 2-dimensional Poisson's equation has to be solved.

Using the GCC model in (8.23) and the velocity-field dependence in (8.24), one can perform dc as well as small-signal ac modeling. This is quite tedious. The interested reader can find the detailed mathematical manipulation and formulae in [8.57]. These formulae can be directly used for Si/SiGe MODFETs and Si δFETs and applied to Si/SiGe PMOSTs after some modifications [8.64].

With respect to HF properties, the parameter of most interest is the cut-off frequency f_T. This is defined as the frequency at which the current gain with shorted output becomes unity. f_T imposes more restriction on the device operation than f_{max}, the maximum available frequency at which the unilateral power gain extrapolates to one, of FETs. The f_T of a device can be either higher or lower than its f_{max}, up to the specific values of the terms governing them [8.59]. In HF modeling, the static approximation (SA) is used. The first-order small-signal equivalent circuit is shown in Fig. 8.11. Although not as accurate as the wave-equation and transmission-line methods in simulating other HF performance, SA gives quite good results for f_T [8.65]. From Fig. 8.11, $f_T \approx g_m[2\pi(C_{\text{gs}} + C_{\text{gd}})]^{-1}$.

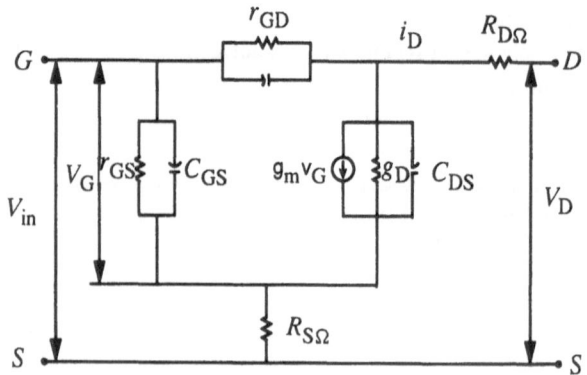

Fig. 8.11. First-order small signal equivalent circuit for a FET

8.3.2 Modeling Results and Comparison with Si n- and p-MOSTs

As theoretically pointed out in [8.13], holes in strained SiGe layers on (100) Si have a higher in-plane mobility over Si. Recently, it has been confirmed experimentally [8.66] that a Si/SiGe p-MOST has a two-fold enhancement (at 300 K) in inversion hole effective mobility over Si for the Ge fraction of 0.3. Such improvement increases with the temperature and reaches about three-fold at 77 K. Using the new type of graded SiGe buffer layers in Si/SiGe n-MODFET structures, it has been found experimentally [8.67] that the mobility enhancement is about two-fold at 300 K and about ten-fold at 77 K. These results indicate a very promising future for Si/SiGe FETs at room temperature and even more at low temperatures. The high mobility of Si/SiGe n-MODFET is helpful for developing high-frequency and high-speed circuits based on Si technology. In this subsection we try to explore the effects of such mobility enhancement on the dc and HF performance of devices using the method mentioned in last section.

Some simulated dc characteristics for Si/SiGe n-MODFETs, Si/SiGe p-MOSTs and Si n-δ FETs at 300 K are shown, respectively, in Fig. 8.12 a, b and c. The Si/SiGe FETs have a Ge fraction of 0.3, an effective gate length of 1 μm (1.2 μm CMOS technology) and the gate barrier thickness of 650 Å (MODFET) or gate oxide thickness of 250 Å (p-MOST). The other technology parameters, such as, junction depth, etc are from [Ref. 8.68, Table 9.6]. Shown by dashed lines in Fig. 8.12 a and b are simulated results for Si n-MOST and p-MOST ($d_{ox} = 250$ Å), respectively, for clear and direct comparisons between Si/SiGe n-MODFET and Si n-MOST, Si/SiGe and Si PMOSTs. The mobility data for Si/SiGe n-MODFET, p-MOST and Si δFET are from experimental results in [8.66, 67, 69], respectively. The saturation velocity at 300 K is 1×10^7 cm/s for electrons and 8×10^6 cm/s for holes.

It is worth noting here that we have used in Fig. 8.12a the same values for the effective gate capacitance C_G of a Si/SiGe MODFET as in Si n-MOST and therefore the same GCC ability, see (8.24), in order to project only the role of mobility enhancement (double). This means that the Si/SiGe MODFET in Fig. 8.12a has a

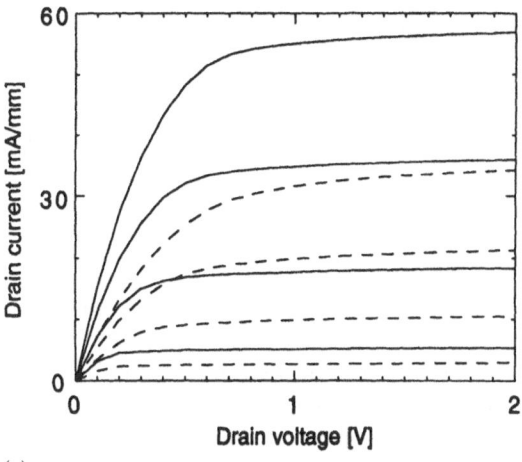

(a)

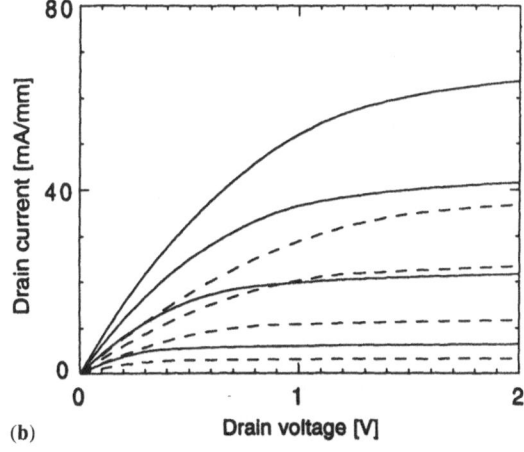

(b)

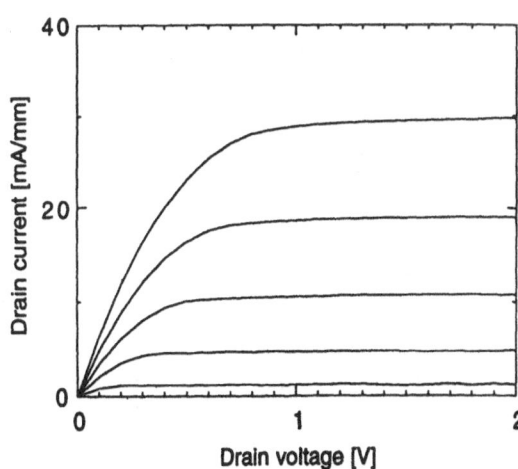

(c)

Fig. 8.12. DC performance in (**a**) the Si/SiGe n-MODFET (d_b = 650 Å, solid line) and Si n-MOST (d_{ox} = 250 Å, dashed line), (**b**) the Si/SiGe p-MOST (d_{ox} = 250 Å, solid line) and the Si p-MOST (d_{ox} = 250 Å, solid line), and (**c**) Si n-δFET (d_{cap} = 300 Å) at 1 μm gate length, the parameters of device structures are the same as in Fig. 8.10 unless said. $V_{gs} - V_{th}$ is in 0.25 V steps from bottom (0.25 V) to top (1 V) lines in (**a**) and (**c**), and in 0.5 V steps from 0.5 V to 2 V in (**b**)

much larger barrier thickness (about $d_b \approx d_{ox}\varepsilon_{SiGe}/\varepsilon_{ox}$) as seen from (8.24). From Fig. 8.12a and b, we see that the improvement does exist, but it is less than double. This is due to the channel-velocity saturation which is important for modern short-gate-length field-effect transistors.

Table 8.2 lists the room temperature modeling results of the drain current I_{ds}, transconductance g_m, cut-off frequency f_T and performance parameter improvement ratio between the Si/SiGe n-MODFETs and Si n-MOSTs at 1 μm and 0.5 μm (0.7 μm CMOS technology, $d_{ox} = 100$ Å [8.68]) effective gate lengths. The first row represents the results of $d_{ox} = d_b = 250$ Å at 1 μm. The improvement is very prominent owing to the higher mobility, and higher channel carrier density available in MODFET. The higher carrier density in MODFET is due to much higher dielectric constant ε_{Si} in Si than ε_{SiO_2} in SiO_2, as seen from (8.23 and 24). Such improvement, however, is much less than the factor $\mu_{SiGe}\varepsilon_{SiGe}/\mu_{Si}\varepsilon_{ox}$ (≈ 6.1) which one would expect from classical long channel FET model. Now one may get a feeling that the circuits built by Si/SiGe n-MODFETs would have a rather high potential in high-frequency and high-speed applications.

In the last two rows of Table 8.2, we list the results under the assumption of the same C_G which is required in the scaling scheme of MOSTs [8.68] at 1 μm and 0.5 μm gate lengths. The GCC ability in Si/SiGe n-MODFETs and Si n-MOSTs is the same. The improvement should be mainly from the double enhancement of the mobility. The improvement is less than double and decreases with the scaling down of the gate length. This is due to the carrier-velocity saturation which implies the reduced averaged carrier mobility with increased electric field. The shorter the gate length is, the stronger the electric field along the channel is and the longer saturation portion of the channel.

Table 8.3 lists the results for Si/SiGe and Si p-MOSTs. The improvement ratio is between 1.3–1.7, depending on the gate length and the parameter under consideration. The improvement ratio is quite remarkable. This means that Si/SiGe PMOST may have a potential in microwave applications if we could continue to increase the Ge content since Ge has the highest intrinsic hole mobility in all

Table 8.2. Comparison between the performance of Si/SiGe n-MODFETs and Si n-MOSTs ($V_{gs} - V_{th} = 1.0$ V, $V_{ds} = 2.0$ V)

Parameter		Si/SiGe n-MODFET	Si n-MOST	Improv. Ratio
$L = 1$ μm	I_{ds} [mA/mm]	135	34.2	3.94
$d_{ox} = d_b = 250$ Å	g_m [mS/mm]	212	55.3	3.83
	f_T [GHz]	11.3	7.5	1.51
$L = 1$ μm	I_{ds} [mA/mm]	56.8	34.2	1.66
$d_{ox} = 250$ Å	g_m [mS/mm]	88.1	55.3	1.59
$d_b = 650$ Å	f_T [GHz]	10.2	7.5	1.36
$L = 0.5$ μm	I_{ds} [mA/mm]	196	143	1.37
$d_{ox} = 100$ Å	g_m [mS/mm]	272	204	1.33
$d_b = 240$ Å	f_T [GHz]	24.8	21.2	1.17

Table 8.3. Comparison between the performance of Si/SiGe and Si p-MOSTs ($V_{gs} - V_{th} = 2.0$ V, $V_{ds} = 2.0$ V)

Parameter		Si/SiGe p-MOST	Si p-MOST	Improv. Ratio
$L = 1$ μm	I_{ds} [mA/mm]	63.7	36.8	1.73
$d_{ox} = 250$ Å	g_m [mS/mm]	45.1	27.9	1.62
	f_T [GHz]	6.9	4.3	1.60
$L = 0.5$ μm	I_{ds} [mA/mm]	247	161	1.53
$d_{ox} = 100$ Å	g_m [mS/mm]	152	111	1.37
	f_T [GHz]	18.8	14.5	1.30

commonly used semiconductors. In addition, the Si/SiGe system would also play a role in improving the speed, integration density and current handling ability of Si CMOS circuits within a frame of the processing technology basically compatible with the well developed Si technology because Si p-MOST is one of the limiting factors in CMOS circuits. The SiGe channel layer is separated by a Si cap layer from the gate oxide. This will reduce the influence of hot carrier stressing and be helpful to solve failure problems of CMOS circuits as the gate length scaling continues.

8.3.3   Experimental Results

Today, only results from dc characterization of MODFETs and MOSFETs have been published. However let us take up one dc parameter, the transconductance, g_m, which directly influences the high frequency properties.

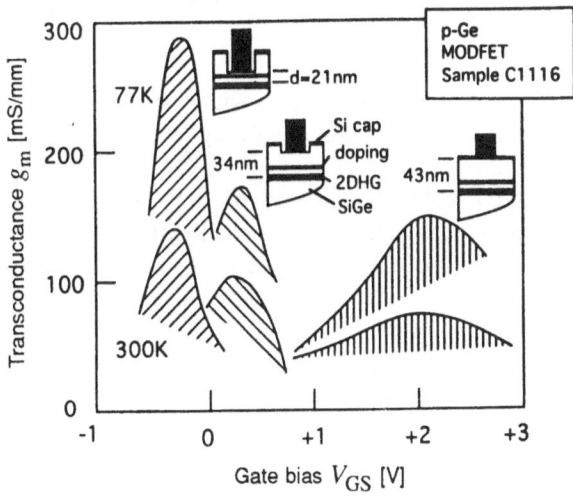

Fig. 8.13. Transconductances of Ge MODFETs. (from U. König)

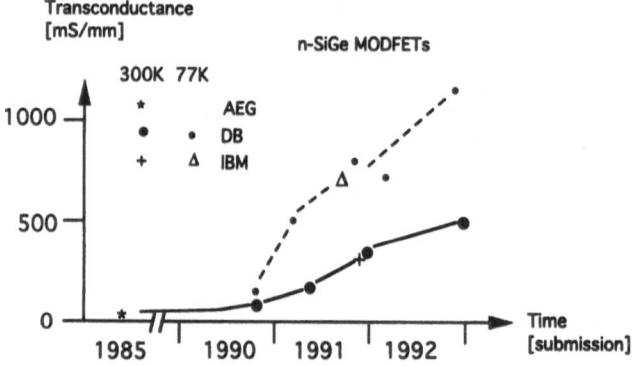

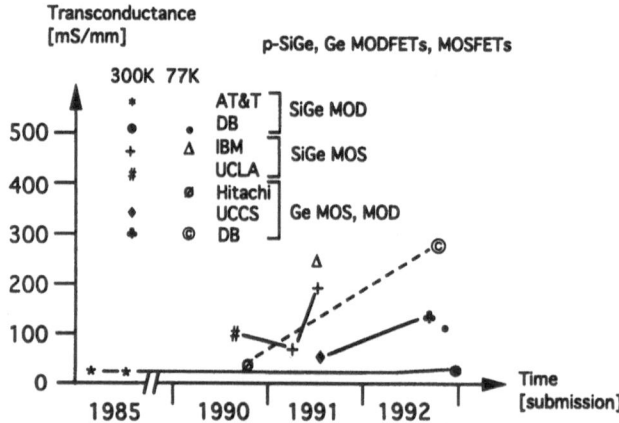

Fig. 8.14. Experimentally obtained transconductances for n-SiGe MODFETs, p-SiGe, Ge MODFETs and MOSFETs. (from U. König)

Theoretically, g_m, could be as high as 100 mS/mm with a distance 10 nm between the gate and the 2DEG and assuming a velocity overshoot up to 2000 mS/mm.

For the first n-Si/SiGe MODFET [8.47,70] demonstrated in 1985, a transconductance of 50 mS/mm was obtained. Today the Daimler Benz Research group has improved their results to 340 mS/mm and 670 mS/mm at RT and 77 K, respectively [8.48]. The transistor, a "normally off" device had a gate-channel distance of 10–15 nm. As also demonstrated by Daimler Benz, the g_m values are sensitive to the rapid thermal anneal (for antimony diffusion), a processing step necessary for activation of the ion implanted contacts (source and drain). The strong improvement of g_m depends mainly on a new type of buffer layer which stops the dislocations before they reach the channel.

In the p-type MODFET, the first (in 1986) demonstrated values from ATT showed a transconductance of 3 mS/mm [8.46]. This value has also been enhanced up to 125 and 290 mS/mm at RT and 75 K, respectively; again by the Daimler Benz group [8.71].

For the p-type MOSFETs the UCLA group obtained a value of 64 mS/mm for g_m [8.49] during 1990.

The transconductance depends also on the transistor's structure in terms of the gate recess, gate length and also on the source resistance. The influence of the gate length for p-Ge MODFET is shown in Fig. 8.13.

Finally, the experimentally obtained results for n-Si/SiGe MODFETs, p-Si/SiGe, Ge MODFETs and MOSFETs are summarized in Fig. 8.14.

8.3.4 Processing Steps: Growth and Post Processing

Si/SiGe layers can be grown by a variety of low-temperature growth techniques, including molecular-beam epitaxy (MBE) [8.41,42] and Chemical Vapour Deposition (CVD) such as in UltraHigh Vacuum (UHV-CVD) [8.44] and in Limited-Reaction Processing (LRP-CVD) [8.43]. Growth of Si/SiGe layers is very critical due to the requirements of low defect densities, in general, and the large lattice mismatch ($\approx$ 4%) between Si and Ge. Epitaxial growth of such strained layer structures requires a balance between two competing temperature requirements. First, the atomic species must have sufficient mobility to obtain high quality epitaxial growth, which requires higher temperatures. However, undesirable islanding growth and strain relaxation are enhanced with the increased temperature. The compromising growth temperature is in the range of $550°-650°C$ [8.72]. The thickness of the desirable Si/SiGe strained layer must be less than the maximum thickness for the pseudomorphic growth (critical thickness) which is an important property of the system [8.45]. As in all epitaxial techniques, it is also crucial to get an atomically clean starting substrate surface. The cleaning process usually consists of a wet etching outside the growth chamber and then an in-situ cleaning.

In MBE, growth is achieved by the direct impingement of the elemental species on a heated substrate under ultrahigh-vacuum conditions. It is a physical vapour deposition process. Doping is carried out by the co-incorporation of the elemental dopant species during epitaxy. Due to the low sticking coefficient of dopants in Si MBE growth, it is often necessary to increase the probability of dopant incorporation. Different doping techniques have been developed, such as doping by secondary implantation [8.73], direct low-energy implantation [8.74] and Solid-Phase Crystallization (SPC) [8.42]. Active doping density of more than 10^{20} cm^{-3} and narrower than 3–4 nm (as required in δFET) have been obtained for n-type doping (Sb) and p-type doping (B).

UHV-CVD is a hot-wall isothermal method of epitaxy. Depositions is carried out by the heterogeneous pyrolysis of gaseous sources, SiH_4, GeH_4 and B_2H_6 in H_2 carrier gas. The growth temperature is between $550°C$ and $630°C$. The Ge fraction in the SiGe layer is determined by the inlet gas composition. Both n- and p-type doping have been achieved. LRP-CVD is a chemical analog of Rapid Thermal Annealing (RTA), and the growth is carried out on a substrate radiant heated in a cold-wall system. The source gases are $SiCl_2H_2$, GeH_4, AsH_3 and B_2H_6. The growth temperature is between 550°C and 650°C.

After growing the desired device structure one starts to process it. The processing steps in a δFET is analogous to conventional MESFET processing techniques. Since there is no high temperature step, the out diffusion of dopants in δ-doping layer can be avoided.

In a Si/SiGe MODFET the key point is to grow a relaxed SiGe layer on Si substrate, without extending the threading dislocation to the Si channel layer. This goal is achieved by employing either a graded buffer layer [8.67], as shown in Fig. 8.9a, or a ramped superlattice buffer layer [8.75] in which the Ge content is increased stepwise. On this graded or ramped layer, a relaxed SiGe layer can be grown; free from the high dislocation density occurring in direct growth on Si substrate. A high electron mobility in the Si channel layer is then obtained. The processing step is trivial except for the annealing procedure required after ion implantation for the source and drain contact (using RTA and $T \leq 850°C$).

It is important in Si/SiGe PMOST to have a cap Si layer (> 5 nm) to get rid of the high density of interface states from direct oxidation of SiGe [8.76]. Even with a Si cap layer, a Plasma-Enhanced CVD (PECVD) gate oxide is used to avoid possible relaxation induced by the direct oxidation of strained structure. The key point in getting a gate quality PECVD oxide is to reduce the oxide deposition rate to several tens of Å/min from the usual high deposition rate (several hundred Å/min) [8.77]. Also, the local oxidation step (LOCOS) used to create a thick field oxide must be replaced by a PECVD oxide deposition of relatively high rate compared with gate oxide deposition.

As a concluding remark, we performed a modeling on Si/SiGe p-MOST, Si/SiGe n-MODFET and Si n-δFET. Comparisons between Si/SiGe and Si p-MOSTs, Si/SiGe n-MODFET and Si n-MOST are made. Also the growth and post-growth processing for these devices are briefly described. Our simulation results show a very promising application future for these FETs in high-frequency and high-speed fields.

Acknowledgement. We like to thank the millimeter-wave group at Chalmers University, Dr. U.König and Prof. E.L. Ivchenko for reading the manuscript.

Appendix 8.A   The Effective-Mass Approximation

The valence band structure is calculated using the 6×6 $k \cdot p$ perturbation Hamiltonian [8.10–14]. We briefly outline the theory.

In a basis of $|x \uparrow>$, $|y \uparrow>$, $|z \uparrow>$, $|x \downarrow>$, and $|z \downarrow>$, where x, y, z denote three orbitals associated with the Γ_{25} representation of the top of the valence band, $\uparrow$ and $\downarrow$ denote spin up and down, and the valence-band Hamiltonian matrix is [8.10,11]:

$$\mathcal{H}_{k \cdot p} = \begin{vmatrix} \mathcal{H}' & 0_{3\times3} \\ 0_{3\times3} & \mathcal{H}' \end{vmatrix} \begin{matrix} x, y, z \uparrow \\ x, y, z \downarrow \end{matrix}, \tag{8A.1}$$

$$\cdot\mathcal{H}' = \begin{vmatrix} Lk_x^2 + M\left(k_y^2 + k_z^2\right) & Nk_xk_y & Nk_xk_z \\ Nk_yk_x & Lk_y^2 + M\left(k_z^2 + k_x^2\right) & Nk_yk_z \\ Nk_zk_x & Nk_zk_y & Lk_z^2 + M\left(k_x^2 + k_y^2\right) \end{vmatrix} \begin{matrix} x \\ y \\ z \end{matrix}.$$

$$(8A.2)$$

The spin-orbit interaction matrix is [8.12]:

$$\mathcal{H}_{SO} = \frac{\Delta}{3} \begin{vmatrix} 0 & -j & 0 & 0 & 0 & 1 \\ j & 0 & 0 & 0 & 0 & -j \\ 0 & 0 & 0 & -1 & j & 0 \\ 0 & 0 & -1 & 0 & j & 0 \\ 0 & 0 & -j & -j & 0 & 0 \\ 1 & i & 0 & 0 & 0 & 0 \end{vmatrix} \begin{matrix} x\uparrow \\ y\uparrow \\ z\uparrow \\ x\downarrow \\ y\downarrow \\ z\downarrow \end{matrix}.$$

$$(8A.3)$$

The effect of strain due to lattice mismatch is included in the band structure by the valence band deformation-potential theory [8.11]

$$\mathcal{H}_\varepsilon = \begin{vmatrix} \mathcal{H}'' & 0_{3\times3} \\ 0_{3\times3} & \mathcal{H}'' \end{vmatrix} \begin{matrix} x, y, z \uparrow \\ x, y, z \downarrow \end{matrix},$$

$$(8A.4)$$

$$\mathcal{H}'' = \begin{vmatrix} l\varepsilon_{xx} + m\left(\varepsilon_{yy} + \varepsilon_{zz}\right) & n\varepsilon_{xy} & n\varepsilon_{xz} \\ n\varepsilon_{yx} & l\varepsilon_{yy} + m\left(\varepsilon_{zz} + \varepsilon_{xx}\right) & n\varepsilon_{yz} \\ n\varepsilon_{zx} & n\varepsilon_{zy} & l\varepsilon_{zz} + m\left(\varepsilon_{xx} + \varepsilon_{yy}\right) \end{vmatrix}.$$

$$(8A.5)$$

For the system under discussion, the epitaxial semiconductor layer is biaxially strained in the xy-plane, by an amount $\varepsilon_1 = \varepsilon_{xx} = \varepsilon_{yy} = a_S/a_L - 1$, where a_S and a_L are the bulk lattice constants of the substrate and the layer material, respectively. The sample is also uniaxially strained in the z-direction, by $\varepsilon_2 = \varepsilon_{zz} = -\varepsilon_1/\sigma$, where σ is the Poisson's ratio. For the growth on a (001) substrate $\sigma = c_{11}/2c_{12}$, where c_{11} and c_{12} are elastic stiffness tensor elements of the layer material.

In [8.13–14], the valence band structures were calculated for the unstrained $Si_{1-x}Ge_x$, biaxially strained $Si_{1-x}Ge_x$ grown on Si, unstrained Si and biaxially strained Si grown on $Si_{1-x}Ge_x$. In the unstrained $Si_{1-x}Ge_x$ and Si, the heavy hole (hh) and light hole (lh) bands are degenerate. However, in the tensile (compressive) strained Si ($Si_{1-x}Ge_x$), the lh band moves upwards (downwards), while the hh band moves downwards (upwards).

Incorporating the interface potential results in an effective Hamiltonian for the envelop function:

$$\mathcal{H}_{\text{eff}} = \mathcal{H}\left(k_x, k_y, k_z\right) + V,$$

$$(8A.6)$$

where

$$\mathcal{H}\left(k_x, k_y, k_z\right) = \mathcal{H}_{k\cdot p} + \mathcal{H}_{SO} + \mathcal{H}_\varepsilon,$$

$$(8A.7)$$

and k_z stands for the operator $(1/j)\partial/\partial z$, and V is the valence band offset. $V = 0.54$ for Si and $V = 0.54(1 - x)$ for $Si_{1-x}Ge_x$, where 0.54 [eV] is the valence band offset between Si and Ge [8.15].

Direct self-consistent calculation of (8A.6 and 7) is very complicated and CPU time consuming, thus special attention must be observed concerning the computer precision problem when numerically handling matrix operations. Simplifications can be approached in the following way. The Fermi level E_F is usually less than 20 meV below the valence band top when the doping levels in the Si substrate and contact layers are less than 5×10^{19} cm^{-3}. Without scattering events (this can be partially assured by the low doping in the barrier and well region), the kinetic energies K_{xy} of tunneling holes in xy-plane are conserved during the z-direction tunneling processes so that are less than E_f. However, the kinetic energies in z-direction, K_z are about eV_{ex} (V_{ex} is the external voltage). Since V_{ex} is usually larger than 0.1 V for an observable tunneling current in experimental I-V characteristics, K_{xy} can be neglected compared with K_z. Thus, the z-direction motions of holes are separated from those in the xy-plane and H' of (8A.2) and H" of (8A.5) become

$$\mathcal{H}' = \begin{vmatrix} Mk_z^2 & 0 & 0 \\ 0 & Mk_z^2 & 0 \\ 0 & 0 & Lk_z^2 \end{vmatrix}, \tag{8A.8a}$$

$$\mathcal{H}'' = \begin{vmatrix} \mu & 0 & 0 \\ 0 & \mu & 0 \\ 0 & 0 & \lambda \end{vmatrix}. \tag{8A.8b}$$

where $\mu = (1 + m)\varepsilon_1 + m\varepsilon_2$ and $\lambda = 2m\varepsilon_1 + 1\varepsilon_2$. It is easy to show that in the following basis

$$|\Gamma_8, 3/2 > = \frac{1}{\sqrt{2}}(|x \uparrow> + j|y \uparrow>), \tag{8A.9a}$$

$$|\Gamma_8, 1/2 > = \frac{1}{\sqrt{6}}(2|z \uparrow> - |x \downarrow> - j|y \downarrow>), \tag{8A.9b}$$

$$|\Gamma_8, -1/2 > = \frac{1}{\sqrt{6}}(|x \uparrow> - j|y \uparrow> + 2|z \downarrow>), \tag{8A.9c}$$

$$|\Gamma_8, -3/2 > = \frac{1}{\sqrt{2}}(|x \downarrow> - j|y \downarrow>), \tag{8A.9d}$$

$$|\Gamma_7, 1/2 > = \frac{1}{\sqrt{3}}(|z \uparrow> + |x \downarrow> + j|y \downarrow>), \tag{8A.9e}$$

$$|\Gamma_7, -1/2 > = \frac{1}{\sqrt{3}}(-|x \uparrow> + j|y \uparrow> + |z \downarrow>), \tag{8A.9f}$$

the Hamiltonian (8A.7) becomes

$$\mathcal{H}(\hat{k}_z) = \begin{vmatrix} \mathcal{H}_{hh} & 0 & 0 & 0 & 0 & 0 \\ 0 & \mathcal{H}_{lh} & 0 & 0 & \mathcal{H}_{lh-sso} & 0 \\ 0 & 0 & \mathcal{H}_{lh} & 0 & 0 & \mathcal{H}_{lh-sso} \\ 0 & 0 & 0 & \mathcal{H}_{hh} & 0 & 0 \\ 0 & \mathcal{H}_{lh-sso} & 0 & 0 & \mathcal{H}_{sso} & 0 \\ 0 & 0 & \mathcal{H}_{lh-sso} & 0 & 0 & \mathcal{H}_{sso} \end{vmatrix} \qquad (8A.10)$$

where

$$\mathcal{H}_{hh} = Mk_z^2 + \mu, \qquad (8A.11a)$$

$$\mathcal{H}_{lh} = \frac{2L + M}{3}k_z^2 + \frac{2\lambda + \mu}{3}, \qquad (8A.11b)$$

$$\mathcal{H}_{sso} = -\Delta + \frac{L + 2M}{3}k_z^2 + \frac{\lambda + 2\mu}{3}, \qquad (8A.11c)$$

$$\mathcal{H}_{lh-sso} = \frac{\sqrt{2}}{3}\left[(L - M)k_z^2 + \lambda - \mu\right]. \qquad (8A.11d)$$

It is then easy to see that the $|\Gamma_8, 3/2 >$ and $|\Gamma_8, -3/2 >$ states are the two degenerate hh states (one spin up, the other spin down) which do not interact with other states, see (8A.10). The $|\Gamma_8, 1/2 >$ and $|\Gamma_8, -1/2 >$ are two lh states, while $|\Gamma_7, 1/2 >$ and $|\Gamma_7, -1/2 >$ are spin-split-off (sso) states. The $\mathcal{H}_{lh-sso}$ describes the interaction between lh and sso states. Further renormalization gives us the following non-interactive lh and sso states:

$$\mathcal{H}_{lh,sso} = \frac{1}{2}\left\{\mathcal{H}_{lh} + \mathcal{H}_{sso}\left[(\mathcal{H}_{lh} - \mathcal{H}_{sso})^2 + 4\mathcal{H}_{lh-sso}^2\right]^{1/2}\right\}. \qquad (8A.12)$$

Instead of the complicated matrix operation (8A.6), we now have three separate scalar equations; heavy hole (8A.11a), lh and sso (8A.12). The hh band of (8A.11a) is parabolic; for lh and sso bands, we can assume an energy-dependent effective mass so that (8.1) agrees with (8A.12). This is similar to the definition of the density of states effective mass and carrier concentration effective mass when calculating the density of states and carrier concentration [8.18].

Parameters used in the calculations are listed in Table 8.1.

Appendix 8.B Maximum Oscillation Frequency and Power Generation

We will here give an outline for the derivation of (8.16 and 18) following E. Kollberg et al. [8.38].

The large signal admittance Y of the drift region can be written

$$Y = G + jB, \qquad (8B.1a)$$

$$G \approx \frac{A\sigma_e}{W}\left(1 + \frac{\sigma_e W}{2\varepsilon V_s}\right)^{-1}, \qquad (8B.1b)$$

$$jB \approx \frac{A\omega\varepsilon}{W}\left(1 + \frac{\sigma_e W}{2\varepsilon V_s}\right)^{-1}, \tag{8B.1c}$$

where σ_e is the effective (large-signal) negative conductance related to the well region. A is the diode area.

When the oscillator starts oscillating, the effective negative conductance G_{eff} is

$$G_{eff} = \frac{1}{2(R_l + R_s)}\left\{1 - \sqrt{1 - [2(R_l + R_s)\omega C_O]^2}\right\} \tag{8B.2a}$$

with

$$f_{max} = \frac{1}{2\pi}\frac{1}{2R_s C_0}. \tag{8B.2b}$$

Hence, σ_e can be determined from $|G| = G_{eff}$. At maximum frequency of oscillation, $R_l = 0$ and $|G| = R_s/2$, we have $j\omega C_O = jB$ and therefore

$$f_{max} \approx \frac{1}{4\pi R_s C_O} = \frac{W}{4\pi R_s A_\varepsilon}\left(1 + \frac{W^2}{4A\varepsilon V_s R_s}\right)^{-1}. \tag{8B.3}$$

Taking into account the negative conductance $-G_{qw}$ and the susceptance B_{qw} of the quantum well and using $-G_{qw}/B_{qw} = -G/B$, (8B.3) can be corrected to

$$f_{max} \approx \frac{W + l_{qw}}{4\pi A\varepsilon(R_s + R_{spch}/2)} \tag{8B.4}$$

with the space-charge resistance $R_{spch} = \frac{W}{2A\varepsilon V_s}$. Eq. (8B.4) is now equal to (8.16).

Let us go over to calculate the maximum power delivered to the load $P_{del.max}$. From (8B.2a), with $G_l = G_{eff}$, we get

$$P_{del} = \frac{\Delta I^2}{8}\frac{1}{G_l}\frac{R_l}{R_l + R_s} = \frac{\Delta I^2 R_l/4}{1 - \sqrt{1 - [2(R_l + R_s)\omega C_O]^2}}. \tag{8B.5}$$

If i_{RF} is the RF current amplitude we have $\Delta I/2 = i_{RF}$. P_{del} is maximal for

$$1 - 2(R_l + R_s)\omega C_O = 0, \tag{8B.6}$$

In the same way as (8B.4) is derived we obtain for the oscillation frequency f:

$$f = \frac{W + l_{qw}}{4\pi A\varepsilon(R_l + R_s + R_{spch}/2)}. \tag{8B.7}$$

From (8B.7 and 4) we get the resistance R_{lopt} for $P_{del.max}$ as

$$R_{lopt} = \frac{1}{4\pi}\frac{W + l_{qw}}{A\varepsilon}\left(\frac{1}{f} - \frac{1}{f_{max}}\right). \tag{8B.8}$$

Hence (8B.5, 6 and 8) give

$$P_{\text{del.max}} = \frac{\Delta I^2}{16\pi} \frac{W + l_{\text{qw}}}{A\varepsilon} \left(\frac{1}{f} - \frac{1}{f_{\text{max}}} \right). \tag{8B.9}$$

Eq. (8B.9) is now equal to (8.18).

It should be noted that the model makes assumptions which are unrealistic when f tends to 0.

References

Section 8.0

8.1   J.F. Luy, K.M. Strohm, J. Buechler: Coplanar silicon monolithic millimeter-wave integrated circuits, in *Monolithic Microwave Integrated Circuits for Sensors, Radar and Communication Systems*, Proc. SPIE **1475**, 129 (1991)

8.2   K. Ismail, B.S. Meyerson, P.J. Wang: "Electron resonant tunneling in Si/SiGe double barrier diodes". Appl. Phys. Lett. **59**, 973 (1991)

Section 8.1

8.3   R. Wessel, M. Altarelli: Resonant tunneling of holes in double-barrier heterostructures in the envelope-function approximation. Phys. Rev. **B39**. 12802; (1989); also Analytic solutions of the effective-mass equation in strained $Si\text{-}Si_{l-x}Ge_x$ heterostructures applied to resonant tunneling. **B40**, 12457 (1989)

8.4   H.C. Liu, D. Landheer, M. Buchanan, D.C. Houghton, M. D'Iorio, S. Kechang: Hole resonant tunneling in Si/SiGe heterostructures. Superlattice and Microstructures. **5**, 213 (1989)

8.5   S.S. Rhee, J.S. Park, R.P.G. Karunasiri, Q. Ye, K.L. Wang: Resonant tunneling through a $Si/Ge_xSi_{l-x}/Si$ heterostructure on a GeSi buffer layer. Appl. Phys. Lett. **53**, 204 (1988)

8.6   G. Schuberth, G. Abstreiter, E. Gornik, F. Schäffler, J.F. Luy: Resonant tunneling of holes in $Si/Si_{l-x}Ge_x$ quantum well structures. Phys. Rev. **B43**, 2280 (1991)

8.7   H.C. Liu, SiGe resonant tunneling devices. Proc. 1st Topical Symp. on Si based heterostructure, ed. by S.S. Iyer, D.C. Houghton, M.L. Green (Am. Vacuum Soc., Baltimore, MD 1990) p. 130

8.8   V.P. Kesan, U. Gennser, S.S. Iyer, T.J. Bucelot: Temperature dependent transport measurements on strained $Si/Si_{l-x}Ge_x$ resonant tunneling devices, Proc. 1st Topical Symp. on Si-Based Heterostructure, ed. by S.S. Iyer, D.C. Houghton, M.L. Green, (Am. Vacuum Society, Baltimore, MD 1990) p. 134

8.9   R.P.G. Karunasiri, K.L. Wang: Quantum devices using SiGe/Si heterostructures. Proc. 1st Topical Symp. on Si based heterostructure, ed. by S.S. Iyer, D.C. Houghton, M.L. Green (Am. Vacuum Society, Baltimore, MD 1990) p. 145

8.10 G. Dresselhaus, A.F. Kip, C. Kittel: Cyclotron resonance of electrons and holes in silicon and germanium crystals. Phys. Rev. **98**, 368 (1955)

8.11 G.E. Pikus, G.L. Bir: Effect of deformation on the hole energy spectrum of germanium and silicon. Sov. Phys.-Solid State **1**, 1502 (1959)

8.12 M. Tiersten: IBM J. Res. Dev. **5**, 122 (1961)

8.13 J.M. Hinckley, J. Singh: Hole transport theory in $Si_{l-x}Ge_x$ alloys grown on (001) Si. Phys. Rev. **B41**, 2912 (1990)

8.14 Y. Fu, Q. Chen, M. Willander: Resonant tunneling of holes in Si/Ge_xSi_{l-x}. J. Appl. Phys. **70**, 7468 (1991)

8.15 C.G. Van de Walle, R.M. Martin: Theoretical calculations of heterojunction discontinuities in the Si/Ge system. Phys. Rev. B34, 5621 (1986)

8.16 J.M. Luttinger: Quantum theory of cyclotron resonance in semiconductors: General theory. Phys. Rev. 102, 1030 (1956)

8.17 D.A. Broido, L.J. Sham: Valence-band coupling and Fano-resonance effects in the excitonic spectrum in undoped quantum wells. Phys. Rev. B34, 3917 (1986)

8.18 F.L. Madarasz, J.E. Lang, P.M. Hemeger: Effective masses for nonparabolic bands in p-type silicon. J. Appl. Phys. 52, 4646 (1981)

8.19 S.M. Sze Physics of Semiconductor Devices (Wiley, New York 1981) p. 850

8.20 J.O. Sofo, C.A. Balseiro: Intrinsic bistability in resonant-tunneling structures. Phys. Rev. B42, 7292 (1990)

8.21 H. Yoshimura, J.N. Schulman, H. Sakaki: Charge accumulation in a double-barrier resonant-tunneling structure studied by photoluminescence and photoluminescence-excitation spectroscopy. Phys. Rev. Lett. 64, 2422 (1990)

8.22 O. Boric, T.J. Tolmunen, M.A. Freking, J.B. Hacker, D.B. Rutledge: S-parameter measurements of quantum-well devices, Proc. 15th Int'l Conf. on Infrared and Micrometer Waves, Orlando, FL (1990), ed. by R.J. Temkin

8.23 Y. Fu, M. Willander: Charge accumulation and band edge in the double barrier tunneling structure. J. Appl. Phys. 71, 3877 (1992)

8.24 Y. Fu, M. Willander: Lateral-nonuniformity effect on the IV spectrum in a double-barrier resonant-tunneling structure. Phys. Rev. B44, 13631 (1991)

8.25 U. Gennser: Private communication (1992)

8.26 S.M. Sze: Microwave diodes, in High-Speed Semiconductor Devices, ed. by S.M. Sze (Wiley, New York 1990) Chap. 9, p. 521

8.27 M. Jonson: Tunneling times in quantum mechanical tunneling, in Quantum Transport in Semiconductor, ed. by D. Ferry, C. Jacoboi (Plenum, New York 1991)

8.28 H.C. Liu: Photoconductive gain mechanism of quantum well intersubband infrared detectors. Appl. Phys. Lett. 60, 1507 (1992)

8.29 S. Collins, D. Lowe, J.R. Barker: A dynamic analysis of resonant tunneling. J. Phys. C20, 6233 (1987)

8.30 Yaotian Fu: Response time in high-frequency quantum transport, in Quantum Coherence in Mesoscopic Systems, ed. by B. Kramer (Plenum, New York 1991) p. 333

8.31 Yaotian Fu, A. Ramaswami: Transient response in quantum transport of noninteracting electrons in nanostructures. Phys. Rev. B44, 10884 (1991)

8.32 Y. Fu, M. Willander: Response of a semiconductor tunneling structure to a time-dependent perturbation. J. Appl. Phys. 72, 3593 (1992)

8.33 T.C.L.G. Sollner, E.R. Brown, W.D. Goodhue, H.Q. Le: Microwave and millimeter-wave resonant tunneling devices, in Physics of quantum electron devices, ed. by F. Capasso, Springer Ser. Electron. Photon., Vol. 26 (Springer Berlin, Heidelberg 1990) p. 147

8.34 F. Stern, W.E. Howard: Properties of semiconductor surface inversion layers in the electric quantum limit. Phys. Rev. 163, 816 (1967)

8.35 N.S. Wingreen, K.W. Jacobsen, J.W. Wilkins: Resonant tunneling with electron-phonon interaction: An exactly solvable model. Phys. Rev. Lett. 61, 1396 (1988)

8.36 W. Cai, T.F. Zheng, P Hu, B. Yudanin, M. Lax: Model of phonon-associated electron tunneling through a semiconductor double barrier. Phys. Rev. Lett. 63, 418 (1989)

8.37 Y. Fu, M. Willander: Evanescent channels in calculation of phonon-assisted tunneling spectrum of a semiconductor tunneling structure. J. Appl. Phys. 73, 1848 (1993)

8.38 E. Kollberg, H. Grönqvist, A. Rydberg: Quantum well and quantum barrier diodes for generating sub-millimeter wave power in Coherent Detection Techniques at Millimeter Wavelengths and Their Applications, ed. by P. Encrenaz, C. Laurent, S. Gulkis, E. Kollberg, Q. Winneser (Nova, New York 1991) p. 143

8.39 J.F. Luy, K.M. Strohm, J. Buechler: A 91 GHz Si/SiGe resonant tunneling detector. Arch. Elektrübutrg, 46, 370 (1992)

Section 8.2

8.40 E. Kollberg, A. Rydberg: Quantum-barrier-varactor diodes for high-efficiency millimetre-wave multipliers. Electron. Lett. **25**, 1696 (1989)

Section 8.3

8.41 S.M. Sze, (ed.): *High-Speed Semiconductor Devices* (Wiley, New York 1990) see also M. Shur: *GaAs Devices and Circuits*, (Plenum, New York 1987)

8.42 E. Kasper, J.C. Bean, (eds.): *Silicon Molecular Beam Epitaxy* (CRC, Boca Raton, FL 1988)

8.43 J.F. Gibbons, C.M. Gronet, K.E. Williams: Limited reaction processing: Silicon epitaxy. Appl. Phys. Lett. **47**, 721 (1985)

8.44 B.S. Meyerson: Low-temperature silicon epitaxy by ultrahigh vacuum/chemical vapor deposition. Appl. Phys. Lett. **48**, 797 (1986)

8.45 R. People: Physics and applications of $Ge_x Si_{1-x}$/Si strained layer heterostructures. IEEE J. QE-**22**, 1696 (1986)

8.46 T.P. Pearsall, J.C. Bean: Enhancement- and depletion-mode Si/SiGe modulation-doped FET's. IEEE EDL-**7**, 308 (1986)

8.47 H. Däembkes, H.J. Herzog, H. Jorke, H. Kibbel, E.H. Kasper: The n-channel SiGe/Si modulation-doped field-effect transistor. IEEE Trans. ED-**33**, 633 (1986)

8.48 U. König, A.J. Boers, F. Schäffler, E. Kasper: Enhancement mode n-channel Si/SiGe MODFET with high intrinsic transconductance. Electron. Lett. **28**, 160 (1992)
      K. Ismail, B.S. Meyerson, S. Rishton, J. Chu, S. Nelson, J. Nocera. High-transconductance n-type Si/SiGe modulation-doped field-effect transistors. IEEE EDL-**13**, 299 (1992)

8.49 D.K. Nayak, J.C.S. Woo, J.S. Park, K.L. Wang, K.P. MacWilliams: Enhancement-mode quantum-well $Ge_x Si_{1-x}$ PMOS. IEEE EDL-**12**, 154 (1991)

8.50 S. Verdonckt-Vanderbroek, E.F. Crabbe, B.S. Meyerson, D.L. Harame, P.J. Restie, J.M.C. Stork, A.C. Megdanis, C.L. Stanis, A.A. Bright: High-mobility modulation-doped SiGe-channel p-MOSFETs. IEEE EDL-**12**, 447 (1991)

8.51 P.M. Garone, V. Venkataraman, J.C. Sturm: Hole-confinement in MOS-gated $Ge_x Si_{1-x}$/Si heterostructures. IEEE EDL-**12**, 230 (1991)

8.52 Q. Chen, M. Willander, J. Carter, C.H. Thaki E.R.A. Evans: Fabrication and performance of delta-doped Si n-MESFET grown by MBE. Electron. Lett. **29**, 671 (1993)

8.53 G.L. Patton, S.S. Iyer, S.L. Delage, S. Tiwari, J.M.C. Stork: Silicon-germanium base heterojunction bipolar transistors by molecular beam epitaxy. IEEE EDL-**9**, 165 (1988)

8.54 A. Gruhle, H. Kibbel, U. König, U. Erben, E. Kasper: MBE- grown Si/SiGe HBT's with high β, f_T, and f_{max}. IEEE EDL-**13**, 206 (1992)

8.55 E.F. Crabbe, J.H. Comfort, W. Lee, J.D. Cressler, B.S. Mayerson, A.C. Megdanis, J. Sun, J.M.C. Stork: 73 GHz self-aligned SiGe base bipolar transistors with phosphorous-doped polysilicon emitters. IEEE EDL-**13**, 259 (1992)

8.56 H.C. Liu, D. Landheer, M. Buchanan, D.C. Houghton: Resonant tunneling in $Si/Si_{1-x}Ge_x$ double-barrier structures. Appl. Phys. Lett. **52**, 1809 (1988)

8.57 Q. Chen, M. Willander: Analytical dc and small signal ac modeling of Si delta-doped field-effect transistors. Solid-St. Electron. **35**, 687 (1992)

8.58 Q. Chen, M. Willander: Analysis of charge control in Si delta-doped field effect transistors. J. Appl. Phys. **69**, 8233 (1991)

8.59 W.-K. Chen: *Active Network and Feedback Amplifier Theory* (McGraw-Hill, New York 1980) Chap. 3

8.60 R. Braunstein, A.R. Moore, F. Herman: Intrinsic optical absorption in germanium-silicon alloys. Phys. Rev. **109**, 695 (1958)

8.61 Y.H. Byun, K. Lee, M. Shur: Unified charge control model and subthreshold current in heterostructure field-effect transistors. IEEE EDL-**11**, 50 (1990)

8.62   C.K. Park, C.Y. Lee, K. Lee, B.J. Moon, Y.H. Byun, M. Shur: A unified current-voltage model for long-channel nMOSFETs. IEEE Trans. ED-**38**, 399 (1991)

8.63   C.G. Sodini, P.K. Ko, J.L. Moll: The effects of high fields on MOS devices and circuits performance. IEEE Trans. ED-**31**, 1386 (1984)

8.64   Q. Chen, M. Willander: to be published

8.65   Y. Tsividis: *Operation and Modeling of the MOS Transistor* (McGraw-Hill, New York 1987)

8.66   P.M. Garone, V. Venkataraman, J.C. Sturm: Hole mobility enhancement in MOS-gated GeSi/Si heterostructure inversion layers. IEEE EDL-**13**, 56 (1992)

8.67   F. Schäffler, D. Többen, H.J. Herzog, G. Abstreiter, B. Holländer: High-electron-mobility Si/SiGe heterostructures: influence of the relaxed SiGe buffer layer. Semicond. Sci. Technol. **7**, 260 (1992)

8.68   H.B. Bakoglu: *Circuits, Interconnections and Packaging for VLSI* (Addison-Wesley, Reading 1990)

8.69   H.M. Li, W.X. Ni, M. Willander, K.F. Berggren, B.E. Sernelius, G.V. Hansson: Electrical characterization and subband structures in antimony δ-doped molecular-beam epitaxy-Silicon layers. Thin Solid Films **183**, 331 (1989)

8.70   H. Däembkes, H.J. Herzog, H. Jorke, H. Kilbbel, E. Kasper: IEDM Tech. Digest IEEE, New York (1985) p. 768

8.71   U. König, F. Schäffler: submitted to IEEE Electron Device Lett. **14**, 205 (1993)

8.72   J.C. Bean, L.C. Feldman, A.T. Fiory, S. Nakahara, I.K. Robinson: Ge_xSi_{1-x}/Si strained-layer superlattice grown by molecular-beam epitaxy. J. Vac. Sci. Technol. A **2**, 436 (1984)

8.73   H. Jorke, H.J. Herzog, H. Kibbel: Secondary implantation of Sb into Si molecular-beam epitaxy layers. Appl. Phys. Lett. **47**, 511 (1985)

8.74   Y. Ota: Silicon molecular beam epitaxy with simultaneous ion implant doping. J. Appl. Phys. **51**, 1102 (1980)

8.75   F.K. LeGoues, B.S. Meyerson, J. Morar: Anomalous strain relaxation in SiGe thin films and superlattices. Phys. Rev. Lett. **66**, 2903 (1991)

8.76   S.S. Iyer, P.M. Solomon, V.P. Kesan, A.A. Bright, J.L. Freeouf, T.N. Nguyen, A.C. Warren: A gate-quality dielectric system for SiGe metal-oxide-semiconductor devices. IEEE EDL-**12**, 246 (1991)

8.77   J. Batey, E. Tierney: Low-temperature deposition of high-quality silicon oxide by plasma-enhanced chemical vapor deposition, J. Appl. Phys. **60**, 3136 (1986)

9 Future Applications

W. Menzel

University of Ulm, Department of Microwave Techniques, 89069 Ulm, Germany

For system applications, the mm-wave frequency range overlaps, on the one hand, with standard microwave frequencies and, on the other hand, with the optical or IR range. Compared to microwave frequencies, mm-waves offer

- additional wide frequency bands for new applications.
- Smaller antenna dimensions for a given gain or higher gain for fixed antenna dimensions.
- Small component size combined with low weight.
- High data rates or spread-spectrum transmission in communication systems.
- High resolution of angle, range and Doppler frequencies in radar systems.

In relation to optical systems, especially the superior transmission properties under adverse weather conditions or in industrial environments with a lot of dust, smoke or dirt should be mentioned.

On the other hand, increasing atmospheric attenuation as well as the influence especially of rain [9.1] limit the application of mm-wave systems mostly to medium or short range. While the minima of atmospheric attenuation around 35, 94, and 140 GHz can be utilized for radar and communication applications up to distances of a few km, the maxima at 60 and 120 GHz allows only short-range systems (up to several hundred meters). This limitation, however, offers an inherent protection against detection, interference, or interception of such links from distances greater than one or two km. Furthermore, these frequencies may be reused in systems only some kms away.

A major precondition for the wide-spread application of mm-wave systems, especially for civil applications, is the availability of semiconductor power-generating elements together with highly integrated (planar) circuits [9.2–5]. The lack of cheap and mass producible components was one of the reasons that the first efforts towards mm-wave systems based on metal waveguide, e.g., for collision avoidance for cars, were not successful [9.6–8]. The situation improved with the development of hybrid integrated circuits [9.9–11], and more applications came up using these techniques [9.12–21].

During the last years, three different aspects speeded up the development of commercial mm-wave systems. The first point is the development of *monolithic* mm-wave integrated circuits on GaAs [9.22] as well as on silicon – described in detail in this book. A second aspect concerns the need for mm-wave sensors and communication equipment for control and safety improvement of the individual road traffic in Europe [9.23] as well as in the US [9.24] and Japan [9.25]. Finally,

Springer Series in Electronics and Photonics, Vol. 32
Silicon-Based Millimeter-Wave Devices, Eds.: Luy et al.
© Springer-Verlag Berlin Heidelberg 1994

the recession of the military business due to the recent political changes, too, urged many companies to open up new fields of application for their micro- and mm-wave capacities.

9.1 Sensor Applications

9.1.1 Measurement Principles

Well-known quantities measured with radar systems are signal propagation time as a measure for target distance and Doppler frequency as a measure for target speed. However, much more information can be gained fully utilizing the available information, even using passive sensors like radiometers. A short overview of possible sources of information is given in Table 9.1. Depending on the required information, the design of the sensor and of the signal processing has to be adjusted.

9.1.2 Radiometric Sensors

A radiometer basically is a sensitive receiver of black-body radiation at microwave frequencies. In the microwave and mm-wave ranges, the black-body radiation power detected by an ideal antenna can be approximated by

$$P_A = kT\Delta f, \tag{9.1}$$

where k is the Boltzmann constant, T the absolute temperature and Δf the receiver bandwidth [9.26]. It should be noted that this power on principle is independent of the actual antenna radiation characteristics as long as the body extends over the complete area illuminated by the antenna. As can be seen from (9.1), the received

Table 9.1. Possible measured sensor quantities and related information about the target

	Passive sensor	Active sensor
Signal delay	–	Distance, ε_r of a fluid or massive body when distance is known
Doppler frequency	–	Speed, acceleration, distance
Phase	–	Relative distance
Amplitude	Temperature, surface and interior structure, reflectivity, material	Reflectivity, size, form, surface and interior structure, material, (distance should be known)
Frequency spectrum	Layer structure of stratified medium using a multi-frequency radiometer	cw *operation*: vibration, rotation and movement of (parts of) the target. *Broadband modulation*: information on the form of the target (reflection centers, resonances, etc.)
Polarisation	Surface and interior structure of the target, surface orientation of the target	

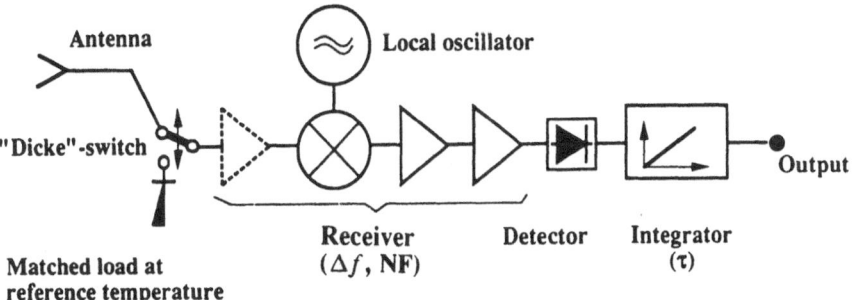

Fig. 9.1. Basic block diagram of a Dicke-type radiometer

power is a measure for the absolute temperature. A real radiometer system, as shown in Fig. 9.1, adds noise to the desired input signal, and the best available temperature resolution is given by

$$\Delta T_{min} = \frac{T_A + (F - 1)T_0}{\sqrt{\tau \times \Delta f}} c, \tag{9.2}$$

where T_A is the temperature of the object of interest ("antenna temperature"), F the noise figure of the receiver, T_0 the ambient temperature (typically ca. 300 K), τ the integration time, Δf the receiver bandwidth, and c the radiometer constant ($c = 1$ for an ideal full power radiometer, $c = 2$ for an ideal Dicke-type radiometer with equal time intervals for target and reference measurements).

This resolution is deteriorated due to limitations of real systems (gain instabilities, antenna sidelobes, etc.).

Even more important is the fact that most materials are not ideal "black" bodies; they reflect part a of the incident radiation and they emit less radiation than according to their temperature. Their detected "temperature" then is composed of two contributions:

$$T_{det} = \left(1 - |r|^2\right) T_B + |r|^2 T_S \tag{9.3}$$

where T_{det} denotes the detected "temperature", T_B the temperature of the object of interest, T_S the temperature of the surroundings of which the radiation is reflected, and r the surface reflection coefficient of the object.

Especially, for metal, $|r| = 1$ is valid. Nevertheless, radiometric sensors can be employed – depending on the circumstances – to directly measure temperature, or indirectly quantities like the absorption of materials or the roughness of surfaces.

For example, *Hetzner* [9.27] has shown that mm-wave radiometers are able to distinguish between dry, wet, and icy road surfaces giving important information for traffic safety systems. The principle of this method is based on measuring the radiometric "temperature" compose of emitted "warm" ground and reflected "cold" sky temperature. A direct temperature measurement, on the other hand, could be employed to detect the overheating of vehicle tyres.

Using SIMMWIC circuits, such radiometers could include a low-noise Schottky-diode mixer, an IMPATT local oscillator, and possibly integrated bipolar transistor IF amplifiers. Special care has to be taken of the IF amplifier design, as gain values of 60–70 dB are required to get a sufficient signal level for the video detector.

9.1.3 cw Radar Sensors

cw radars are well known measuring speed via the Doppler effect. Integrating continuously speed measurements gives the travelled distance, differentiating speed gives acceleration.

A typical Doppler radar set-up is illustrated in Fig. 9.2, but even simpler designs have been realized using the principle of self-oscillating mixers, as it is described in Chaps. 5 and 6.

Typical applications of standard mm-wave Doppler radars are the measurement of true speed over ground for all types of vehicles [9.12,17] or road traffic monitoring [9.15,16]. In the latter case, additional functions can be added to the sensors like length measurements of vehicles to distinguish between personal cars and trucks or buses or even communication facilities for local data exchange.

A detector registrating *moving* targets can be employed, too, as a sensor for the control of illumination, doors, or traffic lights, or to detect rain, snow or hail. Furthermore, such sensors could be used for intruder alarm, especially in indoor applications.

Continuously moving objects are not the only targets of cw radar. They are equally well suited to measure periodic movements, e.g. vibrations [9.28]. This may be very important in a number of industrial applications where excessive vibrations might even destroy that equipment.

An additional example for detecting periodically moving objects is the determination of the upper dead center of the piston in a car engine [9.29]. To this end, a 60 GHz cw signal is fed through the sparking plug, and an autocorrelation function was formed out of the sensor's output signal. Due to the symmetry of this function, the upper dead end could be determined very exactly.

CW-transmitter

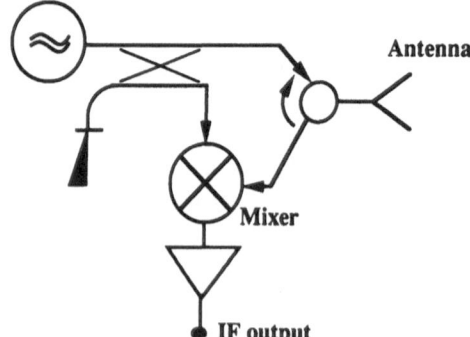

Fig. 9.2. Basic block diagram of a cw Doppler radar

Another area of application is found making use of a quadrature detector for phase measurements. In this way, a very precise relative distance measurement can be achieved down to the range of 1 μm [9.30,31]. Typical applications might be in industrial processes where dust, dirt, or heavy noise prevent the use of optical or acoustic sensors. Although phase is periodic with $\lambda/2$, a wider range can be achieved with continuous measurements, or a (sinusoidal) modulation is applied to the transmitter signal, and the demodulated signal is used for phase comparison. Although the minimum resolution is higher in that case, a higher number of modulation frequencies gives a wide unambiguous range of such systems.

One problem of this type of application is its high sensitivity to unwanted reflections reducing the measurement accuracy. In a cooperative environment, this problem can be overcome by an active reflector [9.31], e.g., some antenna element with an active device modulating the reflected signal. Such an active reflector, too, seems to be a typical application for SIMMWIC circuits.

9.1.4 Frequency Modulated Radar Sensors

The basic set-up of a frequency modulated continuous wave (FM-cw) radar is identical to that of a cw radar (Fig. 9.2), except that the transmitter is (mostly) linearly frequency modulated. Due to the propagation delay, the received signal is frequency shifted compared to the actual transmitted signal resulting in a frequency difference proportional to the distance of the target (Fig. 9.3). This distance can be calculated by

$$R = \frac{c_0 T}{2\Delta f} f_{IF} \tag{9.4}$$

(c_0: speed of light; f_{IF}: detected if frequencies, T: ramp length, Δf: maximum frequency deviation).

Using digital signal processing of the output signal, e.g. a fast Fourier transform, the range resolution can be calculated by

$$\Delta R = \frac{c_0}{2\Delta f}. \tag{9.5}$$

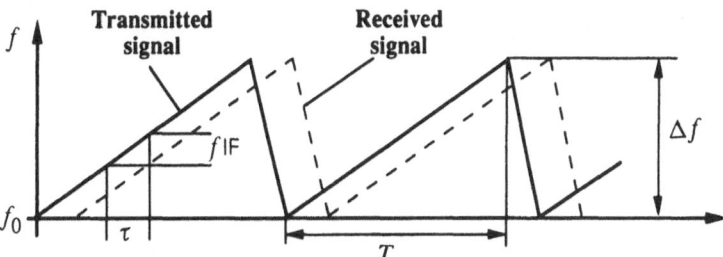

Fig. 9.3. Transmit and receive frequency response of a FM-cw radar

The resolution is solely depending on the maximum frequency deviation. With a wide-band modulation of some GHz, a range resolution of a few cm can be achieved.

Major applications of this type of radar in the mm-wave range are found for short-range distance measurements (a few meters up to some hundred meters). Therefore, a possible mass production is predicted for traffic applications like distance control (as an addition to an automatic speed control), collision warning, electronic rear-view mirror, parking aid at the rear, etc. [9.8,23–25,30,32–34]. On the other hand, this type of sensor is equally well suited for industrial applications like the control of automatically guided transport vehicles, the measurement of the filling height of tanks or furnaces, or distance control in a production process [9.6,30,34].

Evaluating the complete spectrum of a return signal, not only the distance but also the length profile of objects can be measured. Such a profile spectrum contains valuable information about this object and may be used for its recognition independently of its orientation. In this way, a simple system may be built for the identification of parts in a production process [9.30].

9.1.5 Pulse-Modulated Radars

Pulse radars are widely used in many types of applications, especially for medium and long range like air-traffic control. The range is simply determined by the delay time between transmitted and receiver pulse, while the range accuracy depends on the pulse rise time and range resolution on the pulse width.

In the mm-wave range, wide bandwidths are available, thus enabling short pulses with even shorter rise times, allowing the realization of high resolution, but short-range radar sensors. Such systems were investigated for collision avoidance of cars [9.7,35] or helicopters [9.14]. As has been shown in [9.35], a high degree of integration is possible – in that case the complete RF sensor part is integrated in a single monolithic silicon circuit.

For very high resolution, *Lange* et al. [9.37] have presented a 94 GHz pulse radar with extremely low pulse width (about 1 ns) for use as an imaging radar for autonomous vehicles in an industrial environment.

One disadvantage of pulse radars, however, may be the larger if bandwidth compared to FM-cw radars, although the RF bandwidths are comparable for the same resolution. This fact may complicate, to some extent, the realization of a simple and cheap signal processing.

9.2 Communication Applications

9.2.1 General Considerations

One of the major reasons for the development of mm-wave communication systems was the requirement of new services which could not be realized in the crowded

standard frequency bands. Typical applications are (mobile) point-to-point links in
the 30–50 GHz range [9.12,13,18–20,36,38,39] for the distribution of voice, video,
fax, and other data, for Local Area Networks (LAN), or mobile communication
with trains. The transmitter power is generated using Gunn elements or IMPATT
diodes, the power level is in range of some tens to some hundreds of mW, Schottky-
diode mixers are used as receivers, the antenna gain is relatively high due to the
use of parabolic antennas, and frequency modulation or frequency shift keying is
used in most cases. Accordingly, the available hop length amounts up to 10 or
15 km [9.40]. Parallel to these applications, further experiments are conducted in
the 60 GHz frequency range, especially exploiting this frequency range for cellular
radio [9.41] or traffic information services [9.42].

While these systems have been realized in a more conventional technology
like waveguide and hybrid integrated components – although they would make use
of available monolithic integrated circuits – , mass applications for individual use
require much cheaper systems employing a high degree of monolithic integration
combined with much more simple system designs. Once this is available, a high
number of applications in all areas are possible.

9.2.2 Identification Card Systems

The remote identification of persons or objects as well as the remote readout of all
kinds of data has proved to be of extreme importance concerning a wide range of
applications: like

- Access control to private or restricted areas.
- Priority control of public transport vehicles, e.g., at traffic lights.
- Control of the production process, e.g., in the production of cars.
- Identification of trains for display control in railway/underground stations.
- Identification of wagon number, load, and destination of railway wagons.
- Identification of transport containers and their contents/destination.
- Automatic toll debiting systems.
- Automatic parking area management.

Typical ranges of such systems are a few meters to a maximum of some ten
meters. For some of these applications, lower-frequency systems, e.g. at 5.8 GHz,
are on the market since several years. A simplified set-up of such a system is displayed in Fig. 9.4. Typically, a microwave signal is sent to the identification card.
This signal is detected, and the response procedure is triggered. The incident RF
signal is modulated according to the card information and transmitted back to the
stationary unit. The modulation may affect amplitude, phase or frequency of the
incoming signal; in some cases, an additional if frequency is used, or the information is transmitted even at a completely different frequency. For communication
safety, the data usually are transmitted several times. In some cases, the microwave
signal may be used even for the power supply of the digital circuits in the mobile
equipment.

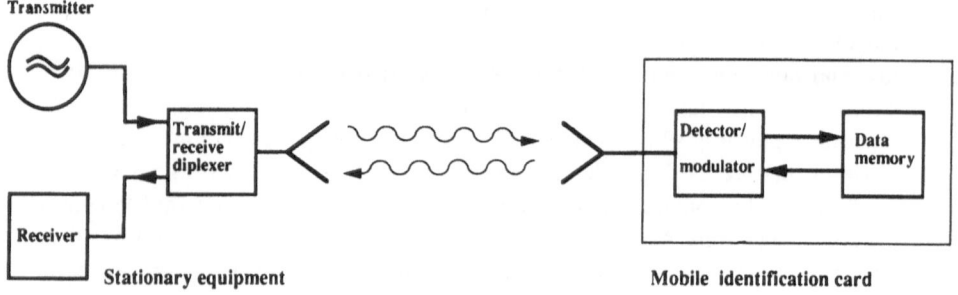

Fig. 9.4. Basic block diagram of an identification card system

While the stationary equipment may be more complex and sophisticated – and therefore more expensive -, the mobile card must be as simple and cheap as possible. For this reason, at low frequencies this card is operated mostly in a passive way, i.e. the incident signal is received by a printed antenna, modulated using simple (beam-lead) diode(s), and returned via the same antenna.

Typically, the RF power received by the mobile antenna is given by

$$P_r = \frac{P_t G_t G_r \lambda^2}{(4\pi R)^2},$$

(9.6)

where P_t is the transmitted power, G_t the gain of transmitter antenna, G_r the gain of receiver antenna, R the distance between receiver and transmitter, and λ the wavelength.

This relation is valid in the opposite direction, too. Additionally, the modulation loss reduces the power budget. The antenna gain values are determined by the required field of view, independent of the frequency. At low frequencies, the power budget allows the use of passive cards; as the budget, however, is strongly dependent on wavelength, this is no longer possible at mm-wave frequencies. Therefore, active cards have to be used including a mm-wave source. This can be done in an economical way only if highly integrated circuits are available. On the other hand, such a source can be used, too, as LO for a receiving mixer improving the sensitivity of the card.

9.2.3 Short-Range Data Transmission

More and more information – numerical and alphanumerical data or graphics – are handled today by computers or computer assisted systems. This does not rise severe problems as long as a stationary environment is concerned, e.g. the offices of some administration. Including mobile partners, however, an extension of the data transmission via radio systems is necessary. One way to cope with these requirements is the use of cellular radio or even satellite systems. While this is effective for medium range up to world-wide coverage, it is not for many close range applications.

A typical example for the latter is the distribution and the road transport of small-piece goods. Storage organization, ordering, delivery preparation, and billing are done using computers. An efficient transportation should include an appropriate computer assistance for route planning, delivery organization, and customer orders in the truck. A data exchange is necessary – in most cases – only when the truck stops at his home company. Although data exchange is possible establishing a cable link at that point or transferring data via diskette, it is much more effective using a short range mm-wave radio link for this purpose.

9.2.4   Information Systems

Further applications of mm-wave systems are related to traffic information [9.16,23,24,33,42]. With increasing traffic density it will become necessary to inform drivers about the traffic situation *relevant for his special route* in contrast to the general wide-area information distributed today by public radio. For example, local mm-wave systems combined with traffic monitoring [9.16] can manage this task. Another example is route guidance in urban areas using electronic maps combined with beacon systems providing information on the actual location and local traffic situation [9.43].

In some scenarios, even communication between cars driving one behind the other is discussed [9.42]. In this way, information on the actual situation some hundred meters in advance – e.g., that some cars in front have started to brake or slow down – could improve safety on the road. Such systems, however, can work successfully only if *all* vehicles are equipped in such a way.

9.2.5   Millimeter-Wave Data Bus

Terminals, PCs, workstations, or peripheral equipment today are widely linked by high speed data buses. In some cases, the flexibility of establishing such a bus link is reduced by the limited access to a computer network. This is especially true thinking of the increasing use of portable PCs (laptops, palmtops, etc.).

There is no reason, however, why a data bus should be restricted to a cable. Using appropriate transmitters and receivers, a data bus can be made available via a radio link. Special applications of such a radio data bus are in larger offices or in fabrication halls where a frequent and flexible rearrangement of data links are required. For the required high data rates as well as with respect to a small system size, mm-waves seem to be the first choice for this type of application. A first system – in this case replacing cables over distances up to a few km – has already been realized and is available commercially [9.44].

9.3   System Requirements

9.3.1   General Systems Aspects

Designing a complete microwave or mm-wave system, there are – from the microwave point of view – three general items which have to be taken into account. The first item refers to the overall system specifications determined by the intended application. This, typically, determines the detailed requirements for the subsystems. However, in the view of simple and cheap systems suitable for a mass production, the availability of special devices, technologies and circuits may modify the overall specs, too. The second point concerns the RF front-end itself, including its general specs like frequency, bandwidth, RF-power, the types of devices or circuits, the choice of materials and technologies, testing, packaging etc. The third one deals with the interfaces and the integration with signal and data processing (up to an accepted man-machine interface), power supply, or general electrical and mechanical construction. Furthermore, devices, components, subsystems as well as the complete system have to be operated in the respective environment including ambient temperature, humidity, presence of a number of agressive chemicals, or incident electromagnetic signals. Additionally, the complete system must not interfere with systems of the same kind or other systems.

In the following, a few of these aspects related to mm-wave circuit technology will be addressed.

9.3.2 Frequency Stability

There are basically two reasons for the requirement of frequency stability – a technical and an administrative one. The technical aspects mostly concern communication applications. To allow a minimum if bandwidth or an easy LO synchronisation, the center frequency of transmitter and receiver should be equal or at least as close together as possible. As this is not possible using a *common* reference source, a frequency stabilization at each location is necessary. In cw radar applications, the RF frequency accuracy directly influences the measurement accuracy, although, in most cases, this accuracy is limited by other factors like the width of the Doppler-frequency spectrum.

A major requirement for frequency stability in all types of applications is given by national and international PTT regulations. Typically, these are given as *absolute* frequency values. With increasing frequency in the mm-wave range, this results in more and more reduced *relative* bandwidths.

Planar circuits, especially monolithic integrated circuits, show a Q-factor much too low for an *internal* frequency stabilization, e.g. with a planar resonator. Therefore, *external* means have to be used, either passive or active. At lower frequencies, dielectric resonators made of high ε_r, temperature compensated materials have proven as adequate means for passive stabilization, [9.45]. This technique basically can be applied to mm-wave sources; increased tolerance requirements and the decreasing size of the resonators, however, do not result in a low-cost design. Some efforts have been made to increase the resonator size using higher order resonances or whispering gallery types of resonators [9.46].

For active stabilization, Phase Lock Loop (PLL) circuits are widely used at microwave as well as mm-wave frequencies. Today, in the microwave area, digital frequency dividers are available to prescale the frequency to a value which can be compared with a reference source, e.g. a crystal oscillator in the 10–100 MHz

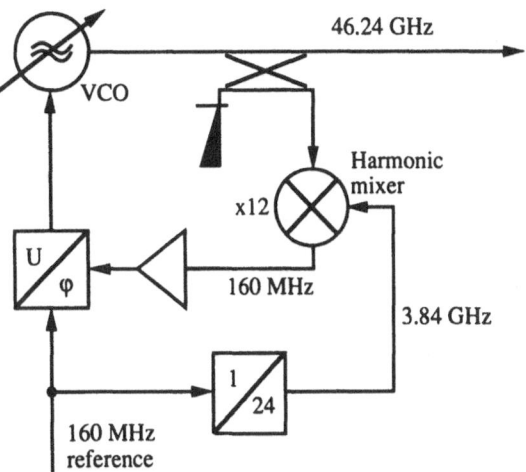

Fig. 9.5. Basic block diagram of a conventional mm-wave PLL stabilization [9.49]

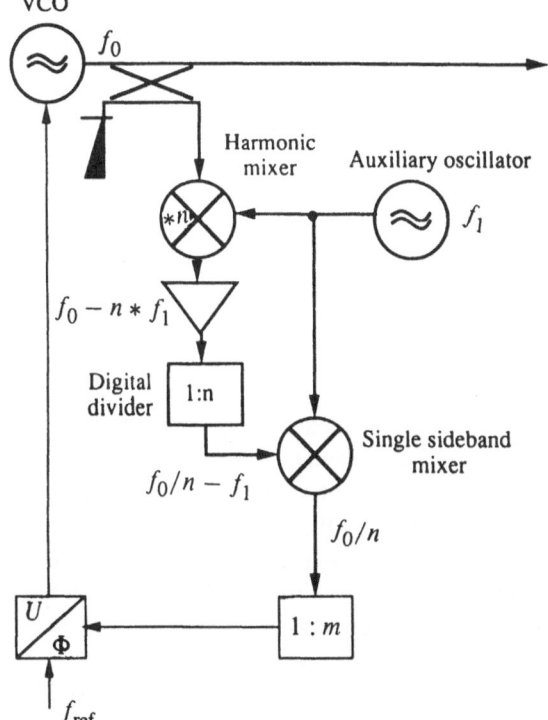

Fig. 9.6. Block diagram of a PLL stabilization using a mm-wave frequency divider [9.50]

range [9.47]. In the mm-wave frequency range, such dividers are not available, and even analogue dividers, as described in [9.48], are not an applicable solution due to the requirement of a mm-wave amplifier. Therefore, up to now, more complex arrangements, as shown in Fig. 9.5, were used [9.49]. While the harmonic mixer can be realized easily, the multiplier chain is a rather complex circuit. At higher frequencies, e.g. in the 77 GHz range for automotive applications, this becomes even more severe. In order to sufficiently suppress unwanted harmonics, the multiplication has to be done in several steps separated by filters and amplifiers.

Recently, a mixed analogue/digital frequency divider suitable for mm-wave frequencies was proposed and tested [9.50]. It again uses a harmonic mixer in the mm-wave range, and an amplifier, a digital divider, a Single Side Band (SSB) mixer and an auxiliary oscillator in the microwave range (Fig. 9.6). Although, at a first glance, this arrangement looks a little bit complicated, all components are suitable for (monolithic) integration leading to a possibly cheap mm-wave (PLL) stabilization, as indicated in Fig. 9.6, too.

9.3.3 Environmental Conditions

Each system has to operate successfully under a wide range of environmental conditions typically for its application. Quantities affecting its performance are temperature, humidity, chemical substances, dirt, dust, mechanical loads (vibration, shock), and incident electromagnetic signals. For indoor applications, these conditions may be less severe than, for example, in outdoor traffic applications.

Environmental conditions influence the design of the overall mechanical housing including a radome or integrated antenna, the packaging of subsystems and of components down to monolithic integrated circuits, and even the passivation of devices and monolithic circuits. A number of solutions to these problems are already known. However, with respect to a cheap mass production of mm-wave systems, many of these methods have to be modified or can not be applied. For example, heating a mm-wave front-end to a constant temperature is not acceptable in most of the planned applications due to an excessive power consumption, a relatively long warm-up time, and the expenses for heater elements, excess power supply, and the control circuiting.

Temperature, on the one hand, affects semiconductor properties and, on the other hand, can lead to mechanical stress as materials with different expansion coefficients have to be used. The latter is of specific importance in conjunction with interconnects between different circuits (MMICs) or between circuits and connectors. To avoid interruptions of such interconnects during temperature cycles, some excess length of bond wire or lead has to be added leading to an arch-type interconnect. This, however, together with end capacitances of the connecting lines, may lead to a low-pass characteristic of the interconnect, critically especially for mm-wave circuits. Only a thorough design with tight tolerances of the interconnect can overcome this problem. Another idea to cope with this problem in conjunction with mm-wave circuits might be the use of electromagnetic coupling [9.51] between different chips or a chip and a base substrate. According to Sects. 9.2 and 3, a great number of planned applications are related to traffic. In this case, the required temperature range goes down to $-40\ldots-30°C$, while the upper limit is at $80\ldots90°C$ or even higher if a sensor has to be mounted close to the motor. High temperatures, too, may occur in industrial applications.

Apart from possible corrosion effects, humidity is especially critical if it condenses on mm-wave circuits. Due to the relatively high dielectric constant and losses, condensed water can seriously affect the performance of devices and circuits up to a complete failure of the performance. These effects are increased if a system or component is undergoing temperature cycles where more and more water may be accumulated inside the equipment. Equally critically are water drops, a water film, ice or snow on the antenna aperture or radome. Inside a system, hermetic sealing might be the best protection against humidity. Concerning the antenna aperture, an optimal choice of material, form and mounting position can minimize the problem.

Dirt and dust may affect circuit performance and wave propagation especially in traffic and industrial applications, although their influence is much less pronounced in mm-wave systems than in optical or IR systems. Chemical substances can lead to corrosion or may affect system components in different other ways. Next to industrial environments, traffic sensors are exposed to a lot of substances, too, like petrol, oil, exhaust gases, or salt (to thaw ice and snow on the road).

9.3.4 Packaging

Regarding the environmental loads as addressed above, a packaging of the mm-wave components will be necessary for most applications. Such a package may contain a single monolithic integrated circuit, but also a number of chips forming a complete functional block. A package has

- to provide a hermetic sealing of the components,
- to withstand the environmental loads,
- to provide good electrical connections for RF and if signals and the dc power supply,
- to remove the excess heat produced by the circuits,
- to provide, if possible, some protection against electromagnetic interferences
- to prevent resonances. (This is a new problem, due to the small wavelengths at mm-wave frequencies.)
- to be designed in a proper way suitable for mass production at a low price.

Hermetically sealed packages are available for MMICs up to 20 GHz [9.52]. Typically, they consist of a gold-plated covar ground plane to which the chip is bonded or soldered, a ceramic frame with coaxial-type electrical feed-through, and a metallic (covar) cover. For the aspired applications, this type of package, however, is not acceptable for different reasons. Although the chosen materials do satisfy many of the requirements listed above, the overall set-up is much too expensive for the civil applications discussed presently. Such a package finally would cost much more than the circuits in it. A second problem refers to feeding mm-wave signals into and out of the package. In this case, coaxial transitions may no longer work sufficiently, and also the interior connections between several chips and between chips and feed-through structures have to be redesigned for mm-wave applications. A number of efforts are going on solving these problems [9.53–55] including flip-chip mounting of MMICs, but these are still strongly influenced by military requirements.

To overcome these problems, a number of new approaches have to be considered.

- Packages made of plastic materials can be mass produced easily by molding techniques. The plastic material has to withstand, however, the required temperature range, humidity, chemicals, etc. Absorbing particles could be included to suppress resonances, and a (partly) metallization could give some protection against electromagnetic interference.
- The number of mm-wave connections into or out of the package should be minimized. This can be done by an optimal *combination of several circuits*, so that, for example, only a single input/output is required. The remaining connections should be made using planar transmission line structures. In some cases, radiating elements can be included in the interior circuits, enabling even a radiation connection through a radome-like package cover.

References

Section 9.0

9.1   A. Van der Forst: MM-Wave atmospheric propagation and system implications. 16th Europ. Microw. Conf., Dublin (1986) pp. 19–30
9.2   H.J. Kumo: Are millimeter-wave systems affordable now? Microwave J. 16–24 (June 1982)
9.3   D.M. Russel: MM-wave snapshot reveals maturing technology. Defense Electronics 113–115 (October 1984)
9.4   N.B. Kramer: Must MM-waves wait again? Microwave J. 24–25 (July 1985)
9.5   A.E. Braun: Progress in millimeter-waves-where is the infrastructure? Defense Electronics 77–82 (December 1982)
9.6   H. Meinel, B. Rembold: Commercial and scientific applications of millimetric and submillimetric waves. The Radio and Electronic Engineer, No. 7/8, 351–360 (1979)
9.7   K. Lindner, W. Wiesbeck: Die Mikrowellenbaugruppen eines 35 GHz Abstandswarnradars für Kraftfahrzeuge (the microwave components of a 35 GHz anticollision radar for cars). Mikrowellen Magazin 398–403 (May 1977)
9.8   G. Neininger: Vehicle Collision Avoidance Radar. Funkschau **49**, 389–393 (September 1977) (in German)
9.9   T.H. Oxley, C. Burnett: MM-Wave (30–110 GHz) hybrid microstrip technology. Microw. J. Part I: 36–44 (March 1986), Part II: 177–185 (May 1986)
9.10   K. Solbach: The status of printed MM-wave E-plane circuits. IEEE Trans. MTT-*31*, 107–121 (1983)
9.11   W. Menzel: Integrated fin-line components for communication, radar, and radiometer applications. *Infrared and Millimeter waves 13*, 77–121 (Academic Press) Orlando 1985
9.12   G. Reinhold, H. Meinel: Verwendung von Millimeterwellen in Bahnsystemen (application of mm-wave in railway systems). NTZ **34**, 352–357 (1981)
9.13   H. Meinel, A. Plattner, G. Reinhold: A 40 GHz railway communication system. IEEE Trans. SAC-*1*, 615–622 (1983)
9.14   B. Rembold, H.G. Wippich, M. Bischoff, W.F.X. Frank: A 60 GHz collision warning sensor for helicopters. Military Microw. Conf., London (1982) pp. 344–351

9.15 W. Linss *et al.*: MM-wave radar sensor for traffic data acquisition systems. Proc. SBMO, Int. Microw. Symp. Sao Paulo, Brazil (1989) pp. 513-521

9.16 H.-J. Fischer: Digital beacon vehicle communication at 61 GHz for interactive dynamic traffic management. 8th Int'l Conf. on Automotive Electronics, London (1991) pp. 120-124

9.17 E. Lissel: Geschwindigkeits- und Wegsensor nach dem Mikrowellen-Doppler-Prinzip (distance meter using the microwave Doppler principle). VDI-Bericht **687**, 257-282 (1988)

9.18 S.A. Mohamed, M. Pilgrim: 29 GHz point-to-point radio systems for local distribution. British Telecom Technology **2**, 29-40 (1984)

9.19 P. Dupuis S. Meyer, M. Goloubkoff, J.J. Guena: Millimeter wave subscriber loops. IEEE Trans. SAC-*1*, 623-632 (1983)

9.20 K. Ogawa, T. Ishizaki, K. Hashimoto, M. Sakarura, T. Uwano: A 50 GHz compact communication system for video link fabricated in MIC. IEEE Intern. Microw. Symp. MTT, New York (1988) pp. 1023-1026

9.21 H. Meinel, W. Menzel: Commercial millimeter-wave applications. Int'l Conf. on IR and MM-Waves, Orlando (1985) T 1.2

9.22 IEEE Microw. and Millimeter-Wave Monolithic Circuits Symp. Dig. (1992)

9.23 H. Meinel: Applications of microwaves and millimeterwaves for vehicle communications and control in Europe. IEEE Int'l Microw. Symp. MTT, Albuquerque (1992) pp. 609-612

9.24 Panel Session on IVHS in America. IEEE Int'l. Microw. Symp. MTT, Albuquerque (1992)

9.25 M. Kotaki, Y. Takimoto, E. Akutsu, Y. Fujita, H. Fukuhara, T. Takahashi.: Development of millimeter wave automotive sensing technology in Japan. IEEE Int'l Microw. Symp. MTT, Albuquerque (1992) pp. 709-712

Section 9.1

9.26 J.H. Rainwater: Radiometers: Electronic eyes that 'see' noise. Microwaves 58-62 (September 1978)

9.27 W. Hetzner: Aktive und passive Straßenzustandserkennung im Millimeterwellenbereich (detection of road surface condition using mm-waves). Frequenz **38**, 179-185 (1984)

9.28 G. Kadel: Radarechos mechanisch schwingender Objekte. Dissertation, TH Darmstadt, VDI Fortschrittsberichte, Reihe 21, Nr. 36

9.29 H.G. Wippich, A. Happe, B. Rembold: Ein 60 GHz Radarsensor zur Bestimmung des oberen Totpunktes bei Verbrennungskraftmaschinen (a 60 GHz radar sensor for the measurement of the upper dead center in combustion engines). VDI Report **509**, 259-261 (1984)

9.30 W. Holpp: High resolution mm-wave sensors converted from military to industrial applications. Proc. Military Microw. Conf., Brighton, England (1992) pp. 13-20

9.31 J. Detlefsen, W.-D. Schuck: Präzisionslängenmessung mit Millimeterwellen (precision length measurements using mm-waves). NTZ **35**, 344-347 (June 1 1982)

9.32 D.A. Williams: Millimeter wave RADARS for automotive applications. IEEE Int'l Microw. Symp. MTT, Albuquerque (1992) pp. 721-725

9.33 Workshop on Advanced Car Electronics and Future Traffic Control Systems Related to Microwaves. 21st Europ. Microw. Conf., Stuttgart (1991)

9.34 W. Holpp: Millimeterwave radar applications in the commercial arena. Proc. Workshop Commercial Applications of Micro- and Millimetre Waves. 22nd Europ. Microw. Conf., Helsinki (1992) pp. 9-17

9.35 N. Haese, M. Benlamlih, D. Cailleur, P.A. Rolland: Low-cost design of a quasi-optical front-end for on-board MM-wave pulse radar. IEEE Int'l Microw. Symp. MTT, Albuquerque (1992) pp. 621-623

9.36 D. Kroll, B. Rembold: Communication with millimeter-waves. MIOP Conf., Wiesbaden, Germany (1987) UE III-2 (in German)

9.37 M. Lange, J. Detlefsen, M. Bockmaier: 94 GHz imaging radar for autonomous vehicles. 18th Europ. Microw. Conf., Stockholm, Sweden (1988) pp. 826-830

Section 9.2

9.38   M. Hata, A. Fukazawa, M. Blesho, S. Makimo, R. Higuchi.: A new 40 GHz digital distribution radio with single local oscillator. IEEE Int'l Microw. Symp. MTT, Ottawa (1978) pp. 236–238

9.39   S. Samejima, S. Kurokawa: Transportable radio equipment using 40 GHz band for video conference system. J. Telecom. Rev. **21**, No. 4, 364–399 (October 1979)

9.40   H.H. Meinel: Millimeter wave system design and application trends in Europe. Alta Frequenza **28**, No. 5–6, 441–456 (September–December 1989)

9.41   F. Baron, M. Liber: 60 GHz communications, technology and applications trends. Proc. Workshop Commercial Applications of Micro- and Millimetre Waves. 22nd Europ. Microw. Conf., Helsinki (1992) pp. 24–28

9.42   H.-J. Fischer: Vehicle communications at 61 GHz, Workshop Commercial Applications of Micro- and Millimetre Waves. Proc. 22nd Europ. Microw. Conf., Helsinki (1992) pp. 35–40

9.43   W. Linss: Futuristic view of the applications of advanced mm-wave integrated circuits in a traffic environment. Proc. 21st Europ. Microw. Conf., Workshop, Stuttgart, Germany (1991) pp. 106–111

9.44   WICOL 38, Product information, Telefunken Sendertechnik, Berlin (1991)

Section 9.3

9.45   U. Güttich, A. Gruhle, J.F. Luy: A Si-SiGe HBT dielectric resonator stabilized microstrip oscillator at X-band frequencies. IEEE Microw. and Guided Wave Lett. **2**, 281–283 (1992)

9.46   D. Cros, P. Guillon: Whispering gallery dielectric resonator modes for W-band devices. IEEE Trans. MTT-*38*, 1667–1674 (1990)

9.47   B. Ress, L. Johnson, D. Apte: Prescaler aids design of fast frequency source. Microwaves & RF, 129 ff (November 1990)

9.48   R.L. Miller: Fractional frequency generators using regenerative modulation. IRE Proc. Vol. 27, 446–457 (1939)

9.49   A. Plattner: A coherent, frequency agile 94 GHz radar with dual polarization capabilities. Proc. 15th Europ. Microw. Conf., Paris (1985) pp. 125–130

9.50   P. Nüchter, W. Menzel: A MM-wave frequency divider. IEEE Int'l Microw. Symp. MTT, Albuquerque (1992) pp. 695–697

9.51   G. Strauss, W.Menzel: A novel concept for mm-wave interconnects and packaging. IEEE Int'l Microw. Symp. MTT, San Diego (1994) pp. 1141–1145

9.52   B. Berson: Strategies for microwave and millimeter wave packaging today. 19th Europ. Microw. Conf., London (1989) pp. 89–95

9.53   G. Jerinic, M. Borkowski: Microwave module packaging. IEEE Int'l Microw. Symp. MTT, Albuquerque (1992) pp. 1503–1506

9.54   H.-J. Kuno, T.A. Midford: Millimeter wave packaging. IEEE Int'l Microw. Symp. MTT, Albuquerque (1992) pp. 1507–1508

9.55   A.K. Agrawal, R.D. Clarz, J.J. Komiaz, R. Browne: Microwave module interconnection and packaging using multilayer thin film/thick film technology. IEEE Int'l Microw. Symp. MTT, Albuquerque (1992) pp. 1509–1511

Subject Index